ORGANIC FARMING FOR SUSTAINABLE DEVELOPMENT

Current Advances in Biodiversity, Conservation, and Environmental Sciences

ORGANIC FARMING FOR SUSTAINABLE DEVELOPMENT

Edited by

Jeyabalan Sangeetha, PhD

Kasem Soytong, PhD

Devarajan Thangadurai, PhD

Abdel Rahman Mohammad Al-Tawaha, PhD

AAP APPLE ACADEMIC PRESS

First edition published 2023

Apple Academic Press Inc.
1265 Goldenrod Circle, NE,
Palm Bay, FL 32905 USA

760 Laurentian Drive, Unit 19,
Burlington, ON L7N 0A4, CANADA

CRC Press
6000 Broken Sound Parkway NW,
Suite 300, Boca Raton, FL 33487-2742 USA

4 Park Square, Milton Park,
Abingdon, Oxon, OX14 4RN UK

Library and Archives Canada Cataloguing in Publication

Title: Organic farming for sustainable development / edited by Jeyabalan Sangeetha, PhD, Kasem Soytong, PhD, Devarajan Thangadurai, PhD, Abdel Rahman Mohammad Al-Tawaha, PhD.
Names: Sangeetha, Jeyabalan, editor. | Soytong, Kasem, editor. | Thangadurai, D. (Devarajan), 1976- editor. | Al-Tawaha, Abdel Rahman Mohammad, editor.
Series: Current advances in biodiversity, conservation, and environmental sciences (Series)
Description: First edition. | Series statement: Current advances in biodiversity, conservation, and environmental sciences | Includes bibliographical references and index.
Identifiers: Canadiana (print) 20220220697 | Canadiana (print) 20220220727 | Canadiana (ebook) 20220220727 | ISBN 9781774910207 (hardcover) | ISBN 9781774910214 (softcover) | ISBN 9781003284055 (ebook)
Subjects: LCSH: Organic farming. | LCSH: Sustainable agriculture.
Classification: LCC S605.5 .O73 2022 | DDC 631.5/84—dc23

Library of Congress Cataloging-in-Publication Data

CIP data on file with US Library of Congress

ISBN: 978-1-77491-020-7 (hbk)
ISBN: 978-1-77491-021-4 (pbk)
ISBN: 978-1-00328-405-5 (ebk)

ABOUT THE CURRENT ADVANCES IN BIODIVERSITY, CONSERVATION AND ENVIRONMENTAL SCIENCES BOOK SERIES

Series Editors

Jeyabalan Sangeetha, PhD
Assistant Professor, Central University of Kerala, Kasaragod, Kerala, India

Devarajan Thangadurai, PhD
Associate Professor, Karnatak University, Dharwad, Karnataka, India

- **Biodiversity and Conservation: Characterization and Utilization of Plants, Microbes, and Natural Resources for Sustainable Development and Ecosystem Management**

  Editors: Jeyabalan Sangeetha, PhD, Devarajan Thangadurai, PhD, Hong Ching Goh, PhD, and Saher Islam, PhD

- **Beneficial Microbes for Sustainable Agriculture and Environmental Management**

  Editors: Jeyabalan Sangeetha, PhD, Devarajan Thangadurai, PhD, and Saher Islam, PhD

- **Algal Genetic Resources: Cosmeceuticals, Nutraceuticals, and Pharmaceuticals**

  Editors: Jeyabalan Sangeetha, PhD, and Devarajan Thangadurai, PhD

- **Organic Farming for Sustainable Development**

  Editors: Jeyabalan Sangeetha, PhD, Kasem Soytong, PhD, Devarajan Thangadurai, PhD, and Abdel Rahman Mohammad Al-Tawaha, PhD

ABOUT THE EDITORS

Jeyabalan Sangeetha, PhD

Assistant Professor, Central University of Kerala, Kasaragod, Kerala, South India

Jeyabalan Sangeetha, PhD, is an Assistant Professor at Central University of Kerala, Kasaragod, South India. She has edited/co-edited several books in her research areas, including environmental toxicology, environmental microbiology, environmental biotechnology, and environmental nanotechnology. She earned her BSc in Microbiology and PhD in Environmental Science from Bharathidasan University, Tiruchirappalli, Tamil Nadu, India. She holds an MSc in Environmental Science from Bharathiar University, Coimbatore, Tamil Nadu, India. She is the recipient of a Tamil Nadu Government Scholarship and a Rajiv Gandhi National Fellowship of the University Grants Commission, Government of India, for her doctoral studies. She served as the Dr. D.S. Kothari Postdoctoral Fellow and UGC Postdoctoral Fellow at Karnatak University, Dharwad, South India, during 2012–2016.

Kasem Soytong, PhD

Associate Professor, King Mongkut's Institute of Technology Ladkrabang, Bangkok, Thailand

Kasem Soytong, PhD, is currently serving as an Associate Professor in the Department of Plant Production Technology in the Faculty of Agricultural Technology, King Mongkut's Institute of Technology Ladkrabang, Bangkok, Thailand. He is the President of the Association of Agricultural Technology in Southeast Asia; Editor-in-Chief of the *International Journal of Agricultural Technology;* Director of Research and Development for BioAgritech Co. Ltd, Vietnam; Executive Research Director at CAS Bio-Agricultural Bio-engineering, China; Director of Research for Shew Khant Kyaw Co. Ltd. (Organic Biofertilizer), Myanmar; Organic Consultant, CGC Lao Coffee, Laos; Organic Consultant, Natural, and Premium Food Co. Ltd, Thailand; and Director, AATSEA Organic Agriculture Model Farm, Thailand. He is an author/co-author of more than 150 scientific publications, including research articles, reviews, book chapters, and books.

He obtained his PhD in Plant Pathology (Mycology) from the University of the Philippines Los Banos (UPLB), Philippines, under the SEMEO-SEARCA Scholarship. His research interests include nanotechnology for agriculture, bioactive metabolites for disease control and immunity, organic agriculture research, and plant pathology. He holds patent rights and product registrations in Thailand, China, Laos, Vietnam, and IFOAM (International Federation of Organic Agriculture Movements). He also has international research collaborations with scientists from Sweden, Russia, Finland, the USA, New Zealand, Australia, Japan, China, Philippines, Cambodia, Myanmar, Vietnam, Lao PDR, Egypt, and Iran.

Also he is the recipient of several awards and honors, including 1981 Gold Medal of First Class Honor in BSc conferred by His Majesty the King of Thailand; 1991 Distinguished Alumnus Award in Research Work conferred by Her Majesty the Princess Sirindhorn of Thailand (RIT-Chantaburi, Thailand); 1994 Distinguished Alumnus Award in Research Work (UPLB, Philippines); 1995 National Research's Inventor Award (National Research Council of Thailand); 1997 Excellency Research Award in Agriculture and Biology (National Research Council of Thailand); 1998 IFS/Silver Jubilee Research Award (IFS, Sweden); 2007 Outstanding Research Award from King Rama IV Foundation by Her Royal Highest Princess Sirindhorn; 2015 Best Research Paper Award at the International Conference in Agriculture and Environment for Sustainable Development, National Research Center, Cairo, Egypt; 2017 Best Research Paper Award at the International Conference on Advanced Technologies and Their Application in Agriculture, Cairo, Egypt; 2017 Distinguished Alumnus Award, University of the Philippines Los Banos (UPLB, the Philippines), 2018 Commemorative Certificate for Research Contribution Award, Faculty of Agricultural Technology, KMITL, Bangkok, Thailand; 2018 India-Guest of Honor Award at the International Conference on Bioproducts Development in Agricultural Inputs for Organic Crop Production, Bharathiyar Arts and Science College for Women, Attur, India; 2019-Recognition Award from China at the International Modern Organic Agriculture Forum, Songyang, Zhejiang, China; 2019 The Second Prize 2019 for the New Entrepreneurial and Innovative Competition in Wuxi, Jiangsu, China by CAS Asian Agriculture.

Devarajan Thangadurai, PhD

Associate Professor, Karnatak University, Dharwad, Karnataka, South India

Devarajan Thangadurai, PhD, is an Associate Professor at Karnatak University in South India and Editor-in-Chief of the international journals *Biotechnology, Bioinformatics, and Bioengineering* and *Acta Biologica Indica*. He has authored/edited over 25 books with national and international publishers. He has visited 24 countries in Asia, Europe, Africa, and the Middle East for academic visits, scientific meetings, and international collaborations. He received his PhD in Botany from Sri Krishnadevaraya University in South India as a CSIR Senior Research Fellow with funding from the Ministry of Science and Technology, Government of India. He served as a Postdoctoral Fellow at the University of Madeira, Portugal; University of Delhi, India; and ICAR National Research Centre for Banana, India. He is the recipient of a Best Young Scientist Award with a Gold Medal from Acharya Nagarjuna University, India, and the VGST-SMYSR Young Scientist Award of the Government of Karnataka, Republic of India.

Abdel Rahman Mohammad Al-Tawaha, PhD

Professor, Al-Hussein Bin Talal University, Jordan

Abdel Rahman Mohammad Al-Tawaha, PhD, is a Professor of Plant Science and former Head of the Department of Biological Sciences as well as former Director of the Planning, Information and Quality Unit at Al Hussein Bin Talal University, Jordan. He is author/coauthor of more than 200 publications in plant science in leading peer-reviewed journals and chapters in edited volumes and books on a broad range of development issues. He is the Founder and Editor-in-Chief of the journal Advances in Environmental Biology. He is also participated in many international conferences as chair or member of the scientific committees. Dr. Al-Tawaha obtained his PhD from McGill University, Montreal, Canada.

CONTENTS

CONTRIBUTORS

Charles Oluwaseun Adetunji
Microbiology, Biotechnology, and Nanotechnology Laboratory,
Department of Microbiology Edo State University Uzairue, Auchi, Nigeria

Hiba Alatrash
General Commission for Scientific Agricultural Research, Syria

Alla Aleksanyan
Institute of Botany aft. A.L. Takhtajyan NAS RA/Department of Geobotany and Plant Eco-Physiology, Yerevan, Armenia

Abdel Rahman M. Al-Tawaha
Department of Biological Sciences, Al-Hussein Bin Talal University, P.O. Box 20, Maan, Jordan

Abdel Razzaq M. Al-Tawaha
Department of Crop Science, Faculty of Agriculture, University Putra Malaysia, Serdang–43400, Selangor, Malaysia

Kangasalam Amala
Department of Microbiology, Sacred Heart College (Autonomous), Tirupattur–635601, Tamil Nadu, India

Amanullah
Department of Agronomy, The University of Agriculture, Peshawar, Pakistan

Osikemekha Anthony Anani
Laboratory of Ecotoxicology and Forensic Biology, Department of Biological Science, Faculty of Science, Edo State University Uzairue, Auchi, Nigeria

Toba Samuel Anjorin
Department of Crop Protection, Faculty of Agriculture, University of Abuja, Nigeria

Adeniran Lateef Ariyo
Department of Veterinary Physiology and Biochemistry, Faculty of Veterinary Medicine, University of Abuja, Nigeria

Mazen A. Ateyya
Faculty of Agricultural Technology, Al Balqa Applied University, Al-Salt–19117, Jordan

Abdul Basit
Department of Plant Pathology, Agriculture College, Guizhou University, Guiyan–550025, P.R. China

Rajan Bhatt
Regional Research Station-Kapurthala, Punjab Agricultural University, Ludhiana, Punjab–144601, India

İsmail Çelik
Çukurova University, Faculty of Agriculture, Department of Soil Science and Plant Nutrition, Adana, Turkey

Apurbo Kumar Chaki
School of Agriculture and Food Sciences, University of Queensland, QLD–4072, Australia; On-Farm Research Division, Bangladesh Agricultural Research Institute (BARI), Gazipur, Dhaka, Bangladesh

Shilpi Das
Bangladesh Institute of Nuclear Agriculture, Mymensingh–2202, Bangladesh; School of Agriculture and Food Sciences, University of Queensland, QLD–4072, Australia

Chibuzor-Onyema Ihuoma Ebere
Department of Microbiology, School of Science and Technology, Babcock University, Ilishan-Remo, Nigeria

Zahra Farrokhi
College of Agriculture and Natural Resources, University of Tehran, Iran

Elif Günal
Gaziosmanpaşa University, Faculty of Agriculture, Department of Soil Science and Plant Nutrition, Tokat, Turkey

Hikmet Günal
Gaziosmanpaşa University, Faculty of Agriculture, Department of Soil Science and Plant Nutrition, Tokat, Turkey

Aparna B. Gunjal
Department of Microbiology, Dr. D. Y. Patil Arts, Commerce, and Science College, Pimpri, Pune–411018, Maharashtra, India

Akbar Hossain
Bangladesh Wheat and Maize Research Institute, Dinajpur – 5200, Bangladesh

Khursheed Hussain
Division of Vegetable Science, SKUAST-Kashmir, Jammu and Kashmir, India

Osebhahiemen Odion Ikhimiukor
Environmental Microbiology and Biotechnology Laboratory, Department of Microbiology, University of Ibadan, Nigeria

Imran
Department of Agronomy, The University of Agriculture, Peshawar, Pakistan

Imranuddin
Department of Horticulture, The University of Agriculture, Peshawar, Pakistan

Mst. Tanjina Islam
Department of Agronomy, Hajee Mohammad Danesh Science and Technology University, Dinajpur–5200, Bangladesh

Saher Islam
Institute of Biochemistry and Biotechnology, Faculty of Biosciences, University of Veterinary and Animal Sciences, Lahore, Pakistan

Tofazzal Islam
Institute of Biotechnology and Genetic Engineering (IBGE), Bangabandhu Sheikh Mujibur Rahman Agricultural University, Gazipur – 1706, Bangladesh

Somdej Kanokmedhakul
Department of Organic Chemistry, Faculty of Science, Khon Khan University, Khon Khan, Thailand

Arun Karnwal
Department of Microbiology, School of Bioengineering and Biosciences, Lovely Professional University, Phagwara, Punjab, India

Malavika Ram Amanthra Keloth
Department of Plant Pathology, School of Agriculture, Lovely Professional University, Phagwara–144411, Punjab, India

Shah Khalid
Department of Agronomy, The University of Agriculture, Peshawar, Pakistan

Samia Khanum
Department of Botany, University of the Punjab, Lahore, Pakistan

Sameena Lone
Division of Vegetable Science, SKUAST-Kashmir, Jammu and Kashmir, India

Debjyoti Majumder
Uttar Banga Krishi Viswavidyalaya, Cooch Behar, West Bengal, India

Oluwadamilola Mathew Makinde
Department of Microbiology, School of Science and Technology, Babcock University, Ilishan-Remo, Nigeria

Nadia
Department of Agronomy, The University of Agriculture, Peshawar, Pakistan

Nidal Odat
Department of Medical Laboratories, Al-Balqa Applied University, Al-Salt–19117, Jordan

Fapohunda Stephen Oyedele
Department of Microbiology, School of Science and Technology, Babcock University, Ilishan-Remo, Nigeria

Mujeeb Ur Rahman
Department of Horticulture, The University of Agriculture, Peshawar, Pakistan

Meenakshi Rana
Department of Plant Pathology, School of Agriculture, Lovely Professional University, Phagwar–144411, Punjab, India

Abdur Rauf
Department of Chemistry, University of Swabi, Anbar, Khyber Pakhtunkhwa, Pakistan

Poonam Roshan
Department of Biotechnology, Guru Nanak Dev University, Amritsar, Punjab–143005, India

Anjorin Toba Samuel
Department of Crop Protection, Faculty of Agriculture, University of Abuja, PMB 117, Abuja, Nigeria

Jeyabalan Sangeetha
Department of Environmental Science, Central University of Kerala, Kasaragod–671316, Kerala, India

Palani Saranraj
Department of Microbiology, Sacred Heart College (Autonomous), Tirupattur–635601, Tamil Nadu, India

Panneerselvam Sivasakthivelan
Department of Agricultural Microbiology, Faculty of Agriculture, Annamalai University, Chidambaram–608002, Tamil Nadu, India

Kasem Soytong
Department of Plant Production Technology, Faculty of Agricultural Technology, King Mongkut's Institute of Technology Ladkrabang, Bangkok, Thailand

Abdulkadir Sürücü
Harran University, Faculty of Agriculture, Department of Soil Science and Plant Nutrition, Şanliurfa, Turkey

Devarajan Thangadurai
Department of Botany, Karnatak University, Dharwad, Karnataka – 580003, India

Ajay Tomer
Department of Plant Pathology, School of Agriculture, Lovely Professional University, Phagwara – 144411, Punjab, India

Munir Turk
Department of Plant Production, Jordan University of Science and Technology, Irbid, Jordan

Ayşe Yeşilayer
Faculty of Agriculture, Tokat Gaziosmanpasa University, Tokat, Turkey

Nandhini Yoga
Department of Agronomy, Tamil Nadu Agricultural University, Coimbatore – 641003, Tamil Nadu, India

ABBREVIATIONS

ABA	abscisic acid
AC	active cells
ACC	aminocyclopropane-1-carboxylic acid
ADD	anaerobically digested dairy
ADIs	acceptable daily intakes
ADP	anaerobically digested pig slurry
ADS	anaerobically digested slurry
AGE	aged garlic extract
Al	aluminum
Al^{3+}	aluminum trication
$AlPO_4$	aluminum phosphate
AM	arbuscular mycorrhizal
AMF	arbuscular mycorrhizal fungi
AMS	agricultural marketing service
AQPs	aquaporins
ARA	acetylene reduction assay
As	arsenic
ASD	anaerobic soil disinfestation
ATP	adenosine triphosphate
ATUs	active taxonomic units
AUDPC	area under disease progress curve
BC	*Botrytis cinerea*
BGA	blue-green algae
BMI	body mass index
BOF	bio-organic fertilizer
BP	bone phosphate
BPB	bromophenol blue
Bt	*Bacillus thuringiensis*
C3	carbon three
C4	carbon four
$C_6H_6Ca_6O_{24}P_6$	calcium phytate
Ca	calcium
Ca^{2+}	calcium dication

$Ca_3(PO_4)_2$	tricalcium phosphate
$CaHPO_4$	dicalcium phosphate
CC	cork compost
Cd	cadmium
CH_4	methane
CL_{50}	concentration limit
Co	cobalt
CO_2	carbon dioxide
Cr (VI)	hexavalent chromium
Cr	chromium
Cu	copper
CV	cultivar
DADS	diallyl disulfide
DAPG	2,4-diacetylphloroglucinol
DAS	diallyl sulfide
DATS	dimethyl trisulfide
DDT	dichlorodiphenyltrichloroethane
DNA	deoxyribonucleic acid
DOR	dry olive residues
DSR	direct-seeded rice
ECV	essential climate variables
EPHs	enhanced plant holobiomes
EU	European Union
FAO	Food and Agriculture Organization
FC	*Fusarium culmorum*
Fe	iron
Fe^{3+}	iron trication
FOL	*Fusarium oxysporum* f. sp. *lycopersici*
FTIR	Fourier transform infrared spectroscopy
FYM	farmyard manure
GHGs	greenhouse gas
GIS	geographic information system
GM	green manure
GMC	grape marc compost
GSPC	global strategy for plant conservation
H_2PO_4	dihydrogen phosphate
HCl	hydrochloride
HCl-P	phosphorus hydrochloride
HKC	heat-killed cells

HMEC	human microvascular endothelial cells
HMs	heavy metals
HPLC	high-pressure liquid chromatography
HPO_{42}-Ca	calcium-hydrogen phosphate
HS	humic substances
HWSC	harvest weed seed control methods
IAA	indole-3-acetic acid
IBA	indole-3-butyric acid
IFOAM	International Federation of Organic Agriculture Movements
IPM	integrated pest management
IPNS	integrated plant nutrition system
ISAC	Institute for Sustainable Agricultural Communities
ISR	induced systemic resistance
ITS	internal transcribed spacer
K	potassium
LC_{50}	lethal concentration
LPS	lipopolysaccharide
LSM	liquid swine manure
M	mycorrhizal
Mabs	monoclonal antibodies
MALDI-TOF	matrix-assisted laser desorption/ionization-time of flight
MAO	monoamine oxidase
MAP kinases	mitogen-activated protein kinase
MDG	millennium development goals
Mg	magnesium
Mg^{2+}	magnesium dication
MMN	mineral nutrients solution
Mn	manganese
MRLs	maximum residue level
MS	mass spectrometry
MSM	mustard seed meal
N	nitrogen
$Na_{15}NO_3$	sodium nitrate
NAA	naphthalene acetic acid
NaCl	sodium chloride
NaOH-P	phosphonium hydroxide
NBRIP	National Botanical Research Institute Phosphate
NH_4	ammonium
Ni	nickel

NM	nonmycorrhizal
NMDS	non-metric multidimensional scaling
NMR	nuclear magnetic resonance
NO_2	nitrous oxide
NO_3	nitrate
NOP	national organic program
NPK	nitrogen, phosphorous, and potassium
OFS	organic farming system
OPB	inorganic plant breeding
OPPs	organophosphorus pesticides
P	phosphate
P_2O_5	iron (III) phosphate
Pb	lead
PCR	polymerase chain reaction
PCR-DGGE	polymerase chain reaction-denaturing gradient gel electrophoresis
PDA	potato dextrose agar
PGPB	plant growth-promoting bacteria
PGPM	plant growth-promoting microorganisms
PGPR	plant growth-promoting rhizobacteria
pH	degree of acid and base
PHB	poly β-hydroxybutyrate
PM	poultry manure
PR-protein	pathogenesis-related proteins
PSB	phosphate solubilizing bacteria
PSI	phosphate solubilizing index
PSM	phosphate solubilizing microorganism
PSRB	phosphate solubilizing rhizospheric bacteria
PVK	Pikovskaya
Rb	rubidium
rDNA	ribosomal DNA
RLSBX	relative length of stem with brown xylem
ROS	reactive oxygen species
RP	rock phosphate
S	sulfur
SAC	S-allycysteine
SANRU	Sari Agricultural Sciences and Natural Resources University
SAR	systemic acquired resistance
SM	*sclerotinia minor*

SMAC	S-allylmercaptocysteine
SOM	soil organic matter
sp.	species
SSA	Sub-Saharan Africa
SSF	solid-state fermentation
SSP	single super phosphorus
TA	titratable acidity
TCP	tricalcium phosphate
TLC	thin layer chromatography
UNCTAD	United Nations Conference on Trade and Development
UNEP	United Nations Environment Program
USA	United States of America
USDA	U.S. Department of Agriculture
UV	ultraviolet
VAM	vesicular arbuscular mycorrhizas
VFA	volatile fatty acid
ZBNF	zero budget natural farming
Zn	zinc
ZnO	zinc oxide

SYMBOLS

%	percentage
β	beta
μm	micrometer
$\mu molL^{-1}$	micro per mole
1°	first degree
3°	third degree
CFU/ml	colony-forming unit per milliliters
dSm^{-1}	decisiemens per meter
g	gram
h	hour
ha	hectare
kg ha^{-1}	kilogram per hectare
kg	kilogram
kg^{-1}	kilo per gram
m	meter
mg kg^{-1}	milligram per kilogram
mg	milligrams
mg/l	milligrams per liter
mg/mL	milligrams per milliliter
min	minute
ml	milliliter
mm	millimeter
mM	millimolar
ng/mL	nanograms per milliliter
°C	degree Celsius
α	alpha
β	beta
μg ml^{-1}	microgram per liter
μm	micrometer

PREFACE

The gravity of organic farming has increased in recent past decades with the concern with the vast deleterious effect of conventional agricultural practices, which employ chemical fertilizers, pesticides, and herbicides for large-scale food production. Organic farming has become an important paradigm for sustainable development as it prevents jeopardizing the ecological balance and secures the natural resources for upcoming generations. In general, organic farming methods make use of natural resources such as compost, bone meals, green/animal manure, microorganisms, and crop rotation to improve soil quality and nurture plants, which enhance the good health and well-being of humans as well as the biological functioning of the soil, water, and the natural niche. The quality of food in terms of nutrition, fertility of the soil, and the surrounding ecosystem such as micro and macroflora has been retained for the long term. In this way, we can efficiently and cost-effectively use natural and available resources to produce healthy food by reducing the lethality of environmental issues.

Compared to conventional farming, which is more output-oriented rather than health concerned, natural farming includes various cultivating rationale that is ecofriendly and aids in tacking the hazardous consequences of chemicals on the environment. Rather than concentrating on the quantity of food production using conventional farming techniques, if we target equitably on both qualities of food and environment, it can bring about the sustainable development of the ecosystem and the socio-economy of any nation. As it is a question of the existence of the whole planet, organic farming has gained in importance and solemnness on a global basis. On average of 2.8 million farmers in around 186 countries worldwide are now following this farming method.

The current status of organic cultivation is at its advancing stages in developing countries. Up date, only a few fractions of total agricultural farmlands are producing organic cultivars. Conversely, suppose we utilize the remaining portions of farmlands for organic farming. In that case, we can expeditiously increase the nutrients content of our food. We would be able to manage the organic wastes coming from other sectors, such as cattle farms, without polluting the surroundings. This is the key benefit of organic farming wherein we can make use of the waste products from one sector as the raw material for another agricultural sector.

As mentioned earlier, one of the significant aspects of this farming method is the maintenance of ecological balance; that is, a prerequisite factor for the conservation of biodiversity. It can also be employed correspondingly on ecological poles and economic levels of national and international markets because of the products' high nutrition value in comparison with traditionally grown food products. Organic farming is a kind of give-and-take approach whereby we can use natural resources for our needs, and we take responsibility to sustain an ecological balance, which is the main agenda of this farming method.

However, one of the widely discussed demerits of organic farming is its low output in bulk quantity, and it is more labor-intensive. Nevertheless, suppose we implement more sustainable methods in organic farming, in that case, that might increase the quantity and quality of crop production, and we can set aside or reduce the environmental repercussions and thereby sustain the ecosystem in and of itself for the future era.

The present book mainly focuses on the application of different natural resources as manure for organic farming. Section I discusses in detail the application of microorganisms such as *Trichoderma* sp., *Azospirillum* sp., endophytic microorganisms, arbuscular mycorrhiza, *Chaetomium* sp., and, and bioactive secondary metabolites thereof in the organic farming practices. Section II explores the potential applications of organic amendments and sustainable practices for plant growth and soil health using garlic products, organic substrates, biochar, organic mulching, and tillage and weed management. In addition, Section III summarizes the impact and prospects of organic crop production technology on health, food safety, and quality.

—Jeyabalan Sangeetha, PhD
Kasem Soytong, PhD
Devarajan Thangadurai, PhD
Abdel Rahman Mohammad Al-Tawaha, PhD

PART I

Agrofriendly Microbes and Their Metabolites for Sustainable Agriculture

CHAPTER 1

TRICHODERMA: AN ECO-FRIENDLY BIOPESTICIDE FOR SUSTAINABLE AGRICULTURE

APARNA B. GUNJAL

Department of Microbiology, Dr. D. Y. Patil Arts, Commerce, and Science College, Pimpri, Pune–411018, Maharashtra, India

ABSTRACT

The use of chemical pesticides to kill the insects and pests in agriculture is toxic, costly, and causes harm to the environment. This is a serious issue and needs attention. The biological approach is necessary to be applied. The introduction, category, and market demand of biopesticides are taken into account. *Trichoderma*-based biopesticide is gaining importance. This chapter focuses on features of *Trichoderma* biopesticide, formulations, and mechanism of action of *Trichoderma* biopesticide for sustainable agriculture. The production of *Trichoderma*-based biopesticide, spore production of *Trichoderma*, methods for application in agriculture, and how *Trichoderma* biopesticides are effective against diseases caused by phytopathogens are also highlighted. The root colonization and interaction of *Trichoderma* with other microorganisms is also highlighted in this chapter. The application of *Trichoderma*-based biopesticide has importance as it is eco-friendly, easy to use, economical, and safe to the environment. Biopesticides now are alternatives to chemical pesticides and have emerged as novel tools to control insects and pests in agriculture. The use of *Trichoderma* sp. in agriculture will increase the yield of various plants. This will also lead to organic farming and organic products, which will be healthy and safe. This chapter thus highlights the development of *Trichoderma*-based biopesticide.

1.1 INTRODUCTION

Agriculture is the most essential aspect. The uses of chemical fertilizers and pesticides are toxic, costly, and harmful to humans as well as the environment (Prabha et al., 2016). The use of chemical pesticides disturbs the balance of the ecosystem. Biopesticides are made from living microorganisms (bacteria, fungi, and viruses) and plants. They protect the plants and crops against insects, pests, and nematodes (Senthil-Nathan, 2015). These are of three types, viz., plant-incorporated protectants, microbial, and biochemical pesticides. The biopesticides have advantages, viz., target-specific, very eco-friendly, economical, non-toxic, and can be used in trace amounts (Kumar, 2012). Due to their advantages are gaining immense importance in comparison to synthetic chemical pesticides (Chandrasekaran et al., 2012). Biopesticides, in combination with integrated pest management (IPM) programs, help to reduce the use of chemical pesticides and increase the growth of plants (Sharma and Malik, 2012). They do not harm the ecosystem and are easily biodegradable. They have gained immense value in the market due to many advantages (Prabha et al., 2016). The demand for biopesticides has increased all over the world. They also have an excellent role to play in IPM. Biopesticides are safe to be used by farmers and maintain a sustainable approach (Kumar et al., 2014a). The health consciousness of humans has immensely increased the need for organic food. This, in turn, has increased the scope for biopesticides. The use of biopesticides has increased by the farmers in agriculture (Kandpal, 2014). The major biopesticides are *Bacillus thuringiensis*, *Pseudomonas fluorescens*, *Fusarium*, *Pythium*, *Penicillium*, *Verticillium*, *Trichoderma harzianum*, *T. viride, and Beauveria bassiana* (Gupta and Dikshit, 2010; Kachhawa, 2017). There are 15 biopesticides registered in India. In the United States of America (USA) and European countries, about 200 and 60 biopesticide products are available respectively in the market (Kumar and Singh, 2015). The use of biopesticides increases by 10% every year (Kumar and Singh, 2015).

1.2 CATEGORIES OF BIOPESTICIDES

1.2.1 MICROBIAL PESTICIDES

These consist of microorganisms used for the control of insects, nematodes, and pests. The most common biopesticide is *Bacillus thuringiensis* (Bt). *B. thuringiensis* produces Bt toxin, which when consumed by insects, pests, or nematodes, breaks the gut cells (Chandler et al., 2011) and kills the insects or pests. The Bt toxin causes the death of insects or pests in about 48 h.

1.2.2 BIOCHEMICAL PESTICIDES

Biochemical pesticides are also termed herbal pesticides (Pal and Kumar, 2013). They control insects and pests by a non-toxic mechanism.

1.2.3 PLANT-INCORPORATED-PROTECTANTS

Plant-Incorporated-Protectants are the biopesticides produced from plants the way of genetic engineering (Tijjani et al., 2016).

1.3 FORMULATIONS OF BIOPESTICIDES

The biopesticides are available in three forms, viz., powder, emulsion, and granules (Singh et al., 2014).

1.3.1 DRY FORMULATIONS

1.3.1.1 DUSTABLE POWDERS

The dustable powder formulation of biopesticide is prepared by adsorption of active ingredient on talc or clay (Tijjani et al., 2016). The concentration of active ingredients used is 10%.

1.3.1.2 GRANULES

The concentration of active ingredient used is 2–20% which either coat or adsorb on the granules. The granules size range between 100 and 600 microns and can be made using kaolin, silica, and starch (Tijjani et al., 2016).

1.3.1.3 SEED DRESSING

In this biopesticide formulation, the carrier is coated onto the seeds for adherence. In some cases, some coloring agent may also be used along with the carrier to ensure safety (Woods, 2003).

1.3.1.4 WETTABLE POWDERS

The wettable powders are prepared by mixing active ingredients with surfactants and inert fillers. These can be applied after suspension in water and can be stored for a long period (Tijjani et al., 2016).

1.3.2 LIQUID FORMULATIONS

1.3.2.1 EMULSION

Emulsion formulations can be oil in water or water in oil. The loss due to evaporation is minimized (Brar et al., 2006).

1.3.2.2 SUSPENSION CONCENTRATE

This biopesticide formulation is prepared by proper mixing of active ingredient in water. The active ingredients become readily available to the plants due to the minimum size of the particles (1–10 μm).

1.3.2.3 SUSPO-EMULSION

This formulation is a combination of emulsion and suspension concentrate. This formulation of biopesticides has more demand in the market (Tijjani et al., 2016).

1.3.2.4 OIL DISPERSION

This formulation is prepared similar to suspension concentrate. The active ingredients should be properly selected.

1.3.2.5 CAPSULE SUSPENSION

In capsule suspension, the active ingredients are enclosed in capsules. The capsules can be prepared using gelatin, starch, or cellulose. This formulation of biopesticides can be applied by interfacial polymerization.

1.4 SCOPE AND MARKET DEMAND OF BIOPESTICIDES

There is worldwide immense scope for biopesticides to control the insects and pests in the field due to their advantages (Oguh et al., 2019). The market demand for biopesticides has tremendously increased in the last few years. The reason behind this is conventional chemicals are toxic, costly, and cause pollution (Glare et al., 2012). The biopesticides are very safe to use. The world biopesticides market is expected to be around $7.7 billion by 2021 (Ruiu, 2018). In the USA, the USA and European market around 200 and 60 biopesticide products, respectively, are there in the market (Kumar and Singh, 2015). The biopesticide use increases by 10% each year, which is expected to increase more (Kumar and Singh, 2015). *Trichoderma* biopesticide will have good market globally.

1.5 MECHANISM OF ACTION OF BIOPESTICIDES

The mechanisms of action of biopesticides are antibiosis, where the microorganisms produce antibiotics to inhibit the pathogens (Rikita and Utpal, 2014); competition where the biopesticides colonize the substrates and control the growth of harmful microorganisms. There can be competition for the nutrients and minerals among the microorganisms, where the most potent microorganism will survive in competition. Antibiotics are reported to have some role in biocontrol activity (Hamid and Mohiddin, 2018). The antibiotics secreted by some fungi can control the growth of harmful fungi. *Trichoderma* biopesticide has been reported to control the pathogen *Erysiphe pisi* (Patel et al., 2016). *Trichoderma* produces toxic metabolites, viz., viridin, tricholin, and alamethicins (Gajera et al., 2013). Also, there is a report on *Trichoderma* for control of various plant diseases (Kumar et al., 2017). Hyper parasitism is another mechanism of action of biopesticide. In hyper parasitism, *Trichoderma* sp. due to chitinase enzyme, degrades chitin and enters the cell wall of the host (Hamid and Mohiddin, 2018). *T. lignorum* usually parasitizes the hyphae of *Rhizoctonia solani*. Due to this, *Trichoderma* spores can be applied to prevent damping off disease in *Citrus* plants (Rikita and Utpal, 2014). Synergism is also the mechanism of action of biopesticides which is a combination of hydrolytic enzymes and antibiotics (Tijjani et al., 2016). *T. asperellum* also stimulates induced systemic resistance (ISR), which helps in the control of plant diseases (Yedidia et al., 2003). There is a report where *Trichoderma harzianum* T-39 showed ISR against *Botrytis cinerea* (BC) and

Podosphaera xanthil, where the *Cucumis sativus*, *Solanum lycopersicum*, *Phaseolus vulgaris* and *Fragaria* × *ananassa* plants were protected from foliar diseases by stimulation of beneficial microorganisms (Levy et al., 2015). The mechanism of *Trichoderma*-based biopesticide is represented in Figure 1.1.

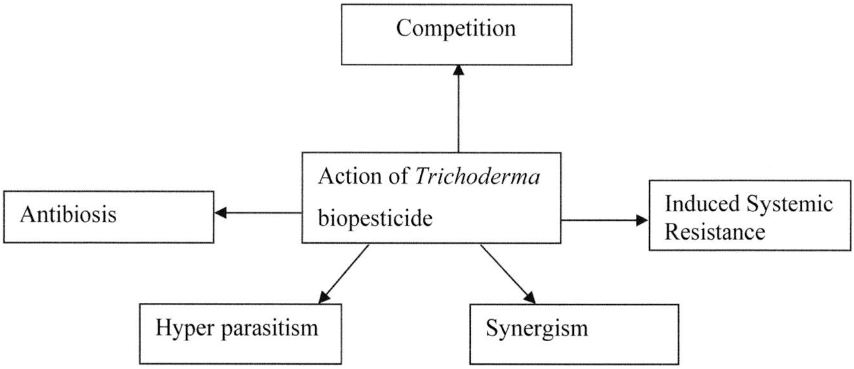

FIGURE 1.1 Mechanism of action of Trichoderma-based biopesticide.

1.6 VARIOUS FUNGI AS BIOPESTICIDES

The various fungi as biopesticides are represented in Figure 1.2.

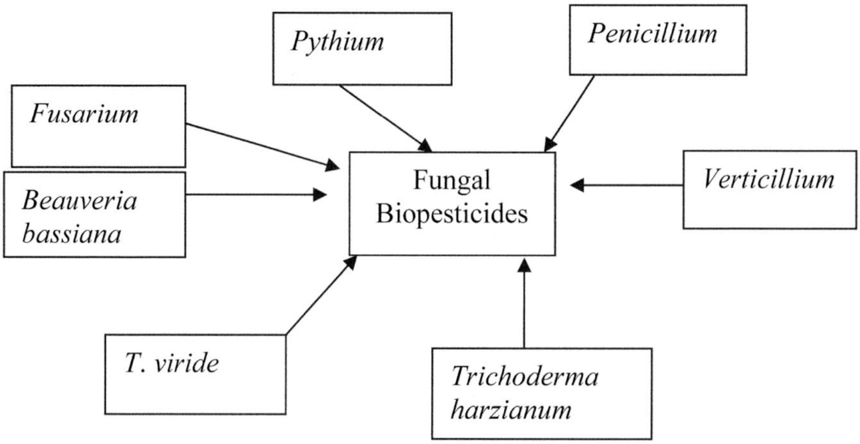

FIGURE 1.2 Various fungi as biopesticides.

1.7 *TRICHODERMA* SP.

Various *Trichoderma* spp. are used as biocontrol agents against plant pathogens (Sharma et al., 2014). *Trichoderma* sp. is free-living fungi and interacts highly with the roots of the plants. *Trichoderma* belongs to the subdivision of Deuteromycotina, classes Deuteromycetes, and familia Tuberculariaceae (Suparno et al., 2016). They have been found effective to control plant diseases caused by *Rhizoctonia solanii* (Gal-Hemed et al., 2011). The species of *Trichoderma* are *T. piluliferum*, *T. polysporum*, *T. hamatum*, *T. koningii*, *T. aureoviride*, *T. harzianum*, *T. longibrachyatum*, *T. pseudokoningii*, *T. reessei*, *T. asperellum*, *T. parareesei*, *T. gamsii*, *T. virens*, and *T. viride* (Waghunde et al., 2016; Zeilinger et al., 2016). There are more than 100 species of *Trichoderma* (Thapa et al., 2020). The mode of reproduction in *Trichoderma* sp. is asexual. *Trichoderma* sp. is found dominantly in soil and their characteristic feature is they produce spores. *Trichoderma* sp. is a strong opportunistic invader and produces antibiotics. *Trichoderma* has antagonistic properties (Reyes et al., 2012) and are known to produce biomass, primary metabolites (e.g., enzymes like cellulase, chitinase, organic acids, antibiotics, etc.); secondary metabolite (e.g., 6 pentyl-alpha-pyrone) and; spores. These are found to play a good role in the control of plant diseases. Also, the study is done on *Trichoderma* sp. as biocontrol of *Colletotrichum* sp. and *F. oxysporum* causing disease in chili (Su et al., 2018; Utami et al., 2019).

Trichoderma grows fast without much need of nutrients. Conidiophores are branched, either loosely or compactly tufted or formed in concentric rings. The main branch produces paired or unpaired branches. The phialides are seen arising near the tip of the main branch. The primary branches produce secondary branches either paired or unpaired. The longest secondary branch is seen close near the tip of the primary branch. This gives *Trichoderma* conidiophores a pyramid look. Phialides are held at 90° clustered or solitary, either cylindrical or globose in shape. Conidia are dry, smooth, ellipsoidal, or tuberculate and measure 3–5×2–4 μm. The conidia are green in color. Chlamydospores are unicellular or multicellular, subglobose in shape and present inside the hyphal cells. Stromata are present which may be brown, yellow, or orange in color and are pulvinate in shape. There are several reports where *Trichoderma* is used as biofertilizer (Gu et al., 2016; Kamal, 2018; Tawfeeq Al-Ani, 2018; Kumar et al., 2019). *Trichoderma* as biofertilizer has shown an increase in the yield of *Triticum* plant (Mahato et al., 2018).

1.7.1 SPORE PRODUCTION OF TRICHODERMA

For the preparation of biological control agents on a large scale by solid-state fermentation (SSF), spore biomass is necessary. SSF process occurs in the absence of water by using a variety of economical and non-toxic substrates, viz., rice, and wheat husk, bagasse, corn cob, pressmud, sawdust, coconut husk, etc. *T. harzianum* can be used for spore production by SSF (Ming et al., 2019). The spore production of *T. harzianum* is represented in Figure 1.3. Glucose, starch, fructose inorganic salt, magnesium, phosphorus, iron, and rice bran help in good sporulation of *Trichoderma* (Ming et al., 2019). There is a report on production of *T. asperellum* spores by SSF using mango waste (Sala et al., 2019; Santos-Villalobos et al., 2012). Study is done *T. harzianum* spore production on rice husk by SSF (Sala et al., 2020). Mycelial growth favors the spore production (Hiba et al., 2019). In the case of liquid-state fermentation, *Trichoderma* is grown in potato dextrose broth on a rotary shaker at 28°C for 10–12 days for the production of biomass of *Trichoderma* (Hamid and Mohiddin, 2018).

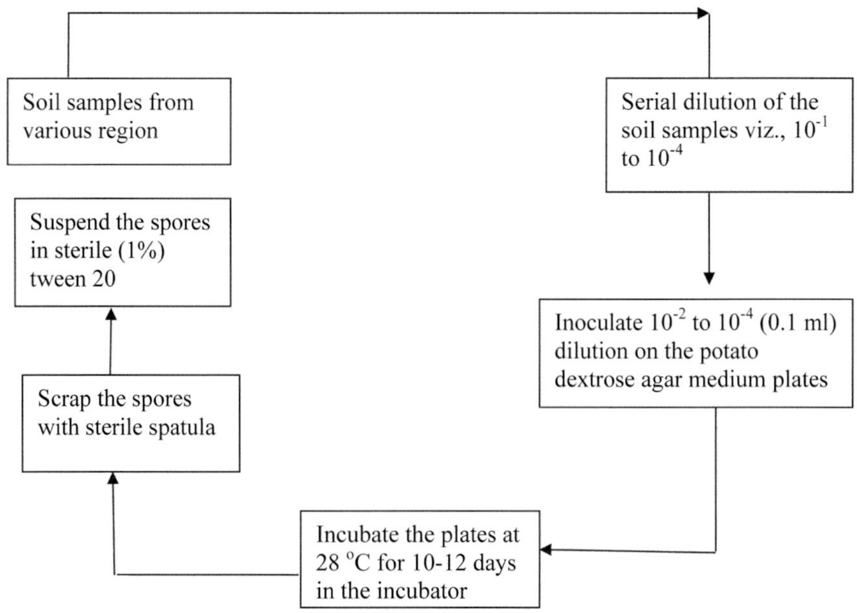

FIGURE 1.3 Spore production of *T. harzianum*.

1.8 ROOT COLONIZATION BY TRICHODERMA

Trichoderma colonizes roots of plants which alters plant metabolism (Brotman et al., 2012). They help the plants to acquire nutrients (Khatabi et al., 2012). *T. harzianum* penetrates the roots of plants by the help of enzymes. The enzyme endopolygalacturonase (ThPG1) from *T. harzianum* helps to penetrate the roots of plants. *Trichoderma* takes help of swollenin protein which has a cellulose-binding module, disrupts the cell wall of plants and helps in root colonization of plants (Druzhinina et al., 2011). The molecular mechanism in root colonization of plants by *Trichoderma* is still not known clearly.

1.8.1 INTERACTION OF TRICHODERMA WITH OTHER MICROORGANISMS

Trichoderma sp. has applications in agriculture to improve the plant yield. *Trichoderma* interacts mostly with pathogenic fungi by various mechanisms. There can be competition for nutrients, minerals, and space which can suppress the activity of pathogenic fungi (Thapa et al., 2020). There are studies done on the interaction of *Trichoderma viride* with *Azotobacter* and *Rhizobium*. *T. hamatum* interaction with *Rhizobium* sp. has been reported to improve the productivity of *V. mungo* (Badar and Qureshi, 2012). *Trichoderma* spp. on interaction has also offered too many fungi in the soil (Khan et al., 2017; Fiorentino et al., 2018). The interaction of *Trichoderma* sp. with other microorganisms leads to the secretion of antibiotics, which can kill other pathogenic fungi. Sometimes interaction of *Trichoderma* sp. with other fungi leads to the secretion of enzymes which can be helpful for biological control and provide resistance to the plants against many diseases.

1.8.2 TRICHODERMA AS A BIOPESTICIDE FOR SUSTAINABLE AGRICULTURE

Trichoderma-based biofungicides have immense importance in the agricultural market. More than 50 formulations are registered products of *Trichoderma* sp. (Waghunde et al., 2016). *Trichoderma* as biopesticide has been studied by Kachhawa (2017). These products have application as strong biocontrol agents. *Trichoderma*-based biocontrol agents are more effective as compared to other fungi. *T. harzianum* is widely used biopesticide in agriculture (Lorito et al., 2010; Prakash et al., 2019). *Trichoderma* sp. is antagonistic against various phytopathogens and this minimizes the diseases in crops (Monte, 2001). The antagonistic activity of *Trichoderma* sp. is due to 'trichodermin' which makes them strong biocontrol agents (Tvetdyukov et al., 1994). The secondary metabolites produced by *Trichoderma* sp. have immense applications (Ramteke, 2019). There is a report on the use of *Trichoderma* in sustainable agriculture by plant disease management (Tawfeeq Al-Ani, 2018). There is a report on the use of these secondary metabolites from *Trichoderma* for the plant growth. Also, there is a report on *Trichoderma* sp. effective against fungal phytopathogens, viz., *Clerotinia sclerotiorum* and *Sclerotinia minor* (SM), which causes major loss to many crops (Ibarra-Medina et al., 2010). There is a report on role of *Trichoderma* as effective biopesticide (Khandelwal et al., 2012; Ramteke, 2019). The efficacy of *T. viride* to control insects and pests is 60–90%. There is a report on the use of *Trichoderma* in agriculture (Sachdev and Singh, 2020). *Trichoderma brevicompactum* has been reported to show activity against *Sclerotium rolfsii*, *Colletotrichum gloesporioides*, *Verticillium dahliae*, *Fusarium oxysporum*, and *Cylindrocladium* sp. (Marques et al., 2018).

1.8.3 FEATURES OF TRICHODERMA BIOPESTICIDE

The features of *Trichoderma* biopesticide are, viz., effective, contains essential nutrients and minerals, good quality, non-toxic, easy to use, eco-friendly, and has good shelf-life.

1.8.4 MECHANISM OF ACTION OF T. HARZIANUM AS BIOCONTROL AGENT

T. harzianum establishes in the rhizosphere region of the plants or crops. The rhizosphere is the region where abundant beneficial microorganisms are present near the roots of the plants. There are different mechanisms of action which makes *T. harzianum* strong biocontrol agent. The growth of *T. harzianum* forms a barrier and prevents the growth of fungal phytopathogens. *T. harzianum* also utilizes the excess nutrients available in the root system, due to which the growth of fungal phytopathogens is inhibited. The second reason is *T. harzianum* produces chitinase enzyme which degrades chitin and ruptures the cell wall. This allows the entry of other microorganisms. These mechanisms help *T. harzianum* to prevent the growth of other fungal phytopathogens and thus help to control diseases in the plants (Gajera et al., 2013). This also enables to increase the growth and yield of the plants.

1.8.5 PRODUCTION OF TRICHODERMA-BASED BIOPESTICIDE

There is a report on production of biopesticide of *T. viride* (Arora et al., 2017). The production of *Trichoderma*-based biopesticide is represented in Figure 1.4. The fungal biopesticides are produced by submerged and SSF. Mostly SSF is usually used for the production of fungal biopesticides. SSF is the fermentation in the absence of water using beneficial microorganisms to get value-added product. For the production of biopesticide, *Trichoderma* is isolated from the rhizosphere region of the plants. It is grown on solid agar medium viz., potato dextrose agar (PDA). A pure is obtained which can be used for further production of the spores. The spores can be mixed with a suitable carrier such as talc and then applied in the field on a large scale.

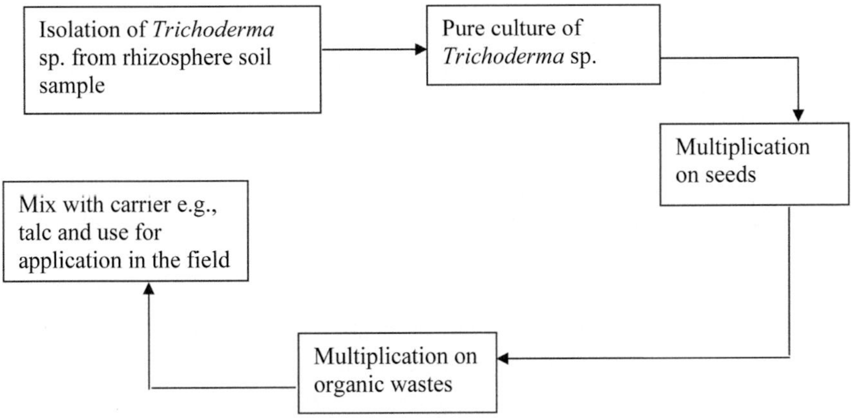

FIGURE 1.4 Production of *Trichoderma*-based biopesticide.

1.8.6 METHODS OF APPLICATION OF TRICHODERMA FOR BIOLOGICAL CONTROL IN AGRICULTURE

In seed treatment, 10 g *Trichoderma* formulation per lit cow dung slurry for the treatment of 1 kg of seed prior to sowing is used. The second method is cutting and seedling root dip where 10 g L^{-1} *Trichoderma* formulation is used to treat the seedlings for 10 min prior to planting. In the case of soil treatment method of application, 1 kg of *Trichoderma* formulation + 100 kg of farmyard manure (FYM) is covered for a week with polythene and mixed properly to apply in the field. The foliar spray application is also used where *Trichoderma* sp. can be applied as foliar spray to control the plant diseases (Mishra et al., 2018). The application of *Trichoderma* in agriculture is represented in Figure 1.5 (Kumar et al., 2014b).

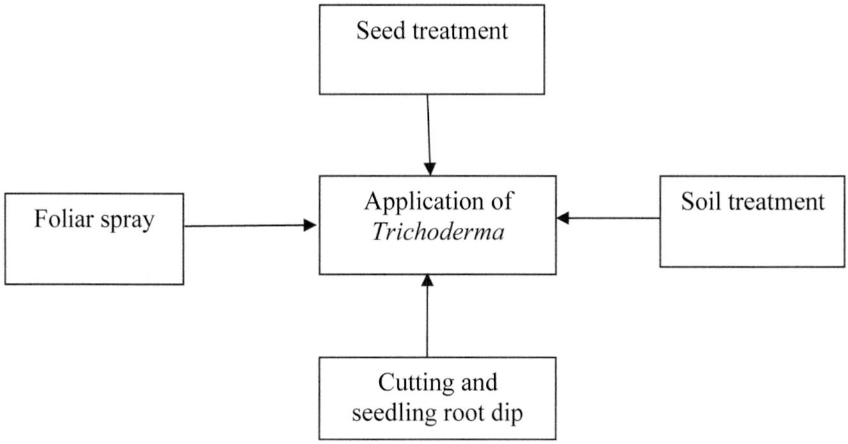

FIGURE 1.5 Application of *Trichoderma* in agriculture.

1.8.7 APPLICATION OF TRICHODERMA SP. BIOPESTICIDES AGAINST VARIOUS DISEASES

The application of *Trichoderma* sp. biopesticides against various diseases is represented in Table 1.1. There is a report where *T. harzianum* and *T. hamatum* are biocontrol agents against *Fusarium oxysporum* which causes lentil wilt (El-Hassan et al., 2013). *Trichoderma* biopesticide is used to control wilt and root rot diseases caused by *Fusarium* sp., *R. solani*, *R. bataticola*, *Sclerotium rolfsii* and *Sclerotium sclerotiorum* in crops (Chaudhary et al., 2004; Mishra et al., 2015). There is a report on eco-friendly management of red rot disease of *Saccharum officinarum* with *Trichoderma* strains (Singh et al., 2008). Also, there is a report of *Trichoderma* from mangrove sediments as biocontrol against *Fusarium* strains (Filizola et al., 2019). *Trichoderma* is reported to exhibit various mechanisms, viz., mycoparasitism, antibiosis, and competition which will be essential for use by farmers in agriculture to control diseases caused by pests and insects (Patil et al., 2016). The control of plant diseases by *Trichoderma* is shown in Figure 1.6.

TABLE 1.1 Application of *Trichoderma* sp. Biopesticides Against Various Diseases

Bioagent	Pathogen	Crop	References
T. harzianum	*Puccinia sorghi*	*Oryza sativa*	Dey et al. (2013)
	Pyricularia oryzae		
	Alternaria alternata	*Nicotina tabacum*	Gveroska and Ziberoski (2012)
	Phytophthora capsici	Red pepper	Savitha and Sriram (2015)
	Fusarium oxysporum	*Cicer arietinum*	Shabir-U-Rehman (2013)
T. viride	*Colletotrichum capsici*	*Capsicum frutescens*	Jagtap et al. (2013)
	Phytophthora capsici	*Piper nigrum*	Mathew et al. (2011)
Trichoderma sp.	*Botrytis cinerea*	*Solanum lycopersicum*	Tucci et al. (2011)
	–	*Fragaria* × *ananassa*	Mutiya and Prilya (2017)
	R. solani	*Oryza sativa*	Chakravarthy and Nagamani (2007)
	Fusarium and *Phoma* sp.	*Solanum tuberosum*	Gogoi et al. (2007)
	F. oxysporum and *F. solani*	*Phaseolus vulgaris*	Abd-El-Elshahawy et al. (2019)
T. atroviride	*R. solani*	*Cucumis sativus*	Yobo (2005)

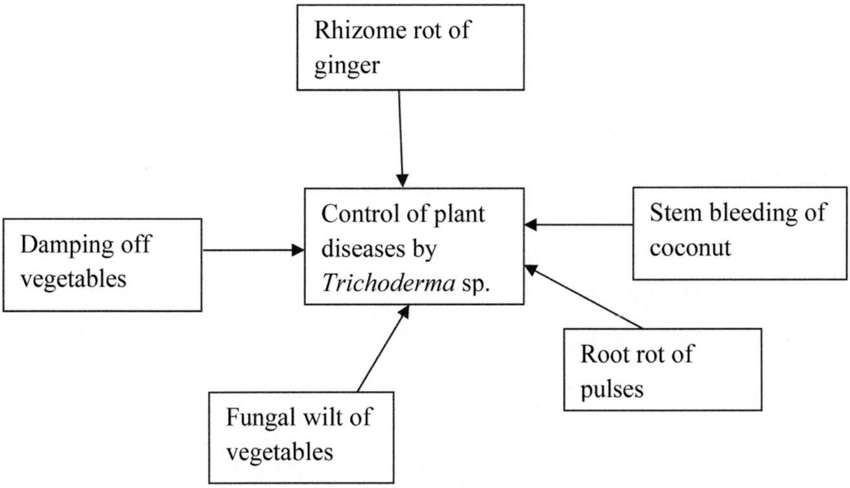

FIGURE 1.6 Control of plant diseases by *Trichoderma*.

1.9 CONCLUSION AND FUTURE PROSPECTS

Trichoderma is excellent biopesticide and will have immense use in agriculture. This will be very easy, eco-friendly, and economical. The use of *Trichoderma* as biopesticide is safe to the humans and to the environment. The *Trichoderma*-based biopesticide is an organic product as it is prepared using microbial approach. It will also help in the growth of various plants and crops. *Trichoderma*-based biopesticide will also protect many plants and crops against various phytopathogens. The farmers will benefit due to the use of *Trichoderma*-based biopesticide. The biopesticides will have an immense advantage in the global market. The use of biopesticides will be definitely very fruitful in the control of insects, nematodes, and pests. The biological approach is thus always interesting and safe as compared to the use of hazardous and toxic chemicals. The ban of chemicals is safe for the human and environment. The research on biopesticides should be carried on large-scale, and the molecular aspect related to the exact mechanisms of biopesticides needs to be studied. New biopesticides should also be developed for the control of insects, pests, and nematodes on crops and other plants. The availability of good quality *Trichoderma* biopesticides at an economical price should be maintained. The awareness regarding biopesticide enterprises should be increased.

KEYWORDS

- ***Bacillus thuringiensis***
- **induced systemic resistance**
- **integrated pest management**
- **pesticides**
- **potato dextrose agar**
- **solid state fermentation**
- **sustainable agriculture**

REFERENCES

Abd-El-Khair, H., Elshahawy, I. E., & Haggag, H. K., (2019). Field application of *Trichoderma* spp. combined with thiophanate-methyl for controlling *Fusarium solani* and *Fusarium oxysporum* in dry bean. *Bull. Natl. Res. Cent., 43*, 1–9.

Al-Ani, L. T., (2018). *Trichoderma*: Beneficial role in sustainable agriculture by plant disease management. In: *Plant Microbiome: Stress Response* (pp. 105–126). Springer.

Arora, A., Kaur, P., Kumar, M., & Saini, V., (2017). Production of biopesticides namely *Trichoderma viride* and *Beauveria bassiana*. *Ind. J. Sci. Technol., 10*, 1–7.

Badar, R., & Qureshi, S. A., (2012). Comparative effect of *Trichoderma hamatum* and host-specific *Rhizobium* species on growth of *Vigna mungo*. *J. Appl. Pharm. Sci., 2*, 128–132.

Brar, S. K., Verma, M., Tyagi, R. D., & Valero, J. R., (2006). Recent advances in downstream processing and formulations of *Bacillus thuringiensis* based biopesticides. *Proc. Biochem., 41*, 323–342.

Brotman, Y., Lisec, J., Méret, M., Chet, I., Willmitzer, L., & Viterbo, A., (2012). Transcript and metabolite analysis of the *Trichoderma*-induced systemic resistance response to *Pseudomonas syringae* in *Arabidopsis thaliana*. *Microbiol.*, 139–146.

Chakravarthy, S. K., & Nagamani, A., (2007). Efficacy of non-volatile and volatile compounds of *Trichoderma* species on *Rhizoctonia solani*. *J. Mycol. Plant Pathol., 37*, 82–86.

Chandler, D., Bailey, A., Tatchell, G. M., Davidson, G., Greaves, J., & Grant, W. P., (2011). The development, regulation and use of biopesticides for integrated pest management. *Phil. Trans. Royal Soc. Bull., 386*, 2–13.

Chandrasekaran, R., Revathi, K., Nisha, S., Kirubakaran, S. A., Sathish-Narayanan, S., & Senthil-Nathan, S., (2012). Physiological effect of chitinase purified from *Bacillus subtilis* against the tobacco cutworm *Spodoptera litura* fab. *Pestic. Biochem. Physiol., 104*, 65–71.

Chaudhary, R. G., Shukla, N., & Prajapati, R. K., (2004). Biological control of soil-borne diseases: An update in pulse crops. In: Shahid, & Narain, (eds.), *Eco-friendly Management of Plant Diseases* (pp. 178–200). Daya Publishing House, New Delhi.

Dey, U., Harlapur, S. L., & Dhutraj, D. N., (2013). Bioassay of fungicides, botanicals, bioagents and indigenous technology knowledge against *Pyricularia oryzae* Cav. causal agent of blast of rice. *J. Plant Dis. Sci., 8*, 102–106.

Druzhinina, I., Seidl-Seiboth, V., Herrera-Estrella, A., Horwitz, B., Kenerley, C., Enrique, M. E., Mukherjee, P., et al., (2011). *Trichoderma*: The genomics of opportunistic success. *Nat. Rev. Microbiol., 9*, 749–759.

El-Hassan, S. A., Gowen, S. R., & Pembrok, B., (2013). Use of *Trichoderma hamatum* for biocontrol of lentil wilt diseases. *J. Plant Prot. Res., 53*, 12–17.

Filizola, P., Luna, M., de Souza, A., Coelho, I., Laranjeira, D., & Campos-Takaki, G., (2019). Biodiversity and phylogeny of novel *Trichoderma* isolates from mangrove sediments and potential of biocontrol against *Fusarium* strains. *Microbial Cell Fact., 18*, 1–14.

Fiorentino, N., Ventorino, V., Woo, S. L., Pepe, O., De Rosa, A., Gioia, L., Romano, I., et al., (2018). *Trichoderma*-based biostimulants modulate rhizosphere microbial populations and improve N uptake efficiency, yield, and nutritional quality of leafy vegetables. *Front. Plant Sci., 9*, 1–15.

Gajera, H., Domadiya, R., Patel, S., Kapopara, M., & Golakiya, B., (2013). Molecular mechanism of *Trichoderma* as biocontrol agents against phytopathogen system: A review. *Curr. Res. Microbiol. Biotechnol., 1*, 133–142.

Gal-Hemed, I., Atanasova, L., Komon-Zelazowska, M., Druzhinina, I. S., Viterbo, A., & Yarden, O., (2011). Marine isolates of *Trichoderma* spp. as potential halotolerant agents of biological control for arid-zone agriculture. *Appl. Environ. Microbiol., 77*, 5100–5109.

Glare, T., Caradus, J., Gelernter, W., Jackson, T., Keyhani, N., Kohl, J., Marrone, P., Morin, L., & Stewart, A., (2012). Have biopesticides come of age? *Trends Biotechnol., 30*, 250–258.

Gogoi, R., Saikia, M., Helim, R., & Ullah, Z., (2007). Management of potato diseases using *Trichoderma viride* formulations. *J. Mycol. Plant Pathol., 37*, 227–230.

Gu, X., Chen, W., Cai, F., Pang, G., & Li, R., (2016). Effect of *Trichoderma* biofertilizer on continuous cropping cucumber cultivation with reduced rates of chemical fertilizer application. *Acta Pedol. Sin., 53*, 1296–1305.

Gupta, S., & Dikshit, A. K., (2010). Biopesticides: An ecofriendly approach for pest control. *J. Biopest., 3*, 186–188.

Gveroska, B., & Ziberoski, J., (2012). *Trichodema harzianum* as a biocontrol agent against *Alternaria altermata* on tobacco. *Appl. Technol. Innov., 7*, 67–76.

Hamid, B., & Mohiddin, F. A., (2018). *Trichoderma* as a potential biocontrol agent. *Int. J. Adv. Res. Sci. Eng., 7*, 68–76.

Hiba, A., Laid, D., Omrane, Z., Sabri, A., & Thonart, P., (2019). Production of *Trichoderma harzianum* (127a and 127b) spores by fermentation (LF and SSF). *Int. J. Innov. Appr. Agric. Res., 3*, 376–384.

Ibarra-Medina, V. A., Ferrera-Cerrato, R., Alarcon, A., Lara-Hernandez, M. E., & Valdez-Carrasco, J. M., (2010). Isolation and screening of *Trichoderma* strains antagonistic to *Sclerotinia sclerotiorum* and *Sclerotinia minor*. *Revista Mexicana de Micologia Impresa en Mexico, 31*, 53–63.

Jagtap, G. P., Mali, A. K., & Utpal, D., (2013). Bioefficacy of fungicides, bio-control agents and botanicals against leaf spot of turmeric incited by *Colletotrichum capsici*. *Afr. J. Microbiol. Res., 7*, 1865–1873.

Kachhawa, D., (2017). Microorganisms as biopesticides. *J. Entomol. Zool. Stud., 5*, 468–473.

Kamal, R. K., (2018). *Trichoderma*: A most common biofertilizer with multiple roles in agriculture. *Biomed. J. Sci. Tech. Res., 4*, 4136–4137.

Kandpal, V., (2014). Biopesticides. *Int. J. Environ. Res. Develop., 4*, 191–196.

Khan, M. Y., Haque, M. M., Molla, A. H., Rahman, M. M., & Alam, M. Z., (2017). Antioxidant compounds and minerals in tomatoes by *Trichoderma*-enriched biofertilizer and their relationship with the soil environments. *J. Integr. Agric., 16*, 691–703.

Khandelwal, M., Datta, S., Mehta, J., Naruka, R., Makhijani, K., Sharma, G., Kumar, R., & Chandra, S., (2012). Isolation, characterization and biomass production of *Trichoderma viride* using various agro products: A biocontrol agent. *Pelagia Res. Lib., 3*, 3950–3955.

Khatabi, B., Molitor, A., Lindermayr, C., Pfiffi, S., Durner, J., Von, W. D., Kogel, K., & Schafer, P., (2012). Ethylene supports colonization of plant roots by the mutualistic fungus *Piriformospora indica*. *PLoS One, 7*, 1–8.

Kumar, A., Patel, A., Singh, S., & Tiwari, R., (2019). Effect of *Trichoderma* spp. in plant growth promotion in chilli. *Int. J. Curr. Microbiol. Appl. Sci., 8*, 1574–1581.

Kumar, G., Maharshi, A., Patel, J., Mukherjee, A., Singh, H., & Sarma, B., (2017). *Trichoderma*: A potential fungal antagonist to control plant diseases. *SATSA Mukhapatra, 21*, 206–218.

Kumar, S., & Singh, A., (2015). Biopesticides: Present status and the future prospects. *J. Fertil. Pestic., 6*, 1–2.

Kumar, S., (2012). Biopesticides: A need for food and environmental safety. *J. Biofertil. Biopestic., 3*, 4.

Kumar, S., Palanisamy, V., & Pandian, R., (2014b). *Trichoderma: A Biocontrol Weapon for Sustainable Agriculture* (Vol. 2, pp. 143–148). Popular Kheti.

Kumar, S., Thakur, M., & Rani, A., (2014a). *Trichoderma*: Mass production, formulation, quality control, delivery and its scope in commercialization in India for the management of plant diseases. *Afr. J. Agric. Res., 9*, 3838–3852.

Levy, N. O., Meller, H. Y., Haile, Z. M., Elad, Y., David, E., Jurkevitch, E., & Katan, J., (2015). Induced resistance to foliar diseases by soil solarization and *Trichoderma harzianum*. *Plant Pathol., 64*, 365–374.

Lorito, M., Woo, S. L., Harman, G. E., & Monte, E., (2010). Translational research on *Trichoderma*: From omics to the field. *Annu. Rev. Phytopathol., 48*, 395–417.

Mahato, S., Bhuju, S., & Shrestha, J., (2018). Effect of *Trichoderma viride* as biofertilizer on growth and yield of wheat. *Malaysian J. Sust. Agric., 2*, 1–5.

Marques, E., Martins, I., & Marques De, M. S., (2018). Antifungal potential of crude extracts of *Trichoderma* spp. *Biota Neotrop., 18*, 1–5.

Mathew, S. K., Mary, C. F., Gleena, G. K., & Surendra, G. D., (2011). Antagonistic activity of endophytic *Trichoderma* against *Phytophthora* rot of black pepper (*Piper nigrum* L.). *J. Biol. Cont., 25*, 48–50.

Ming, S., Rong, J., Zhang, C., Li, C., Zhang, C., Zhang, Y., Zhou, R., & Li, G., (2019). The solid fermentation state's optimization of *Trichoderma harzianum* M1. *IOP Conference Series: Materials Science and Engineering,* 022111.

Mishra, R. K., Bohra, A., Kamaal, N., Kumar, K., Gandhi, K., Sujayanand, G., Saabale, P., et al., (2018). Utilization of biopesticides as sustainable solutions for management of pests in legume crops: Achievements and prospects. *Egyptian J. Biol. Pest Cont., 28*, 1–11.

Mishra, R. K., Saabale, P. R., Naimuddin, Jagadeeswaran, R., & Mishra, O., (2015). *Potential Trichoderma sp. from Pulses Rhizosphere* (p. 3). Pulses Newsletter.

Monte, E., (2001). Understanding *Trichoderma*: Between biotechnology and microbial ecology. *Int. Microbiol., 4*, 1–41.

Mutia, D., & Prilya, F., (2017). Exploration of *Trichoderma* spp. and fungal pathogen that causes a strawberry anthracnose and examination of *in-vitro* antagonistic activity. *Biotika, 5*, 58–68.

Oguh, C. E., Okpaka, C. O., Ubani, C. S., Okekeaji, U., Joseph, P. S., & Amadi, E. U., (2019). Natural pesticides (Biopesticides) and uses in pest management: A critical review. *Asian J. Biotechnol. Genet. Eng., 2*, 1–18.

Ortuno, N., Castillo, J., Miranda, C., Claros, M., & Soto, X., (2017). The use of secondary metabolites extracted from *Trichoderma* for plant growth promotion in the Andean highlands. *Renew Agric. Food Syst., 32*, 366–375.

Pal, G. K., & Kumar, B., (2013). Antifungal activity of some common weed extracts against wilt causing fungi, *Fusarium oxysporum*. *Curr. Discov., 2*, 62–67.

Patel, J., Sarma, B., Singh, H., Upadhyay, R., Kharwar, R., & Ahmed, M., (2016). *Pseudomonas fluorescens* and *Trichoderma asperellum* enhance expression of Gα subunits of the pea heterotrimeric G-protein during *Erysiphe pisi* infection. *Front. Plant Sci., 6*, 1–13.

Patil, A. S., Patil, S. R., & Paikrao, H. M., (2016). *Trichoderma* secondary metabolites: Their biochemistry and possible role in disease management. In: Choudhary, D., & Varma, A., (eds.), *Microbial-Mediated Induced Systemic Resistance in Plants* (pp. 69–102). Springer, Singapore.

Prabha, S., Yadav, A., Kumar, A., Yadav, A., Yadav, H., Kumar, S., Yadav, R., & Kumar, R., (2016). Biopesticides: An alternative and eco-friendly source for the control of pests in agricultural crops. *Plant Arch., 16*, 902–906.

Prakash, N., Kumari, B., & Karn, S., (2019). Biopesticides: Introduction and its prospects. *Int. J. Curr. Microbiol. Appl. Sci., 8*, 2960–2964.

Ramteke, A. S., (2019). *Trichoderma* species: Isolation, characterization, cultivation and their role as effective biopesticides. *Int. J. Res. Rev., 6*, 315–320.

Reyes, Y., Infante, D., Garcia-Borrego, J., Del, P. E., Cruz, A., & Martínez, B., (2012). Compatibilidad de *Trichoderma asperellum* Samuels con herbicidas de mayor uso en el cultivo del arroz. *Proteccion Veg., 27*, 45–53.

Rikita, B., & Utpal, D., (2014). An overview of fungal and bacterial biopesticides to control plant pathogens/diseases. *Afr. J. Microbiol. Res., 8*, 1749–1762.

Ruiu, L., (2018). Microbial biopesticides in agroecosystems. *Agron., 8*, 1–12.

Sachdev, S., & Singh, R. P., (2020). *Trichoderma*: A multifaceted fungus for sustainable agriculture. In: Bauddh, K., Kumar, S., Singh, R., & Korstad, J., (eds.), *Ecological and Practical Applications for Sustainable Agriculture*. Springer, Singapore.

Sala, A., Artola, A., Sanchez, A., & Barrena, R., (2020). Rice husk as a source for fungal biopesticide production by solid-state fermentation using *B. bassiana* and *T. harzianum. Bioresour. Technol., 296*, 122322.

Sala, A., Barrena, R., Artola, A., & Sanchez, A., (2019). Current developments in the production of fungal biological control agents by solid-state fermentation using organic solid waste. *Crit. Rev. Environ. Sci. Technol., 49*, 655–694.

Santos-Villalobos, S., Hernandez-Rodriguez, L., Villasenor-Ortega, F., & Pena-Cabriales, J., (2012). Production of *Trichoderma asperellum* T8a spores by a "home-made" solid-state fermentation of mango industrial wastes. *Bioresour., 7*, 4938–4951.

Savitha, M. J., & Sriram, S., (2015). Morphological and molecular identification of *Trichoderma* isolates with biocontrol potential against *Phytophthora* blight in red pepper. *Pest Manage. Horti. Ecosyst., 21*, 194–202.

Senthil-Nathan, S., (2015). A review of biopesticides and their mode of action against insect pests. In: Thangavel, P., & Sridevi, G., (eds.), *Environmental Sustainability* (49–63). Springer, India.

Shabir-U-Rehman, Dar, W. A., Ganie, S. A., Javid, A. B., Mir, H. G., Lawrence, R., Sumati, N., & Singh, P. K., (2013). Comparative efficacy of *Trichoderma viride* and *Trichoderma harzianum* against *Fusarium oxysporum* f. sp. *ciceris* causing wilt of chickpea. *Afr. J. Microbiol., Res., 7*, 5731–5736.

Sharma, P., Sharma, M., Raja, M., & Shanmugam, V., (2014). Status of *Trichoderma* research in India: A review. *Ind. Phytopath., 67*, 1–19.

Sharma, S., & Malik, P., (2012). Biopesticides: Types and applications. *Int. J. Adv. Pharm. Biol. Chem., 1*, 508–515.

Singh, A., Jain, A., Sarma, B. K., Upadhyay, R. S., & Singh, H. B., (2014). Rhizosphere competent microbial consortium mediates rapid changes in phenolic profiles in chickpea during *Sclerotium rolfsii* infection. *Microbiol. Res., 169*, 353–360.

Singh, V., Joshi, B. B., Awasthi, S. K., & Srivastava, S. N., (2008). Eco-friendly management of red rot disease of sugarcane with *Trichoderma* strains. *Sugar Tech., 10*, 158–161.

Su, D., Ding, L., & He, S., (2018). Marine-derived *Trichoderma* species as a promising source of bioactive secondary metabolites. *Mini-Rev. Med. Chem., 18*, 1702–1713.

Suparno, Sukoso, Hakim, L., & Aidawati, N., (2016). *Trichoderma* spp. as agent of biocontrol in local rice diseases in Tidal Swamplands in South Kalimantan, Indonesia. *J. Agric. Veter. Sci., 9*, 1–6.

Thapa, S., Sotang, N., Limbu, A., & Joshi, A., (2020). Impact of *Trichoderma* sp. in agriculture: A mini-review. *J. Biol. Today's World, 9*, 1–5.

Tijjani, A., Bashir, K. A., Mohammed, I., Muhammad, A., Gambo, A., & Musa, H., (2016). Biopesticides for pests control: A review. *J. Biopest. Agric., 3*, 6–13.

Tucci, M., Ruocco, M., De Masi, L., De Palma, M., & Lorito, M., (2011). The beneficial effect of *Trichoderma* spp. on modulated by the plant genotype. *J. Mol. Plant Pathol., 12*, 341–354.

Tvetdyukov, A. P., Nikonov, P. V., & Yushchenko, N. P., (1994). *Trichoderma. Rev. Plant Pathol., 739*, 273.

Utami, U., Nisa, C., Putri, A., & Rahmawati, E., (2019). The potency of secondary metabolites endophytic fungi *Trichoderma* sp. as biocontrol of *Colletotrichum* sp. and *Fusarium oxysporum* causing disease in chili. *International Conference on Biology and Applied Science (ICOBAS), AIP Conference Proceedings 2120* (pp. 1–6). AIP Publisher.

Waghunde, R., Shelake, R., & Sabalpara, A., (2016). *Trichoderma*: A significant fungus for agriculture and environment. *Afr. J. Agric. Res., 11*, 1952–1965.

Woods, T. S., (2003). Pesticide formulations. In: *AGR 185 in Encyclopedia of Agrochemicals* (pp. 1–11). John Wiley and Sons, New York.

Yedidia, I., Shoresh, M., Kerem, Z., Benhamou, N., Kapulnik, Y., & Chet, I., (2003). Concomitant induction of systemic resistance to *Pseudomonas syringae* pv. lachrymans in cucumber by *Trichoderma asperellum* (T-203) and accumulation of phytoalexins. *Appl. Environ. Microbiol., 69*, 7343–7353.

Yobo, K. S., (2005). *Biological Control and Plant Growth Promotion by Selected Trichoderma and Bacillus Species.* PhD Thesis, KwaZulu-Natal University, South Africa.

Zeilinger, S., Gruber, S., Bansal, R., & Mukherjee, P., (2016). Secondary metabolism in *Trichoderma*-chemistry meets genomics. *Brit. Mycol. Soc., 30*, 74–90.

CHAPTER 2

ARBUSCULAR MYCORRHIZAS: APPLICATIONS IN ORGANIC AGRICULTURE AND BEYOND

CHARLES OLUWASEUN ADETUNJI,[1]
OSIKEMEKHA ANTHONY ANANI,[2] DEVARAJAN THANGADURAI,[3]
and SAHER ISLAM[4]

[1]*Microbiology, Biotechnology, and Nanotechnology Laboratory, Department of Microbiology Edo State University Uzairue, Auchi, Nigeria*

[2]*Laboratory of Ecotoxicology and Forensic Biology, Department of Biological Science, Faculty of Science, Edo State University Uzairue, Auchi, Nigeria*

[3]*Department of Botany, Karnatak University, Dharwad, Karnataka, India*

[4]*Institute of Biochemistry and Biotechnology, Faculty of Biosciences, University of Veterinary and Animal Sciences, Lahore, Pakistan*

ABSTRACT

The role of beneficial microorganisms, including arbuscular mycorrhizal fungi (AMF), has been recognized as a feasible, economical, and reliable biotechnological tool that could contribute towards the achievement of sustainable organic agriculture. Some of the important functions of AMF in sustainable organic agriculture include: (i) improvement of plant nutrition through enhancing nutrient uptake; (ii) improvement of soil texture and structure; (iii) ecorestoration of heavily polluted soil; (iv) mitigation of biotic and abiotic stresses; and (v) application as biopesticides in place of synthetic pesticides. This chapter reviews the modes of action employed by AMF and provides some recommendations on areas where further studies are needed to be able to fully harness the potential of these products.

2.1 INTRODUCTION

The significant role of soil microorganisms for the improvement of plant health, soil fertility, inhibition of plant pathogenic microorganisms, increase in the biomass of crops and their effect against the effect of climate change have severely been demonstrated. Such beneficial microorganisms include arbuscular mycorrhizal fungi (AMF). These organisms play a crucial role in enhancing the uptake of nutrients by crops. This they do by enhancing the absorptive surface area of the host plant root systems. The hyphae of these symbiotic fungi also possess the capacity to increase the surface area for enhanced interaction with other beneficial microorganisms. Additionally, they possess the capability to improve the process of translocation in plants (Gianinazzi et al., 2002).

Several anthropogenic activities have been observed to affect the natural functions of mycorrhizosphere organisms. These include intensive agricultural practices such as tillage (McGonigle and Miller, 1996; Sturz, et al., 1997) and application of synthetic pesticides (Gianinazzi et al., 1994, 2002; Adetunji et al., 2019a–d; Adetunji, 2019; Adetunji and Ugbenyen, 2019). These activities have over time led to a drastic reduction of beneficial microbial diversity and activity (Meader et al., 2002). There is therefore need to adopt agricultural practices that could encourage the shift from conventional intensive management to those that could support sustainable farming. AMF have been reported to develop a strong symbiotic relationship with several plant species (Read et al., 1992; Richardson et al., 2000; Sramek et al., 2000). Specifically, about 6,000 fungal species have the capacity to develop mycorrhizas with about 240,000 plant species (Bonfante and Perotto, 1995). The application of AMF has the capacity to be utilized for several developmental programs, and most especially for low-input agriculture (Schreiner and Bethlenfalvay, 1996).

It has been reported that the occurrence of AMF in plants varies depending on the types of crops, which may be 79% of monocotyledonous and 83% of a dicotyledonous plant. They also occur in about 1,000 genera of plants from 200 families. Some other benefits of AMF in sustainable agriculture include enhanced plant production, increase in the production of growth-regulating substances (Danneberg et al., 1992), facilitating rapid uptake of nutrients-most especially micronutrients and phosphorus (Douds and Millner, 1999), development of high resistance to pests and diseases (Ho, 1998), enhancement of soil aggregate stability (Azizah and Idris, 1996), remediation of toxic metals (Bonifacio et al., 1999), absorption of water (Jastrow et al., 1998), improvement of N_2-fixation through increased P supply (Kucey and Paul, 1982), absorption of essential micronutrients (Marschner and Dell, 1994), improvement of water uptake and, osmotic adjustment when exposed to drought stress (Masri, 1997). Comprehensive knowledge about the role of AMF is essential in helping institute measures for a sustainable enhancement of soil fertility and an increase in crop production.

This chapter provides a comprehensive review of the application of AMF in organic agriculture, their modes of action, and the role they play in the ecorestoration of heavily polluted soils. Future perspectives of AMF are also elaborated, with the mechanisms of action described in detail. The practical application of AMF is also elaborated using specific examples.

2.2 APPLICATIONS OF ARBUSCULAR MYCORRHIZAL FUNGI (AMF)

Arbuscular mycorrhiza is one of a group of fungi that colonizes the roots of higher plants. It infiltrates the cortical chambers of the roots and forms arbuscles (networks of hyphae in certain fungi) in the vascular tissues. These arbuscles aid the fungi to absorb nutrients such as micronutrients [P (Phosphate), S (Sulphur), N (Nitrogen)], and trace-nutrients from the soil. One of the benefits of AM is therefore its utilization as agro-fertilizer (Timmer and Leyden, 1980; Grant, 2005; Berruti et al., 2016).

2.3 APPLICATION AMF AS BIOFERTILIZER

Laila et al. (2019) investigated the effect of AMF on *Allium cepa* L. as a permanent replacement to phosphorus synthetic fertilizer. Maximum yields were recorded in the plant treated with AMF, attesting to its potential for use as a fertilizer. Carina et al. (2016) evaluated the effect of co-inoculation of *Azospirillum* and AMF on micro-propagated banana seedlings. The result obtained indicates that 1.5×10^8 CFU/ml of *Azospirillum* and 200 g of AMF had the highest growth and yield output when compared to the control plant. Their study showed a synergetic effect between the fungus-bacteria inoculum used during this experiment. The high doses showed an enhanced nutritional improvement. After all, it gives that plant a better nitrogen and phosphorus absorption because it enables the microorganism to solubilize the nutrients available for the plant nutrient uptake when compared to the treatment without microbial inoculum.

Igiehon and Babalola (2017) did a review on the sustainable utilization of AMF as biofertilizers. They noted that AMF could help in the reduction of wide usage of inorganic chemical fertilizers. AM aids in the uptake of chemical nutrients in the soil and transports them to various vascular tissues in the plant with the aid of a phosphate trans-mobilizer. They stressed the cons facing the possible massive production of AM because of the inhibition of plant-host microbes. However, the utilization of AMF in several biological controlled settings (greenhouse and field assay trials) indicated that it has persisted and portends the capability to colonize some tested plants like soybeans while they could also enhance the growth and development in all ramifications. The authors stated that AM has been proposed as a first-class new generation bio-fertilizer because it could contribute to the control and upkeep of agronomic expansion, aid in the controlling of some soil degradation factors, augmenting phytoremediation and elimination of other microbes that might hinder the growth of the root system.

Sadhana (2014) wrote an extensive review of the utilization of AM as a bio-fertilizer. The author demystifies those major microbes (cyanobacteria, bacteria, and fungi) are used as bio-fertilizers to boosting the soil quality and vigor for speedy growth and development. That plant plays some symbiotic association with most of these microbes. Apart from the other microorganisms and their related strains, a species of fungi AM (arbuscular mycorrhizal), has an extensive and unique potential feature above the others. The author stated that the sporulation of AMF is controlled by certain soil edaphic and climatic conditions. That soil-based pot assay is the most widely biological controlled method in the fungal inoculum production of AM. The authors recommend that a plant's growth and nutritional requirements solely reside on the role AM plays in forestry and agricultural purposes. To boost the plants' efficiency towards adverse edaphic and climatic conditions, AMF can be utilized in dual inoculum forms; Rhizobium, and other bacteria species in different field trials to test its efficacy and future advances.

Junior et al. (2019) tested and evaluated how agroecological coffee management influences the positive utilization and diversity of AMF. The authors stressed the ecosystem services' potential and biological importance of AMF in modern agricultural practices to reduce the impacts of nutrient deficiency and boost plant growth and development. The biological controlled experiments were done during the phenological phases (flowering, grain filling, harvesting) of the coffee plant. Different root and soil samples were obtained from three controlling systems and areas. The asexual reproductive bodies (Spores) from the fungi community were removed, identified, and were assayed using a PCR-DGGE (polymerase chain reaction denaturing gradient gel electrophoresis). The results from the dendrograms and bio-informatics indicated no differences between the management systems and the richness of the morphospecies, the duration, and areas. However, a little information about some tropical species was noted. The PCR-DGGE (molecular analysis) also indicated that the organic forest was similar to the agroecological controlling structure and a high variety of indices were observed in conservative management. The findings from this study revealed that there is a need to relate traditional taxonomic valuations with genetic methods because diverse methods can result in dissimilar results. In conclusion, the authors provided suggestions for sustainable agricultural management via agroecological coffee management.

Baslam et al. (2011) tested and evaluated the impact and association of AMF on the development and growth of lettuce in a greenhouse setting. The results of the tested biological controlled experiment indicated that the AM elicited the growth of the lettuce plant, therefore, boosting other macro and micronutrients (carotenoids, Ca, anthocyanins, Fe, phenolics, Mn, and Cu) in the tested plant than a none AM symbiotic plants.

Oladele and Awodun (2014) tested and evaluated the reaction of lowland *Oryza sativa* to bio-fertilizers inoculated with AMF and its impacts on plant growth and yield. The results revealed a significant difference in the impact of the bio-fertilizer, showing positive development and growth on the reproductive and vegetative parts of the rice plant.

Casazza et al. (2017) tested and evaluated the richness and variety of AMF and how they are related to the soil interaction talus gradient in terms of plant sustenance and tension tolerance in Florida greeneyes plant. Samples of the plants were examined and analyzed via molecular assay to determine the community structure of AM in the plant. The results of the biological controlled experiment revealed that the AM in Florida greeneyes roots was subjugated by Endomycorrhizal fungus common to three talus gradient. The septate endophytes were totally different from the ones formed by AM as well as the vegetation coverage and the percentage (%) root colonization did not impact the plant community positively. The results of the abundance of AM in the root system of the Florida green eyes showed that there was a positive correlation between the physical and chemical properties of the soil, therefore indicating the effect of AM on the quality of the soil. Hence, the degree of this influence was also correlated with the NMDS (non-metric multidimensional scaling) initiation of AM showing strong diversity and community composition at various regions both on the soil quality and the talus gradient. In conclusion, the authors opined that gradient prompted the physical disruption of the mountain taluses, which may also result in the assortment of disorders in the AM group and finally led to different fungi-plant congregations.

Yeasmin (2017) tested and evaluated the relationship between AM as a bio-fertilizer and plant growth. The author stressed the need for sustainable farming using organic fertilizer; bio-fertilizer to replace the widely used inorganic ones and to combat the food insecurity caused by soil infertility, nutrient instability, pest resistance, and soil degradation. The introduction of beneficial microorganisms such as AM fungi can remediate the dire need for food security. The results of the symbiotic association of the AM in the root mottled between 10% and 90% with 85% of the plants surveyed. It was observed that the AM fungi played a significant role in the growth rate of the plant as well as protection. The rate of the soil phosphorus greatly increased, and there was a positive correlation between the AM fungi with potassium, calcium, total nitrogen, soil texture, water holding capacity, phosphate, and soil pH which as well greatly influences the plant growth. The spores of the vesicular AM fungi were later abstracted using the wet sieving technique; slope centrifugation. The plant garden croton was used as a stock inoculum of AM fungi for bio-fertilizer. The results indicated a sharp improvement in plant growth, rapid productivity, nutrients stability, and crop yield. In conclusion, the author recommended AM fungi as a future sustainable green bio-fertilizer candidate for reducing edaphic and environmental problems related to inorganic chemicals.

Andrade et al. (2009) did a review on the relationship between AM and coffee plants. The authors stressed the need to look at the relationship between the subject areas because of the paucity of literature. However, because of the economic importance of coffee species, AM association or utilized as a bio-fertilizer can improve the dietary status and lessen environmental stress commonly faced by the plant species in the tropics. The authors also elucidate and evaluate the natural manifestation of AM fungi in varied soil types in cultivated coffee farms and propose an advance survey on the prospective benefits it tends to portend to related plant species.

Itelima et al. (2018) did a review on the major role of endo and ecto AM fungi in eliciting and boosting soil fertility and plant productivity. The authors heralded that plant nutrient is significant in the sustenance of the ever-teeming growing population. That of recent, chemical fertilizers have resulted in serious degradation of essential environmental parameters in the ecosystem, and humans are on the verge of ecological and health risks. A sustainable agricultural practice of using an ecosystem friendly, eco-safety, non-toxic, and economical bio-fertilizer for farming are not negotiable at this extant food security era! The authors heralded that bio-fertilizers can serve as essential ingredients in nutrient management. In conclusion, the authors propose AM as a potential bio-fertilizer in improving certain non-living (abiotic) and living (biotic) factors that militating against plant growth. That it will help farmers to manage the cost of procuring inorganic chemicals, thus protecting the ecosystem, and bringing sustainable eco-balance and productivity of farm plants.

Mycorrhiza formation is recognized to change numerous features of plant physiology, including hormonal balance, mineral nutrient constituents, and carbon distribution patterns. Demür (2004) evaluated the influence of AMF *Glomus intraradices* Schenck and Smith on the physiological development of pepper. For the proper description of the physiological growth of the target plant to be carried out, some physiological growth parameters were evaluated in the leaves and the shoots of nonmycorrhizal (NM) and mycorrhizal (M). It was observed that all the evaluated physiological parameters were enhanced in the M pepper plants (by 12%–47%) when compared to NM plants. It was also discovered that the level of phosphorus concentration showed a positive correlation with sugar and chlorophyll contents. Their study showed that enhanced phosphorus concentration might be linked to the mycorrhizal symbioses with positively influence the physiological activity of pepper plants.

Berruti et al. (2016) reviewed the past achievements of AMF as an organic bio-fertilizer. AMF is one of the obligate biotrophic fungi that are symbiotic to about 80% of higher plants. AMF is considered a bio-fertilizer because of the roles it plays in nutrients and water uptake in plants, anabolic processes such as photosynthesis, and protecting plants from disease-causing organisms. In other words, it plays self-ecosystem services in re-instituting and balancing the missing nutrient links and an alternative to the conventional inorganic fertilizer in agriculture. That, the key approach to this method is by the introduction and reintroduction of the inoculates into the targeted soil. The authors stressed the need for a genetic engineering procedure in advancing the genomes of AM to interact with plant-soil host microorganisms that can inhabit its functional role in the ecosystem. The authors, in conclusion, recommend the use of AM in field trials because it has the potential to correct the environmental degradation caused by the wildly used inorganic fertilizers.

Binh et al. (2019) tested and evaluated the effects of AM fungus in the development and sustenance of 15 agronomic plant species (vegetables, legumes, and cereals). The authors reconnoitered varied plant types (C4, C3, dicots, monocots, and non-nitrogen fixer) on how they respond to AM colonization. The results of the study showed that leek plants had more colonization potentials, wet, and dry (biomass) responses, and high nutrient contents compared to the rest of the test plants. More so, the metal (Zn and Cu), phosphate, and sulfate contents were generally enhanced by the inoculation of the AM fungus and the ionomes (inorganic chemical nutrients of cellular organisms in trace composition) fluctuated between the AM fungus and the plant species. In conclusion, the authors propose a further study of ionomics in AM fungus to better elucidate its potential role in prospective agricultural practices to standardize growth and development in crops.

Knerr et al. (2018) tested and evaluated AM fungi associated with natural and conventional *Allium cepa* grown in the Pacific Northwest USA. In this study, the authors compared the AM fungi associations with the root colonization in field trials fumigated with metam-sodium vs. non-fumigated fields. The results of the biological controlled experiment revealed that the AM fungi colonize the roots of *Allium cepa* in the midsummer in the conservative fields (67 vs. 51%) than the non-fumigated conservative fields (45 vs. 67%). A molecular assay using a pyro-sequencing denoted 4 AM fungi of different orders: Diversispora, Claroideoglomus, Glomus, and Paraglomus, and four main active taxonomic units (ATUs); [Glomus (MO_G17), *Funneliformis mosseae*, Whitfield type 17 Glomus, and *Claroideoglomus lamellosum*]. The results of the AM fungi gotten from plants at the midsummer were richer in organic acid than the conventional form, and the diversity was also greater. There was no significant influence of organic *vs*. conservative plants on main ATUs. However, little richness of ATUs was observed in the organic fields, but not in conventional fields. The findings of the study indicated that there was no significant impact of metam-sodium chemigation on the AM fungi groups found in the *Allium cepa* crops. The authors recommend sustainable agricultural practices by the utilization of natural and conservative methods in cultivating *Allium cepa* and the usage of metam sodium to fumigate soil does not seem to be the chief teamsters of AM fungi populations.

2.4 AMF USED IN THE ECORESTORATION OF HEAVILY POLLUTED SOIL, DISEASE, AND ENVIRONMENTAL STRESS

Nurbaity et al. (2016) tested and evaluated the application of Endomycorrhizal fungus (*Glomus* sp.) and Gammaproteobacteria (*Pseudomonas diminuta*) to lessen the utilization of inorganic fertilizers in the cultivation of *Ipomoea batatas*. A biological controlled greenhouse trial was conducted to test the efficacy of the Endomycorrhizal and Gammaproteobacteria on potato cultivated on different soil types (Inceptisols and Andisols) mixed with N (Nitrogen), P (Phosphorus), and K (Potassium) fertilizer of concentration; 0, 25, 50, 75, and 100%. The results of the biological controlled experiment revealed that the application of the Endomycorrhizal and Gammaproteobacteria microbes lessen the inorganic fertilizer (NPK) to 50%, increase the uptake of nutrients, and elicited rapid growth in the *Ipomoea batatas* plant. The results of the responses of the different soils used indicated that Inceptisols had a better positive reaction compared to Andisols. The findings in this study revealed that bio-fertilizers can lessen the impact of inorganic chemical fertilizers. The authors proposed the sustainable utilization of bio-fertilizers to forestall and economically viable plant production and a greener ecosystem for *Ipomoea batatas* production.

Pozo et al. (2002) tested and evaluated the local impact of AM fungi on the security of tomato crops against plant infection and total cleanses of it. Two AM fungi (*G. intraradices* and *G. mosseae*) were used to provoke local resistance to the tomato root disease (*P. parasitica*). The results of the study revealed that *G. mosseae* was able to effectively reduce the disease signs caused by *P. parasitica* contagion. The facts indicated a mixture of both systemic and local mechanisms elicited by bio-protector (AM fungi) impact from the biochemical analysis. These effects were new hydrolytic enzymic defense iso-forms (superoxide dismutase, chitinase, and β-1,3-glucanase, chitosanase) generated by the AM strain to induce fortification against oxidative stress in the tomato plant cell. However, a systemic modification of the action of some of the defense iso-forms was also noticed in some root parts of non-mycorrhizal plants. The results of the root protein analysis also supported the systemic impact of the symbiotic AM fungi on the effective resistance of the tomato plant to the *P. parasitica* disease.

Barea et al. (1998) tested and evaluated the biocontrol impact of the strain of AM fungi in the control of soil-borne plant disease. In this study, 3 (three) strains of AM fungi [F113, F113G22, and F113 (pCU203)] were tested. The result of the study showed that strain F113 which is the wild-form and strain F113G22 which is then transformed type elicited the development of mycelial and spores from *G. mosseae* growing in a colonized tomato. While strain F113 (pCU203) had no negative impact on the mycelial growth, but not to the spore development of *G. mosseae* but, sensitive to a concentration (10 μM) of DAPG, that can cause the production of the rhizosphere. The findings of the study showed that F113 had the ability to produce DAPG (2,4-diacetylphloroglucinol) which is capable of controlling soil-borne plant diseases.

Garg and Chandel (2010) did a review on the sustainable advances of AM fungi in the biocontrol of environmental stressors. The authors opined those environmental variations caused by natural and anthropogenic activities have intensified air, water, and soil pollution, which have led to the contamination of agricultural lands and results in the global reduction of farm yields, capital returns, and food insecurity. The use of novel technology like bio-fertilizer derived from microbes to combat agro-chemicals like heavy metals (HMs) is a restoration model for a sustainable ecosystem. AM fungi play a pivotal role in soil fertility, crop productivity, and the bio-control of environmental contaminants and plant pathogens, the authors stated. They do this by secreting specific enzymes that will induce the plant cell to be resistant to infection and the soil microbes to mutual degrade some chemical structure therein. The authors recounted that in-depth scientific information on the biological composition of AM fungi is still lacking. However, recommend AM as a first-class bio-control organism against environmental stress, heavy metal control, and for the control of plant pathogens because it was environmentally friendly, non-toxic, and cheap.

Hussain et al. (2017) did a review on AM fungi as a biocontrol agent in contemporary sustainable agro practices. The authors condemned the indiscriminate use of inorganic chemicals, which has resulted in various deteriorations of the quality of the environment. The use of inorganic chemicals such as fertilizers have changed the natural settings of the soil and water contents and caused serious ecological and health impacts on the ecosystem. To bring about environmental and food safety, a sustainable approach to agricultural fertilizers; using AM fungi as a bio-fertilizer and bio-control of environmental pollutants and pathogen have been recommended by many schools of scientific thought because it is environmentally safe, non-toxic, and cheap. The authors opined that vesicular AM fungi have the potentials beyond normal usage and portend to be one of the bio-control tools for future sustainable management of problems encountered in modern agriculture. The mode of action of this fungus is via integrated pattern with similar fungi strains and soil bacteria, colonize the roots of the plant and affect the physiological activities of the host plant, thereby accelerating plant nourishment, and improve the soil structure, quality, and health status. The authors, in conclusion, recommend vesicular AM fungi as a special candidate for large scale use as a bio-control agent. However, still, at its nascent stage, it has the potential to correct the anomalies caused by the conversion of inorganic chemicals in the environment.

Cabral et al. (2015) did a review of the phytoremediation potentials of AM fungi on sites contaminated with HMs. The authors recounted the complex role microbes play in decontaminating pollutants in the environment. The role of AM has been stressed by numerous researchers in the bio-control of pathogens, chemicals, and nutrient stabilization in deficient agricultural soils. AM has been considered as the most viable organism because of its broad-spectrum presence in the soil and roots of higher plants and the symbiotic role it plays in their bio-mechanism. The authors remarked that one of the phytoremediation potentials of AM is the possession of natural contrivances in resisting and tolerating heavy and trace metals in the media they habit. Because of this, the authors recommend the utilization of AM fungi because of the benefits it tends to portend for a sustainable phytoremediation purpose.

Garg et al. (2017) did a review on the molecular and physiological mechanism of AM fungi and its ability to tolerate HM in the plant. The authors recounted that AM a symbiotic fungus has been recorded as one of the plant root fungi to reduce HM stress in crops. This is because of its biochemical, physiological, and molecular variations of the rhizosphere (root micro-biome) in the plant they inhabit. The authors backed these facts by establishing that the node of the actions of AM fungi; chelating of noxious metal ions to the wall of the plant cell, secretion of bio-chemical organic compounds (enzymes) such as oxalic and citric acids, glomalin, and glycoprotein have aided it to bind and reduce the HMs absorption in plants. Also, AM fungi likewise, augment the antioxidant defense reactions to offset HMs oxidative pressure. More so, it regulates the genome responsible for nutrient absorption of phytochelatins and metallothioneins, thus stimulating the sequestration of HMs in the mycorrhizal as well in the plant host plant structures. On this ground, the authors recommend AM fungi as a potential cleaning agent of HMs in agricultural soils.

Medina and Azcón (2010) did a review on the efficacy of the utilization of AM fungi as a natural cleanser to advance the performance and quality of plant under serious environmental stress. The authors recounted the stressed plant undergoes serious HMs contamination from polluted sites and adverse weather conditions. That to improve this dire state, AM fungi which have been documented and proposed to be one of the bio-control microbes can be utilized to buffer such conditions, creating a mechanism that will enable the plant to be HMs tolerant and drought resistant. In the same vein, the application of this natural buffer can aid, alter the degradation of the soil characteristics. Some agricultural wastes (rock phosphorus, sugar beet, and dry lime cake), can be used as a mixture or bio-sorption substrate with AM to increase its vigor and bio-control abilities; increase the enzyme production, nutrient availability, tolerance to arid and HMs conditions of the plants. In conclusion, the author proposed an integrated treatment of soil with AM and agricultural waste residues as a re-vegetation method for plants recital in phosphate deficiency soil under serious arid region and the treatment of inorganic waste residues in agricultural land. This will serve as an effective bio-tech apparatus for the recovery of HMs polluted soils.

Hashem et al. (2019) tested and evaluated tolerance of chickpea to arid conditions by using AM fungi as a biocontrol agent. The authors recounted the utilization of AM with biochar to improve the quality of plants to ameliorate some environmental stressors. The biological controlled experiment involved two treatments (AM and bio-char). They were applied as individual and combined treatment(s) of chickpea crop plants. The results of the biological study revealed that the treatments provoked the growth and development of the root and shoot length and other sprouting parts of the plant when inoculated. It was also noticed that the resistance rate of the arid condition also increased, therefore stabilizing the water content and membrane stability index of the plant vascular systems. When both treatments were combined, the rate of anabolism specifically, photosynthesis increased and led to net photosynthetic effectiveness. Individual treatment with AM fungi and bio-char respectively, or in amalgamation, alleviated the poisonous impacts to a significant amount and initiated an important improvement in water content and membrane stability index of the plant vascular systems under standard settings. More so, alterations with AM fungi and bio-char inoculation augmented the N-fixation qualities as well as the amount and load of protuberances, leg-hemoglobin substances, and activity of some enzymes is related to the control plants. The findings of their study indicated that chickpea was able to adapt to drought conditions after the treatment with the mixtures (AM fungi and bio-char), able to regulate the uptake or absorption of phosphate and nitrogen and production of chlorophyll for photosynthetic processes. The authors recommend the bio-char and AM as potential bio-control agents for sustainable agricultural purposes.

Agely et al. (2005) tested and evaluated the use of AM fungi as and hyper-accumulator of toxic Arsenic (As) from the Chinese brake fern. The authors recounted the hyper-accumulative potential of Chinese brake fern from previous literature on its role in the uptake of As from contaminated soils in South America. A greenhouse biological controlled three factorial experiment was conducted with a concentration of As (0, 50, and 100 mg kg^{-1}) and phosphate (0, 25, and 50 mg kg^{-1}) and with or without Chinese brake fern contaminated The results of the biological controlled experiment showed that the AM fungi tolerated the altered concentrations of the As, as well as increased the leaf biomass of the fern plant. Besides, the fungi also amplified the absorption of As across an assortment of phosphate concentrations, whereas phosphate uptake was in general enhanced on the condition of no As alteration. The findings from this study revealed that AM fungi have significant hyper-accumulative potentials towards As uptake by Chinese brake fern. In conclusion, the authors recommend the use of AM fungi as a phyto-cleanser to remediate the impact of any polluted soil and propose a commercial production for agricultural purposes.

Sara et al. (2008) tested and evaluated the influence of AM fungi on the growth of *Zea mays* under cadmium stressed environment and phosphate source. The bio-control roles of AM fungi in the mitigation of environmental stressors were documented from various kinds of literature by the authors. The biological controlled experiment was done in a water system devoid of the soil environment (hydroponic) in an arbitrary 2×2×2 factorial setting. The plant *Zea mays* was inoculated with a strain of AM fungi *Glomus macrocarpum* and exposed to 0–20 $\mu molL^{-1}$ and 0–20 $\mu molL^{-1}$ nutrient concentration of Cd and P, respectively. The results of the study showed that AM fungi elicited the growth of the *Zea mays* while Cd lessen the dry mass content of the plant. The accumulation of Cd was strictly found in the root region. While there was no significant impact between Cd in the AM fungi and the non-inhabited AM plants. Further results showed traces of Cd in the root cell wall compared with the cytoplasm of the cell, and 26% in the cell wall of the non-inhabited AM plant compared to the cytoplasm of the same cell. The results of the comparison of the root and shoot system showed more Cd in the AM inhabited plants compared to the non-inhabited AM plants. The length and colonization of the mycorrhizal mycelium were reduced by the addition of Cd addition. This was the result of the high phosphate supply. When guaiacol peroxidase and some protein, nutrients were added to the Cd, the AM fungi mycelium showed no root induction indicating maximum tolerance to Cd reaction. The author concluded in their findings that Cd affected the AM fungi relationship with the plant roots, however, an addition of nutrients and enzyme (peroxidase) to the media, made it more efficient to subdue cadmium stress.

Christophersen et al. (2012) tested and evaluated the extrication and the colonization effects of AM fungi to As (Arsenic) tolerance in barrel clover plants. The authors tested three hypotheses and measured the genetic expression of the plants as well as the impact of As on the plant's genome. The results of the biological controlled experiment revealed that the plants displayed high gene expression (MtPT4) and those injected with *G. mosseae* showed higher selectivity counter to As and much lesser expression (MtPht1:1) and a specific degree of MtPht1:2 expressions to *G. intraradices* injected with non-Mycorrhizal plants. The findings of their study showed active phosphate and/or As selectivity in AM plants specifically to *G. mosseae*. That, the degree to which this selectivity is not clear. However, depended on the uptake of the phosphate and As in their cell wall. The clear, up-regulation of the genomes (PCS and ACR) gene in AM fungi, plants, might also be involved and need an additional inquiry.

Clark and Zeto (1996) isolated and tested Fe (iron) acquirement by AM fungi in *Zea mays* cultivated in a high pH (alkaline) soil. The results of the roots and shoot biomass were higher in AM fungi, plants than what was obtained in non-AM fungi plants. The result of the contents and concentrations of elemental nutrients in the plant leaf was slightly more in AM fungi, plants compared to non-AM fungi plants. There was an improvement of iron contents under severe Fe conditions of the *Zea mays* plant when inoculated with AM strain. Findings from the study indicated that the benefits (booting of Fe deficiency by AM fungi) derived by the plant, cannot be replicated in a field setting because the outputs were extremely low for a rapidly developing plant of such. In conclusion, the authors recommend better host plants and the addition of phosphate to boost the uptake and control of iron by the AM fungi.

Evelin et al. (2009) did a review on the mitigation of salinity stress in soil using AM fungi. The authors recounted the environmental stress faced by plants under serious saline conditions. AM fungi use enhance mechanism in alleviating saline stress in the plant which is nutrient acquirement; Ca, Mg, N, and P, upkeep of the Na^{+}-K^{+} ion potentials of the plant K, biochemical alterations and balance of some plant chemicals; antioxidants, carbohydrates, polyamines, betaines, and prolines, physiological alteration and balance of some plant activities; N-fixation, root nodulation, accumulation of abscisic acid (ABA), the hydro status of the plant, relative penetrability and efficacy of photosynthesis and molecular alteration of gene expression of some plant extreme-structures such as LsP5CS, Lslea, Na(+)/H(+), Lsnced, antiporters, and PIP, antiporters, Lsnced, Lslea, and LsP5CS. The authors in conclusion recommend that extreme structural and molecular alterations of AM fungi should be looked into for further research and large-scale production.

Ferrol et al. (2009) did a review of the survival methods of AM fungi in the copper polluted ecosystem. The authors recounted that the mechanism that AM fungi use in the mitigation of copper in polluted soils are found in their intra-cellular media and cell cytoplasm-cytosol. The copper is therefore trans-located to the several sub-cellular parts via the root vacuoles that may create lesser damage to the plant. The AM fungi have also evolved with time some compartmentalization approaches or strategies with extra-radical reproductive structure and intra-radical vesicles. More so, they have an antagonistic mechanism to reduce the oxidative stress of copper as well as to repair damaged tissues induced by the toxic Cu.

Ferrol et al. (2016) tested and evaluated the HMs absurdity in AM fungi, its mechanism, and prospective utilization. That is a close association of AM fungi with the root cortex of plants; the external hyphae absorb nutrients such as phosphate, manganese, iron, zinc, and copper. Above and beyond, the AM fungi not only improve the nutrient quality of the plant but also aid in the remediation of the HMs impacts as well. This makes AM fungi a potent bio-regulatory organism against nutrient deficiency in the soil and alleviation of soil toxins. In conclusion, the authors proposed an in-depth investigation of the homeostatic response of AM fungi to plant HMs acquirement under lacking and noxious HMs conditions.

Galli et al. (1994) did a review on the potential ability of AM fungi on chelating HMs from the soil to eke plant growth (nutrients) and alleviating eco-stress. The authors stated that this has also been demonstrated by many studies. However, the extra-matrical mycelium of the fungi cell wall has mechanisms such as melanin, cellulose, and chitin that allow the chelating of HMs and enable the fungi to decontaminate the toxic soil. In conclusion, the authors proposed the utilization of multi chains of phosphate granules with mixtures of nitrogen and sulfur as bio-sorption agents for effective soil remediation.

Jambon et al. (2018) wrote a comprehensive review on the application of mycorrhizal fungi for their bioremediation potential of an inorganic and organic pollutant that is available in the groundwater and soil. Also, their effectiveness in promoting plant growth was emphasized. The authors also highlighted the synergetic effects of mycorrhizal fungi with free-living saprotrophs and bacteria for their degradative potential for highly polluted environments. Profiting from improved knowledge of microbial community structure and their gathering gives a proper understanding of how holobiont can be altered to increase their rate of pollutant degradation and drastic improvement of agricultural plant growth. The authors provide a holistic and proper understanding of plant-bacteria-fungi interfaces and the prospects of utilizing these tripartite interfaces to improve the efficiency of phytoremediation of organic pollutants.

Chen et al. (2018) reviewed the ecological role of AMF and its advantageous services in agriculture. The authors hailed AM as one of the most effective symbionts with several ecological functions including plant boosters, stress-tolerant, and resistance, soil pH buffer, soil profile management, and fertility enhancer. The ability of AMF to interact with most plant species especially, tree fruits, vegetables, cereals, and legumes make it a sustainable bio-fertilizer for agriculture. The authors stated several types of research that have stressed the utilization of AM and re-echo its signaling trails as a further intervention for ecosystem reconstruction and management. Still, they stressed the exchange of nutritional benefits the association tends to get during the interactive biological functioning. Because of the benefits, the authors concluded that AM stands the chance to be used in different related areas (landscaping and horticulture) apart from agriculture and recommend it for novel developments in these areas.

2.5 IMPROVEMENT OF SOIL STRUCTURE AND BIODIVERSITY

It has been observed that the effect of AMF on various soil activities is a very complex process to define because there is the various level of variation which depends on types of plants, environmental condition, and the types of fungal genotypes. In view of the aforementioned, Hamel wrote a comprehensive review on the effect of AMF-microbe interactions, mobilization of nitrogen and phosphorus most especially in the root zone, the types of mycorrhizosphere as well as the significance of AMF in the root zone of the phosphorus and nitrogen dynamics. It was observed that the spatial arrangement of the hyphae and the roots in the soil gives a better understanding that AMF may increase the process of reabsorption of nutrients losses mainly through root exudation. These fungi portend the capability to influence other soil microorganisms, which infers that they play a crucial role in various biochemical reactions that occur in the soil, such as nitrification and mineralization of organic matter. They also highlighted that AMF has the potential to utilize organic sources of nitrogen and nitrogen while AMF has been highlighted to have the capability to enhance the capability of the plant to compete with saprotrophs for the absorption of nitrogen and phosphorus. This might be liked by the fact that AMF could utilize amino-acid-N, their capability to produce phosphatase enzymes as well as the production of organic anions. It has been stated that AMF possesses improved competitive capacity of mycorrhizal plants under conditions of low soil Nitrogen and phosphorus level which could minimize the populations of saprotrophic populations, and whenever this occurs, it could regulate soil matter production rate. This shows that adequate management of beneficial soil resources could be led to an effective sustainable cropping system for effective crop production.

Berardia subacaulis Vill has been recognized as a monospecific genus that is endemic to the southwestern Alps where it is cultivated on alpine screes, which are life-threatening environments characterized by limiting growth conditions and soil disturbance. The rate of root colonization in this region by AMF has been recognized as highly significant to these regions because it plays a high impact on the stress tolerance and plant nutrition as well as amendment and restructuring of soil.

Casazza et al. (2017) evaluated the biotic factor and soil features that enhance the community composition and abundance of AMF in the roots of *B. subacaulis*, which had formerly been discovered to be mycorrhizal. Hence, the authors evaluate the effects of environmental and soil properties of AMF community composition and abundance available in the roots of *B. subacaulis*, sampled on three different scree slopes were analyzed. The result indicated that the dark septate endophytes and vegetation coverage did not affect the plant community and the percentage of AMF colonization. The abundance of the AMF in the roots could be linked to the presence of some chemicals mainly inform of potassium and calcium and AMF colonization, physical properties such as field capacity, cation exchange, and electrical conductivity which indicated that AMF could enhance the quality of the soil tested. The NMDS experiment performed on the AMF community indicates that the diversity of AMF present on various sites was stimulated by the slope and the quality of the soil. Their study showed that slope stimulates physical disturbance of alpine screes which might have led to disruption tolerant AMF taxa, which constitutes various plant-fungus assemblages.

Biological invasions have been recognized as a significant Biological invasions alteration that can influence the soil diversity and ecosystem function. It has been acknowledged that soil microorganisms play a crucial role in plant development and their eventual establishment. But there is a need to establish their role in biological inversions. The fastest approach to it is to inquire whether invasive plants host has numerous microbes than their nearby native plant species. Because of the aforementioned, DeBellis et al. (2019) evaluated the variation that occurs in the related microorganism of native exotic Norway maple (*A. platanoides* L.) and sugar maple (*Acer saccharum* Marsh.) obtained from a forested reserve in eastern Canada. The authors utilized microscopy to investigate root fungi and high-throughput sequencing for the molecular characterization of the fungal, bacteria, and AM communities of the maple species. The result obtained indicates that there was variation in the root-related fungal and bacterial community between the host species. Moreover, the operational taxonomic units of the fungal and bacterial found in the Norway maple was greater than the operational taxonomic units while the indicator species evaluation shows that there were three bacterial and nine fungal operational taxonomic units with special reference for sugar maple Also, the highest dominant bacterial phyla discovered on the roots of both maple species were Proteobacteria and Actinobacteria respectively. The following fungal species discovered with the sugar maple were Hypocreales, Agaricales, Helotiales, Capnodiales, Pleosporales, while those found on the Norway maple roots were Trechisporales, Helotiales, Pleosporales, Hypocreales Agaricales, respectively. The highest dominant orders discovered in the sugar maple were Hypocreales and Capnodiales while the level of Dark septate fungi colonization was significant in sugar maple, but the rates of colonization by the arbuscular mycorrhizal fungal communities were not observed. Their study indicated that two congeneric plant species planted at a close distance can harbor specific root microbial communities. Their study provides a baseline for the significance of plant species in the structuring root-related microbe community. Also, the high level of colonization detected in the Norway maple showed that AMF had great compatibility in the introduced range. Their study also showed that plant-related microbial communities could influence host fitness and role in numerous ways. Hence their study indicated that biotic interaction could play a role in the dynamics between invasive and native species.

2.6 BIOLOGICAL CONTROL

Wehner et al. (2009) wrote a comprehensive report on the protective roles of AMF against some pathogens that could affect the root of crops, and the modes of actions of these fungi were also highlighted. It has been stated that there is a tendency that the AMF taxa differ in numerous ways that they could limit the adverse effects of pathogens on the host plants, but the synergetic effect among different members of AMF assemblages and communities might lead to enhanced protection of pathogen. Therefore, the authors described the between AM and pathogenic fungi and cited some specific examples that compare the efficacy of single- and multi-species AM fungal assemblages. The authors also highlight the specific modes of action by which AMF exhibits their protective role against their pathogens, and special interests were placed on the functional diversity among AM fungal taxa in terms of their modes of action. The authors proposed that functional complementarity among AMF fungal taxa in relationship with pathogen might be linked to the reason for the association between plant productivity and AMF fungal diversity.

2.7 MODES OF ACTION OF AMF AGAINST PESTS AND PATHOGENS

AMF has the capability to mitigate against the biotic stress using several approaches such as competition with the pathogens or through indirect means mainly by plant-mediated influence. The direct influence includes completion for nitrogen, carbon, and other growth factors and survival for niches or specific infection sites. It has been highlighted that AMF possesses the capacity to minimize the effect of pathogenic fungi present in the root region of plants (Filion et al., 2003). Apparently, AMF and pathogenic utilized the available resources within the root, containing infection/establishment location, space, and the rate of their photosynthesis (Whipps, 2004). Moreover, it has been documented that a negative correlation exists among the pathogenic microorganisms and AMF that dwells in the soil and the roots (St. Arnaud and Elsen, 2005).

Cordier et al. (1998) also affirmed that arbusculated cells could inhibit pathogenic oomycete *Phytophthora*. Also, AMF has the capacity to cause changes in the root system structure in morphology as well as in the root exudates (Pivato et al., 2008). The alteration may affect the rate of infection against the invading pathogens, and this may also affect the microbial population of the mycorrhizosphere which empowers the capability of the microbiota to inhibit many pathogens (Badri and Vivanco, 2009).

Vos et al. (2011) also affirmed that alteration in the level of root exudation may affect the nematodes and microbial pathogens. Recently it was documented that AMF has the capacity to regulate the capability of AMF to reorganize or regulate the rate of expression available in plants gene (Campos-Soriano et al., 2012). This may also influence the rates of secondary or primary metabolism of the plant which are all utilized in plant defense (Lopez-Raez et al., 2010a, b).

Moreover, it has been observed that AMF has the capability to stimulate plant defense response at the beginning of attachment to their plant host just like other biotrophs (Paszkowski, 2006). Hence, after the successful colonization, the AMF has the capability to endure the plant counter-reaction which it could overcome as time goes on and energetically modify plant defense rejoinders. This then led to stimulation of the plant defense after a challenger attack which is referred to as priming (Pozo and Azcon-Aguilar, 2007). Priming helps the plant to get ready for any attack by forging a body that may be faster and/or stronger compared to a plant that has not been previously exposed to prime stimulus thereby enhancing the plant resistance (Walters and Heil, 2007).

2.8 REGULATION OF THE PLANT HOST IMMUNE SYSTEM

The stimulation of resistance around the root of the pant by AMF does not require direct stimulation of defense mechanisms but can be obtained can result from sensitization of the tissue upon suitable prompt to show basal defense mechanisms more proficiently after successive pathogen attack (Jung et al., 2012). The stimulation of resistance that occurs whenever AMF colonizes a plant might not need direct stimulation of the defense mechanism but could be from the triggering of the tissue by necessary stimulation to show dense action more enhanced after the pathogen attack (Jung et al., 2012). The priming of the immune system of the plant has been documented to be common whenever beneficial microorganisms come in contact with the plant host system in comparison to direct stimulation of the defense (Conrath, 2009). Several modes of action have been highlighted to be involved in the stimulation of the primed state as a moderate build-up of defense-associated molecules such as chromatin modifications and transcription factors or MAP kinases (mitogen-activated protein kinase) (Pastor et al., 2012). Benhamou et al. (1994) stated that there was a strong defense mechanism that was triggered as the Mycorrhizal-transformed carrot roots showed a powerful defense reaction as the location when attacked by *Fusarium*.

Moreover, it has been observed that in tomato, AM colonization systemically guilds the root of tomato plant against the action of *Phytophthora parasitica* infection but the AMF develop a papilla-like structure around the location of pathogen infection through the dropping of non-esterified callose and pectins that could avert the action pathogens from spreading around the plant hosts. They also prompt the build-up of β-1,3 glucanases and mycorrhizal-transformed carrot roots displayed stronger defense reactions at sites challenged by *Fusarium* (Benhamou et al., 1994). In tomato, AM colonization systemically protected roots against *Phytophthora parasitica* infection. Only mycorrhizal plants formed papilla-like structures around the sites of pathogen infection through deposition of non-esterified pectins and callose, preventing the pathogen from spreading further, and they accumulated significantly more PR-1a than those that are not mycorrhizal plants upon Phytophthora attack (Pozo et al., 1999, 2002).

Also, the build-up of the following compounds such as solavetivone, phytoalexins, and rishitin were discovered inside mycorrhizal potatoes when *Rhizoctonia* infection but interestingly AMF does not have any influence on the level of these compounds (Yao et al., 2003). The build-up of several phenolic compounds could be linked to the protective role against *F. oxysporum* by date palm trees (Jaiti et al., 2007), while the action of the priming has also been confirmed as a protective mechanism against nematode (Hao et al., 2012). Also, the primed response is not limited to plant roots but could also be found in the shoot region (Pozo et al., 2010). Moreover, it was observed that a high level of induced systemic resistance (ISR) was observed in the tomato plant against the action of necrotrophic foliar pathogen *Botrytis cinerea* (BC). It was discovered that the amount of the pathogen available in mycorrhizal plants was significantly reduced while there was an enhanced manifestation of some defense-associated, jasmonate-regulated available in the plants (Pozo et al., 2010). The application of tomato mutants impaired in jasmonate acid signaling affirmed that jasmonate acid is necessary for AM-induced resistance against Botrytis (Jung et al., 2012).

2.9 CONCLUSION AND FUTURE DIRECTION

This chapter has established the significance of AMF in sustainable organic agriculture. This chapter has also validated the sustainable application of AMF for their effectiveness against pest and plant pathogens, their application as a biofertilizer, they restructure of dilapidated soil, and ecorestoration of heavily polluted soil. It was also affirmed that the colonization of AMF with roots of the plant could enhance the host plant physiology. This also plays a crucial role in regulating the capacity of the plants to regulate abiotic and abiotic stress. The modes of action utilized by AMF have been discussed extensively in this chapter. Also, the role of AMF in the maintenance of soil diversity and its functions with other beneficial microorganisms has been elucidated in detail. Furthermore, specific examples have been cited that illustrates the potential of AMF as a plant enhancer for improvement of the plant in term of yield and growth parameters after applying them as an inoculant to various crops.

KEYWORDS

- **abiotic stresses**
- **arbuscular mycorrhizal fungi**
- **biofertilizer**
- **biotic stress**
- **ecorestoration**
- **organic agriculture**

REFERENCES

Adetunji, C. O., & Ugbenyen, A. M., (2019). Mechanism of action of nanopesticide derived from microorganism for the alleviation of abiotic and biotic stress affecting crop productivity. In: *Nanotechnology for Agriculture: Crop Production and Protection*. https://doi.org/10.1007/978-981-32-9374-8_7.

Adetunji, C. O., (2019). Environmental impact and ecotoxicological influence of bio-fabricated and inorganic nanoparticle on soil activity. In: Panpatte, D., & Jhala, Y., (eds.), *Nanotechnology for Agriculture*. Springer, Singapore. https://doi.org/10.1007/978-981-32-9370-0_12.

Adetunji, C. O., Egbuna, C., Tijjani, H., Adom, D., Khalil, L., Al-Ani, T., & Partick-Iwuanuyanwu, K. C., (2019). Pesticides, home made preparation of natural biopesticides and application. In: *Natural Remedy for Pest, Diseases and Weed Control* (pp. 1–12). Elsevier.

Adetunji, C. O., Kumar, D., Raina, M., Arogundade, O., & Sarin, N. M., (2019b). Endophytic microorganisms as biological control agents for plant pathogens: A panacea for sustainable agriculture. In: *Plant Biotic Interactions*. https://doi.org/10.1007/978-3-030-26657-8_1.

Adetunji, C. O., Oloke, J. K., Bello, O. M., Pradeep, M., & Jolly, R. S., (2019d). Isolation, structural elucidation and bioherbicidal activity of an eco-friendly bioactive 2-(hydroxymethyl) phenol, from *Pseudomonas aeruginosa* (C1501) and its ecotoxicological evaluation on soil. *Env. Tech. Innov., 13*(2019), 304–317.

Adetunji, C. O., Panpatte, D., Bello, O. M., & Adekoya, M. A., (2019c). Application of nanoengineered metabolites from beneficial and eco-friendly microorganisms as biological control agents for plant pests and pathogens. In: *Nanotechnology for Agriculture: Crop Production and Protection*. https://doi.org/10.1007/978-981-32-9374-8_13.

Agely, A. L., Sylvia, D. M., & Ma, L. Q., (2005). Mycorrhizae increase arsenic uptake by the hyperaccumulator Chinese brake fern (*Pteris vittata* L.). *J. Env. Qual., 34*(6), 2181–2186.

Andrade, S. A. L., Mazzafera, P., Schiavinato, M. A., & Silveira, A. P. D., (2009). Arbuscular mycorrhizal association in coffee. *The J. Agri. Sci., 147*(2), 105–115. https://doi.org/10.1017/S0021859608008344.

Azizah, H., & Idris, Z. A., (1996). Microbes: How they can do wonders to soil fertility. *Malaysian Technol. Bull. 4*, 92–94.

Badri, D. V., & Vivanco, J. M., (2009). Regulation and function of root exudates. *Pl. Cell Env., 32*, 666–681.

Barea, J. M., Andrade, G., Bianciotto, V., Dowling, D., Lohrke, S., Bonfante, P., O'Gara, F., & Azcon-Aguilar, C., (1998). Impact on *Arbuscular mycorrhiza* formation of *Pseudomonas* strains used as inoculants for biocontrol of soil-borne fungal plant pathogens. *Appl. Env. Microb., 64*(6), 2304–2307.

Baslam, M., Garmendia, I., & Goicoechea, N., (2011). Arbuscular mycorrhizal fungi (AMF) improved growth and nutritional quality of greenhouse-grown lettuce. *J. Agri. Food Chem., 59*(10), 5504–5515. https://doi.org/10.1021/jf200501c.

Benhamou, N., Fortin, J. A., Hamel, C., St Arnaud, M., & Shatilla, A., (1994). Resistance responses of mycorrhizal Ri T-DNA-transformed carrot roots to infection by *Fusarium oxysporum* f. sp. *chrysanthemi*. *Phytopathology, 84*, 958–968.

Berruti, A., Lumini, E., Balestrini, R., & Bianciotto, V., (2016). Arbuscular mycorrhizal fungi as natural biofertilizers: Let's benefit from past successes. *Fron. Microbiol., 6*(426). https://doi.org/ 10.3389/fmicb.2015.01559.

Binh, T. T., Tran, A. B., Stephanie, J., Watts-Williams, A. C., & Cavagnaro, T. R., (2019). Impact of an arbuscular mycorrhizal fungus on the growth and nutrition of fifteen crop and pasture plant species. *Func. Plant Biol., 46*(8), 732–742. https://doi.org/10.1071/FP18327.

Bonfante, P., & Perotto, D. S., (1995). Strategy of arbuscular mycorrhizal fungi when infecting host plants. *New Phytology, 130*, 3–21.

Bonifacio, E., Nicolotti, G., Zanini, E., & Cellerino, G. P., (1999). Heavy metal uptake by mycorrhizae of beech in contaminated and uncontaminated soils. *Fresenius Environ. Bull., 7*, 408–413.

Cabral, L., Soares, C. R., Giachini, A. J., & Siqueira, J. O., (2015). Arbuscular mycorrhizal fungi in phytoremediation of contaminated areas by trace elements: Mechanisms and major benefits of their applications. *World J. Microbiol. Biotechnol., 31*(11), 1655–1664. https://doi.org/10.1007/s11274–015–1918-y.

Carina, V. T., Emilia, M. M., Miguel, U. S., Ortiz, D. R., Manjunatha, B., Thangaswamy, S., & Mulla, S. I., (2016). Effect of Arbuscular mycorrhizal fungi (AMF) and *Azospirillum* on growth and nutrition of banana plantlets during acclimatization phase. *J. Appl. Pharm. Sci., 6*(06), 131–138.

Casazza, G., Lumini, E., Ercole, E., Dovana, F., Guerrina, M., Arnulfo, A., et al., (2017). The abundance and diversity of arbuscular mycorrhizal fungi are linked to the soil chemistry of screes and to slope in the Alpic paleo-endemic *Berardia subacaulis*. *PLoS One, 12*(2), e0171866. https://doi.org/10.1371/journal.pone.0171866.

Chen, M., Arato, M., Borghi, L., Nouri, E., & Reinhardt, D., (2018). Beneficial services of arbuscular mycorrhizal fungi-from ecology to application. *Front. Plant Sci., 4*. |https://doi.org/10.3389/fpls.2018.01270.

Christophersen, H. M., Smith, F. A., & Smith, S. E., (2012). Unraveling the influence of arbuscular mycorrhizal colonization on arsenic tolerance in Medicago: *Glomus mosseae* is more effective than *G. intraradices*, associated with lower expression of root epidermal Pi transporter genes. *Front. Physiol., 13*(3), 91. doi:10.3389/fphys.2012.00091.

Clark, R. B., & Zeto, S. K., (1996). Iron acquisition by mycorrhizal maize grown on alkaline soil. *J. Plant Nutr., 19*, 2.

Conrath, U., (2009). Priming of induced plant defense responses. In: Loon, L. C. V., (ed), *Advances in Botanical Research* (pp. 361–395). Academic Press. https://doi.org/10.1016/S0065-2296(09)51009-9.

Cordier, C., Pozo, M. J., Barea, J. M., Gianinazzi, S., & Gianinazzi-Pearson, V., (1998). Cell defense responses associated with localized and systemic resistance to *Phytophthora parasitica* induced in tomato by an arbuscular mycorrhizal fungus. *Mol. Plant-Microbe Interact., 11*, 1017–1028.

Danneberg, G., Latus, C., Zimmer, W., Hundes-Hagen, B., Schneider-Poetsch, H., & Bothe, H., (1992). Influence of vesicular-arbuscular mycorrhiza on phytohormone balances in maize (*Zea mays* L.). *Journal of Plant Physiology, 141*, 33–39.

DeBellis, T., Kembel, S. W., & Lessard, J. P., (2019). Shared mycorrhizae but distinct communities of other root-associated microbes on co-occurring native and invasive maples. *Peer J.* https://doi.org/10.7717/peerj.7295.

Demür, S., (2004). Influence of arbuscular mycorrhiza on some physiological growth parameters of pepper. *Turk. J. Biol., 28*, 85–90.

Douds, D. D., & Millner, P., (1999). Biodiversity of arbuscular mycorrhizal fungi in agroecosystems. *Agri, Ecosys. Environ., 74*, 77–93.

Evelin, H., Kapoor, R., & Giri, B., (2009). Arbuscular mycorrhizal fungi in alleviation of salt stress: A review. *Annals Botany, 104*(7), 1263–1280. https://doi.org/10.1093/aob/mcp251.

Ferrol, N., Gonzalez-Guerrero, M., Valderas, A., Benabdellah, K., & Azcon-Aguilar, C., (2009). Survival strategies of arbuscular mycorrhizal fungi in Cu-polluted environments. *Phytochem. Rev., 8*(3), 551–559. https://doi.org/10.1007/s11101-009-9133-9.

Ferrol, N., Tamayo, E., Martínez, E., & Paola, V., (2016). The heavy metal paradox in arbuscular mycorrhizas: From mechanisms to biotechnological applications. *J. Exp. Bot., 67*(22), erw403. https://doi.org/10.1093/jxb/erw403.

Filion, M., St Arnaud, M., & Jabaji-Hare, S. H., (2003). Quantification of *Fusarium solani* f. sp. *phaseoli* in mycorrhizal bean plants and surrounding mycorrhizosphere soil using real-time polymerase chain reaction and direct isolations on selective media. *Phytopathology, 93*, 229–235.

Galli, U., Schüepp, H., & Brunold, C., (1994). Heavy metal binding by mycorrhizal fungi. *Physiologia Plantarum, 92*(2), 364–368. https://doi.org/10.1111/j.1399-3054.1994.tb05349.x.

Garg, N., & Chandel, S., (2010). *Sustainable Development Arbuscular Mycorrhizal Networks: Process and Functions. a Review on Agronomy and Sustainable Development* (Vol. 30, No. 3). Springer Verlag/EDP Sciences/INRA.

Garg, N., Singh, S., & Kashyap, L., (2017). Arbuscular mycorrhizal fungi and heavy metal tolerance in plants: An insight into physiological and molecular mechanisms. In: Varma, A.,

Prasad, R., & Tuteja, N., (eds.), *Mycorrhiza-Nutrient Uptake, Biocontrol, Ecorestoration.* Springer, Cham., https://doi.org/10.1007/978-3-319-68867-1_4.

Gianinazzi, S., Schuepp, H., Barea, J. M., & Haselwandter, K., (2002). *Mycorrhizal Technology in Agriculture-From Genes to Bioproducts* (p. 296). Springer, Basel. https://doi.org/10.1007/978-3-0348-8117-3.

Grant, C., Bitman, S., Montreal, M., Plenchette, C., & Morel, C., (2005). Soil and fertilizer phosphorus: Effects on plant supply and mycorrhizal development. *Canadian J. Plant Sci., 85*, 3–14. https://doi.org/doi:10.4141/P03-182.

Hamel, C., (2004). Impact of arbuscular mycorrhizal fungi on N and P cycling in the root zone. *Can. J. Soil Sci., 84*, 383–395.

Hao, Z., Fayolle, L., Van, T. D., Chatagnier, O., Li, X., Gianinazzi, S., & Gianinazzi-Pearson, V., (2012). Local and systemic mycorrhiza-induced protection against the ectoparasitic nematode xiphinema index involves priming of defense gene responses in grapevine. *J. Exp. Bot., 63*, 3657–3672.

Hashem, A., Kumar, A., Al-Dbass, A. M., Abdulaziz, A. A., FahadAl-Arjani, A. B., Garima, S., Farooq, M., & Fathi, A. E., (2019). Arbuscular mycorrhizal fungi and biochar improves drought tolerance in chickpea. *Saudi J. Biol. Sci., 26*(3) 614–624. https://doi.org/10.1016/j.sjbs.2018.11.005.

Hussain, M., Hussain, D. Z. D., & Reshi, A., (2017). Vesicular arbuscular mycorrhizal (VAM) fungi-as a major biocontrol agent in modern sustainable agriculture system. *Russian Agri. Sci., 43*(2), 138–143. https://doi.org/10.3103/S1068367417020057.

Igiehon, N. O., & Babalola, O. O., (2017). Bio-fertilizers and sustainable agriculture: Exploring arbuscular mycorrhizal fungi. *Appl. Microbiol. Biotechnol., 101*(12). https://doi.org/10.1007/s00253-017-8344-z.

Itelima, J. U., Bang, W. J., Onyimba, I. A., & Egbere, O. J., (2018). A review: Biofertilizer; a key player in enhancing soil fertility and crop productivity. *J. Microbiol. Biotechnol. Rep., 2*(1), 22–28.

Jaiti, F., Meddich, A., & El Hadrami, I., (2007). Effectiveness of arbuscular mycorrhizal fungi in the protection of date palm (*Phoenix dactylifera* L.) against bayoud disease. *Physiol. Mol. Plant Pathol., 71*, 166–173.

Jambon, I., Thijs, S., Weyens, N., & Vangronsveld, J., (2018). Harnessing plant-bacteria-fungi interactions to improve plant growth and degradation of organic pollutants. *J. Plant Interactions, 13*(1) 119–130. https://doi.org/10.1080/17429145.2018.1441450.

Jastrow, J. D., Miller, R. M., & Lussenhop, J., (1998). Contributions of interacting biological mechanisms to soil aggregate stabilization in restored prairie. *Soil Biol. Biochem., 30*, 905–916.

Jung, S. C., Martinez-Medina, A., Lopez-Raez, J. A., & Pozo, M. J., (2012). Mycorrhiza-induced resistance and priming of plant defenses. *J. Chem. Ecol., 38*, 651–664.

Junior, P. P., Moreira, B. C., Da Silva, M. C. S., Veloso, T. G. R., Stürmer, S. L., Fernandes, R. B. A., De Sá Mendonça, E., & Kasuya, C. M., (2019). Agroecological coffee management increases arbuscular mycorrhizal fungi diversity. *PLoS One, 14*(1), e0209093. https://doi.org/10.1371/journal.pone.0209093.

Knerr, A. J., Wheeler, D., Schlatter, D., Sharma-Poudyal, D., Du Toit, L. J., & Paulitz, T. C., (2018). Arbuscular mycorrhizal fungal communities in organic and conventional onion crops in the Columbia Basin of the Pacific Northwest United States. *Phytobiomes J., 2*, 194–207.

Kucey, R. M. N., & Paul, E. A., (1982). Carbon flow, photosynthesis and N_2 fixation in mycorrhizal and *Nodulated fababeans* (*Vicia faba* L.). *Soil Biol. Biochem., 14*, 407–412.

Laila, A., Trisnaningrum, N., & Hamawi, M., (2019). Significant potential of arbuscular mycorrhizae fungi to increase on yield of Shallot. *IOP. Conf. Series: Earth Environ. Sci., 292*, 012017. https://doi.org/10.1088/1755-1315/292/1/012017.

Lopez-Raez, J. A., Flors, V., Garcia, J. M., & Pozo, M. J., (2010b). AM symbiosis alters phenolic acid content in tomato roots. *Plant Signal Behav., 5*, 1138–1140.

Lopez-Raez, J. A., Verhage, A., Fernandez, I., García, J. M., Azcón-Aguilar, C., Flors, V., & Pozo, M. J., (2010). Hormonal and transcriptional profiles highlight common and differential host responses to arbuscular mycorrhizal fungi and the regulation of the oxylipin pathway. *J. Exp. Bot., 61*(10), 2589–2601.

Maeder, P., Fliessbach, A., Dubois, D., Gunst, L., Fried, P., & Niggli, U., (2002). Soil fertility and biodiversity in organic farming. *Science, 296*(5573), 1694–1697.

Marschner, H., & Dell, B., (1994). Nutrient uptake in mycorrhizal symbiosis. *Plant Soil, 159*, 89–102.

Masri, B. M., (1997). *Mycorrhizal Inoculation for Growth Enhancement and Improvement of the Water Relations in Mangosteen (Garcinia mangostana L.) Seedlings*. Ph.D. Thesis. Universiti Putra Malaysia, Serdang, Malaysia.

McGonigle, T. P., & Miller, M. H., (1996). Development of fungi below ground in association with plants growing in disturbed and undisturbed soils. *Soil Biol. Biochem., 28*, 263–269.

Medina, A., & Azcón, R., (2010). Effectiveness of the application of arbuscular mycorrhiza fungi and organic amendments to improve soil quality and plant performance under stress conditions. *J. Soil Sci. Plant Nut., 10*(3), 354–372.

Nurbaity, A., Sofyan, E. T., & Hamdani, J. S., (2016). Application of *Glomus* sp. and *Pseudomonas diminuta* reduce the use of chemical fertilizers in the production of potato grown on different soil types. *IOP Conf. Ser. Earth Env. Sci., 41*(1), 012004. https://doi.org/10.1088/1755-1315/41/1/012004.

Oladele, S., & Awodun, M., (2014). Response of lowland rice to bio-fertilizer inoculation and their effects on growth and yield in Southwestern Nigeria. *J. Agric. Environ. Sci., 3*(2), 371–390.

Pastor, V., Luna, E., Mauch-Mani, B., Ton, J., & Flors, V., (2012). Primed plants do not forget. *Environ. Exp. Bot.* http://dx.doi.org/10.1016/j.envexpbot.2012.02.013.

Paszkowski, U., (2006). Mutualism and parasitism: The yin and yang of plant symbioses. *Curr. Opin. Plant Biol., 9*, 364–370.

Pivato, B., Gamalero, E., Lemanceau, P., & Berta, G., (2008). Cell organization of *Pseudomonas fluorescens* C7R12 on adventitious roots of *Medicago truncatula* as affected by arbuscular mycorrhiza. *FEMS Microbiol. Lett., 289*, 173–180.

Pozo, M. J., & Azcon-Aguilar, C., (2007). Unraveling mycorrhiza-induced resistance. *Curr. Opi. Plant Biol., 10*, 393–398.

Pozo, M. J., Azcon-Aguilar, C., Dumas-Gaudot, E., & Barea, J. M., (1999). β-1,3-glucanase activities in tomato roots inoculated with arbuscular mycorrhizal fungi and/or *Phytophthora parasitica* and their possible involvement in bioprotection. *Plant Sci., 141*, 149–157.

Pozo, M. J., Cordier, C., Dumas-Gaudot, E., Gianinazzi, S., Jose, M. B., & Azcón-Aguilar, C., (2002). Localized versus systemic effect of arbuscular mycorrhizal fungi on defense responses to Phytophthora infection in tomato plants. *J. Exp. Bot., 53*(368), 525–534. https://doi.org/10.1093/jexbot/53.368.525.

Pozo, M. J., Jung, S. C., Lopez-Raez, J. A., & Azcon-Aguilar, C., (2010). Impact of arbuscular mycorrhizal symbiosis on plant response to biotic stress: The role of plant defense mechanisms. In: Kapulnick, Y., & Douds, D. D., (eds.), *Arbuscular Mycorrhizas: Physiology and Function* (pp. 193–207). Springer, Dordrecht.

Read, D. J., Lewis, D. H. A., Fitter, H., & Alexander, I. J., (1992). *Mycorrhizas in Ecosystems*. CAB International, Oxford.

Richardson, D. M., Allsopp, N., Antonio, C. M. D., Milton, S. J., & Rejmanek, M., (2000). Plant invasions-the role of mutualisms. *Biol. Reviews Cambridge Philos. Soc., 75*, 65–93.

Sadhana, B., (2014). Arbuscular mycorrhizal fungi (AMF) as biofertilizers: A review. *Int. J. Curr. Microbiol. Appl. Sc., 3*(4), 384–400.

Sara, A. L. D. A., & Adriana, P. D., D. S., (2008). Mycorrhiza influence on maize development under Cd stress and P supply. *Braz. J. Plant Physiol., 20*(1). http://dx.doi.org/10.1590/S1677-04202008000100005.

Schreiner, R. P., & Bethlenfalvay, G. J., (1996). Mycorrhizae, biocides, and biocontrol. 4. Response of a mixed culture of arbuscular mycorrhizal fungi and host plant to three fungicides. *Biol. Fertil. Soils., 23*, 189–195. https://doi.org/10.1007/BF00336062.

Sramek, F., Dubsky, M., & Vosatka, M., (2000). Effect of arbuscular mycorrhizal fungi and *Trichoderma harzianum* on three species of balcony plants. *Rostlinna Vyroba, 146*, 127–131.

St Arnaud, M., & Elsen, A., (2005). Interaction of arbuscular mycorrhizal fungi with soil-borne pathogens and non-pathogenic rhizosphere microorganisms. In: Declerck, S., Fortin, J. A., & Strullu, D. G., (eds.), *In Vitro Culture of Mycorrhizas* (pp. 217–231). Springer, Berlin.

Sturz, A. V., Carter, M. R., & Johnston, H. W., (1997). A review of plant disease, pathogen interactions and microbial antagonism under conservation tillage in temperate humid agriculture. *Soil Tillage Res., 41*, 169–189.

Timmer, L., & Leyden, R., (1980). The relationship of mycorrhizal infection to phosphorus-induced copper deficiency in sour orange seedlings. *New Phytologist, 85*, 15–23. https://doi.org/10.1111/j.1469-8137.1980.tb04443.x.

Vos, C., Claerhout, S., Mkandawire, R., Panis, B., De Waele, D., & Elsen, A., (2011). Arbuscular mycorrhizal fungi reduce root-knot nematode penetration through altered root exudation of their host. *Plant Soil, 354*, 335–345.

Walters, D., & Heil, M., (2007). Costs and trade-offs associated with induced resistance. *Physiol. Mol. Plant Pathol., 71*, 3–17.

Wehner, J., Antunes, P. M., Powell, J. R., Mazukatow, J., & Rillig, M. C., (2009). Plant pathogen protection by arbuscular mycorrhizas: A role for fungal diversity? *Pedobiologia*, https://doi.org/10.1016/j.pedobi.2009.10.002.

Whipps, J. M., (2004). Prospects and limitations for mycorrhizas in biocontrol of root pathogens. *Can J Bot., 82*, 1198–1227.

Yao, M. K., Desilets, H., Charles, M. T., Boulanger, R., & Tweddell, R. J., (2003). Effect of mycorrhization on the accumulation of rishitin and solavetivone in potato plantlets challenged with *Rhizoctonia solani*. *Mycorrhiza, 13*, 333–336.

Yeasmin, T., (2017). Association of arbuscular mycorrhizas in plants: Future perspectives of biofertilizer in Bangladesh. In: *10th International Conference on Agriculture and Horticulture*. London, UK.

CHAPTER 3

AZOSPIRILLUM BIOINOCULANT TECHNOLOGY: PAST TO CURRENT KNOWLEDGE AND FUTURE PROSPECTS

PALANI SARANRAJ,[1] ABDEL RAHMAN M. AL-TAWAHA,[2] PANNEERSELVAM SIVASAKTHIVELAN,[3] ABDEL RAZZAQ M. AL-TAWAHA,[4] KANGASALAM AMALA,[1] DEVARAJAN THANGADURAI,[5] and JEYABALAN SANGEETHA[6]

[1]*Department of Microbiology, Sacred Heart College (Autonomous), Tirupattur–635601, Tamil Nadu, India*

[2]*Department of Biological Sciences, Al-Hussein Bin Talal University, Maan, Jordan*

[3]*Department of Agricultural Microbiology, Faculty of Agriculture, Annamalai University, Chidambaram–608002, Tamil Nadu, India*

[4]*Department of Crop Science, Faculty of Agriculture, University Putra Malaysia, Selangor, Malaysia*

[5]*Department of Botany, Karnatak University, Dharwad, Karnataka–580003, India*

[6]*Department of Environmental Science, Central University of Kerala, Kasaragod–671316, Kerala, India*

ABSTRACT

Azospirillum is a curve-shaped vibroid Gram-negative, measuring about 1–1.5 μm in diameter, motile because of possessing Peritrichous flagella that are capable of fixing atmospheric nitrogen under microaerophilic conditions in association with the roots of cereals. *Azospirillum* was well known for many years as PGPR because it was isolated from the rhizosphere region of many grasses and cereals universally in tropical as well as in temperate climates. *Azospirillum* is considered as the most valuable bioinoculant in the group of plant growth-promoting rhizobacteria (PGPR) because it is not only the microorganism capable of colonizing the roots of agricultural crops, along with root colonization *Azospirillum* sp. have the tendency of producing more beneficial compounds which are highly beneficial to crops. *Azospirillum* biofertilizers have a history going back to the beginning of the 20th century. However, researchers still must find innovative and novel strains and improve production strategies and methods of application and mode of action. *Azospirillum* is one of the widely used biofertilizers in organic farming. *Azospirillum* sp. contribute to increased yields of cereal and grasses by improving root development in properly colonized roots, increasing the rate of water and mineral uptake from the soil, and by biological nitrogen fixation. A better understanding of the basic biology of the *Azospirillum* microbe and plant root interaction may lead to greater efficacy in its application as an effective biological nitrogen-fixing biofertilizer. The products containing *Azospirillum* strains and their use begin to play an important role in recent agriculture and as an alternative to expensive chemical fertilizers and sometimes not environmentally friendly products. This chapter analyzes the genus *Azospirillum*, taxonomy, occurrence, distribution, characters, interactions, nitrogen fixation, growth hormones, siderophore production, and nutrient uptake.

3.1 INTRODUCTION

India is one of the important countries which are playing a major role in bio inoculants production and consumption for agricultural purposes. It was predictable that the current level of biological fertilizer usage is relatively low, and it is expected to increase to 80,000–85,000 tons by 2025 (Bhattacharyya and Kumar, 2000). In order to enhance the productivity of agricultural land, different types of crop nutrients are contributing and playing a vital role in the maximization of growth, yield, and biochemical parameters of agricultural crops. Out of various nutrients required, the highly required nutrient for increasing plant growth is nitrogen, phosphorous, and potassium (NPK), and in the short term, it is referred to as NPK.

Microorganisms are commonly used as bioinoculants and the most commonly used microbial fertilizers are Nitrogen-fixing bacteria (*Rhizobium* sp., *Bradyrhizobium* sp., *Azospirillum* sp. and *Azotobacter* sp.), nitrogen-fixing cyanobacteria or blue-green algae (BGA) (*Anabaena* sp.), phosphate solubilizing bacteria (PSB) or phosphate solubilizers (*Pseudomonas putida*), biocontrol agents (*Pseudomonas fluorescens*, *Bacillus* sp. and *Trichoderma* sp.), microbial insecticides (*Bacillus thuringiensis*, *Beauveria bassiana*, *Metarhizium* sp. and *Verticillium* sp.) and arbuscular mycorrhizal (AM) fungi (Pindi, 2012). When applied to the agricultural field, the microbial inoculants promote the plant growth by various beneficial activities, *viz.*, fixation of atmospheric nitrogen, production of plant growth-promoting phytohormones, solubilization of complex phosphates into simple phosphates, and solubilization of various minerals and nutrients for plant uptake. An efficient microbial inoculant should be viable for a minimum of six months. The major factors responsible for the viability of microbial biofertilizers during mass production, formulation, storage, transportation, and field application are related directly to the plant growth-promoting potential of a biofertilizer formulation.

Plant growth-promoting rhizobacteria (PGPR) which is familiarly called PGPR, plays a key role in cycling of various nutrients, producing plant-growth-promoting hormones, preventing from various phytopathogens, solubilizing inorganic phosphates, maintaining the soil fertility, and establishing the positive interactions with plant roots in agro environment through Rhizosphere and Phyllosphere. In the current research, agricultural scientists are giving more interest in the improvement of the positive interactions and beneficial associations between microorganisms and plants (Bilal et al., 2014). Microbial colonization in plant roots is influenced by various biotic and abiotic factors such as dynamics of microbial population, nature of soil, and characteristics of plants. Biological and chemical alterations of the rhizosphere environment also influence the health of plants through various factors like production of plant growth-promoting hormones, mobilizing various nutrients, and suppressing the phytopathogens through induced systemic resistance (ISR).

Azospirillum is a Gram-negative spiral-shaped bacteria that are capable of fixing atmospheric nitrogen under microaerophilic conditions in association with the roots of cereals, particularly paddy (*Oryza sativa* L.) (Bashan and Levanony, 1990). The genus *Azospirillum* is distributed widely in the rhizosphere of tropical and sub-tropical grasses, particularly cereals (Bashan and Holguin, 1997). The mechanisms by which *Azospirillum* sp. can exert a positive effect on plant growth were probably composed of multiple effects (Pereyra et al., 2007). However, these beneficial effects by *Azospirillum* inoculation are not consistent (Shanon, 2013). In the mid-70s, it was recognized that the strains of *Azospirillum* are abundantly associated with the rhizosphere region in the roots of cereals and millets and helps in fixing substantial amounts of nitrogen. The *Azospirillum* sp. is the most widely studied PGPR than the other rhizosphere microorganisms because of its multiple benefits to the agricultural crops, particularly cereals (Bashan and de-Bashan, 2010).

3.2 THE GENUS *AZOSPIRILLUM* AND ITS TAXONOMY

The soil nitrogen-fixing bacterium *Azospirillum* sp. was first isolated in the year 1925 and described by a Dutch Microbiologist M. W. Beijerinck from nitrogen-poor sandy soils of the Netherlands (Beijerinck, 1925). That was a significant landmark in the contribution in the field of non-symbiotic nitrogen-fixing microorganisms. In 1976, Dobereiner and Day reported a new diazotrophic association between a curved bacterium and the grass *Digitaria decumbens* and named as *Spirillum lipoferum*. At the time of its isolation from soil, the soil bacterium *Azospirillum* was originally named as *Spirillum lipoferum*. The *Spirillum lipoferum* was classified under genus *Spirillum* of the order Pseudomonadales by Breed et al. (1957). Becking (1963) described the *Azospirillum* as *Spirillum* of vibrioid in nature and worked on the taxonomy of the bacterium. After few decades, the genus *Spirillum* was renamed as *Azospirillum*.

Dobereiner et al. (1976) reported that the occurrence of *Azospirillum lipoferum* in tropical soil and they are responsible for the nitrogen fixation in grasses because of their occurrence near the root regions and with the root tissues. Worldwide, *Azospirillum* species has been isolated from the roots of numerous grasses, cereals, legumes, millets, and other non-cereal crop plants and from tropical, subtropical, and temperate soils (Tilak, 1997; Purushothaman, 2002).

Azospirillum is a Gram-negative, motile, and spiralor vibrioid shaped bacilli, which is the common inhabitant of root and soil. Among the various species, the most thoroughly studied *Azospirillum* species are *Azospirillum brasilense* and *Azospirillum lipoferum*. A number of physiological and morphological characteristics differentiates the various species of *Azospirillum*. In Brazil, *Azospirillum amazonense* is an acid-tolerant species, isolated from the grasses and palm trees. In Pakistan, *Azospirillum halopareferens* was isolated from the root surface of *Leptochloa fusca* (Reinhold et al., 1987).

The occurrence of *Azospirillum* in the rhizosphere of cereal crops was varied from 1% to 10% of total rhizosphere population was reported by Okon (1985). More than 100 *Azospirillum* sp. are present in the soil and comparatively the populations of *Azospirillum* species are more in rhizosphere soil than in the non-rhizosphere soils (De Coninck et al., 1988). Michiels et al. (1989) recorded the presence of *Azospirillum* species in different agro-climate zones in crops which are prevalent in all over the universe. Sumner (1990) reported the occurrence of multiple number of *Azospirillum* sp. in various agricultural crops like rice, maize, sorghum, wheat, and pearl millet.

According to Krieg (1976), the *Spirillum lipoferum* belong to the genus *Azospirillum* but its taxonomic position was completely uncertain. *Azospirillum* is having a distinct feature that is the production of Poly-β-hydroxybutyrate granules, in the short term, it is called PHB (Dobereiner and Day, 1976; Okon et al., 1976). *Spirillum lipoferum* was rearranged into three major groups based on its denitrification efficiency when cultures were grown with ammonium nitrate (Neyra and Dobereiner, 1977). Strains with high denitrification rate were placed under Group I. The *Azospirillum* strains which are failed to exhibit the disappearance of nitrate were included in the Group III. Strains with intermediate level of nitrite are coming under the Group II. At present, seven species, viz., *Azospirillum brasilense*, *Azospirillum lipoferum*, *Azospirillum amazonense*, *Azospirillum halopraeferens*, *Azospirillum irakense*, *Azospirillum largimobile* and *Azospirillum doebereinerae* have been described in the genus *Azospirillum*.

3.3 OCCURRENCE, DISTRIBUTION, AND ISOLATION OF *AZOSPIRILLUM*

Azospirillum was well known for many years as PGPR because it was isolated from the rhizosphere region of many grasses and cereals universally in tropical as well as in temperate climates (Steenhoudt and Vanderleyden, 2000). Since the 1970s, the *Azospirillum* sp. has the consistency and was proven as very promising PGPR because this bacterium was the selective microflora for fixing the atmospheric nitrogen in the environment (Bashan et al., 2004).

The global occurrence and distribution of *Azospirillum* from various geographical regions of our universe was reported by Amer et al. (1977). They reported the occurrence of *Azospirillum* sp. in the rhizosphere and rhizoplane of agricultural crops which are belonging to the family Leguminosae,

Cruciferceae, Gramineae, Solanaceae, and Tiliaceae. The occurrence of *Azospirillum* from continuous cultivation of the barely was reported by Idris et al. (1981). Nayak (1981) isolated the *Azospirillum lipoferum* and analyzed that the variations in their nitrogen-fixing efficiencies was wide in the surface of root and leaf of multiple medicinal plants such as *Euphorbia hirta*, *Ipomoea repens*, *Phyllanthus niruri*, *Clerodentron viscosum*, *Pistilla stratiotes*, *Leucas aspera*, *Cyprus* sp. and *Marsilia quadrifolia*. Tilak and Murthy (1981) documented the occurrence of *Azospirillum brasilense* as a microaerophilic because it requires only the minute quantity of oxygen for its survival, and this cannot be able to survive in the presence of high level of oxygen. They also noticed the *Azospirillum* sp. as non-symbiotic nitrogen fixer because of its association with the roots of agricultural crops like rice, sorghum, and maize.

Purusothaman and Oblisami (1985) reported the presence of nitrogen-fixing *Azospirillum* sp. in the rhizosphere region of herbal plants like *Cynodon dactylon*, *Boerhavia diffusa*, *Ipomea* sp. and *Cyperus rotundus* which are growing in the stress-tolerant saline and alkaline soils. Widespread geographic distributions of *Azospirillum* sp. are documented by a number of researchers, and its abundance was particularly observed in the rhizosphere region of tropical soils. Climatically, *Azospirillum* sp. is able to survive and grow in both cold and temperate climates. In the rhizosphere region, the occurrence of *Azospirillum* sp. was varied from 1% to 10% of the total rhizosphere microbial population. According to DeConinck et al. (1988), the population of *Azospirillum* sp. in the rhizosphere soil was comparatively more than the soil of the non-rhizosphere region.

Michiels et al. (1989) isolated the *Azospirillum* sp. globally from the wide variety of plants from tropical region, temperate region, salt-affected soil region, desert soil region to water flooded conditions. Likewise, Sumner (1990) also reported the occurrence of *Azospirillum* sp. in the rhizosphere region of field crops such as rice, maize, wheat, sorghum, and pearl millet. The rare air-borne phase dispersal of *Azospirillum* sp. in air was reported by Bashan (1990). Under diverse agronomic practices, George (1990) detected the *Azospirillum* from the roots of the coconut plant. The presence of *Azospirillum* sp. in the roots of sunflower was recorded in the research of Fages and Lux (1991) and the prevalence of *Azospirillum* sp. in the phyllosphere region of marine mangrove plants was proposed by Chaudhury and Sengupta (1991).

Roots of non-graminaceous crops also act as a source of *Azospirillum brasilense*. In Japan, Gamo and Alm (1991) isolated the *Azospirillum* sp.

from the roots of five non-graminaceous crops including *Spinacia oleracea*, *Brassica chinesis*, *Brassica rapa*, *Glycine max* and *Cucumis sativas*. Singh (1992) isolated the *Azospirillum* sp. within the stem and root nodules and stems of *Aeschynomene indica* and *Aeschynomene aspera*. *Azospirillum* mostly lives in association with the plants, particularly in their rhizosphere region of the root as Ectosymbiont (on the root surface) and to a lesser extent as Endosymbiont (inside the root) (Saha et al., 2001).

In the Tundra and Canadian high arctic region, Nosko et al. (1994) isolated the *Azospirillum* species from the 10 graminoid root species which are adjacent to the soil. Bashan and Holguin (1995) reported that the species of *Azospirillum* are able to colonize an average of 64 plant species, and among that, 18 plant species were belonged to weed species. Baldani et al. (1997) grouped the nitrogen-fixing bacteria into three categories based on their colonizing ability in graminaceous plants. The three categories are rhizosphere diazotrophs, facultative endophytic diazotrophs and obligate endophytic diazotrophs. Nath et al. (1997); Webber et al. (1999); and Hartman and Bashan (2009) isolated the *Azospirillum* sp. from the crop plants including grasses and cereals. Kirchhof et al. (1997) isolated the endophytic *Azospirillum lipoferum* from the rhizosphere of *Muscanthus* sp., *Muscanthus sinensis*, *Pennisetum purpureum* and *Spartina pectinata*.

Kavitha (2000) isolated the *Azospirillum* spp. from the wetland rice field and reported that the 18% of the heterotrophic population of bacteria are belonged to the genus *Azospirillum*. Yu and Mohan (2001) isolated the thermotolerant strains of *Azospirillum* from an aerated lagoon, where pulp and paper mill effluent are treated. Ravi Kumar et al. (2002) isolated the high-density saline tolerant *Azospirillum* strain from the roots of *Avicennia marina* which was more prevalent in the alkaline soils.

Around 300 isolates of *Azospirillum* sp. was isolated from the root tissues of cashew and its characteristics were completely studied by Purushothaman (2002). *Azospirillum* sp. has a tendency to survive in both saline and non-saline soils. The work of Saleena et al. (2002) was an evidence for that. They studied the diversity of *Azospirillum* in the rhizosphere of paddy plants which are grown in the soil of coastal agro-ecosystem with saline and non-saline nature. They also reported the predominance of two different *Azospirillum* sp. viz., *Azospirillum brasilense* and *Azospirillum lipoferum* on the coastal soil. Rao and Charyulu (2003) stated the rhizosphere region soil of foxtail millet contains associative symbionts like *Azospirillum* spp.

Yasmin et al. (2004) isolated and characterized different plant growth-promoting bacteria (PGPB) *Azospirillum* sp. in four soils and evaluated their potential use as biofertilizers for rice. *Azospirillum* sp. were isolated from the rhizosphere and rhizoplane of certain medicinal herbs, viz., *Catharanthus roseus*, *Ocimum sanctum*, *Phyllanthus amarus*, *Coleus forskholii* and *Aloe vera* and arid mine tailings.

Usha and Kanimozhi (2011) isolated the 10 strains of *Azospirillum* from the paddy field soil and characterized based on its morphological characters, cultural characters, and biochemical characters. Salt tolerance of the identified isolates was studied and it was observed that out of 10 strains, four strains were highly tolerant up to 70 mM. They concluded that the coastal environment frequently using the *Azospirillum* species for the cultivation of paddy plants.

Somayeh et al. (2012) tested the effectiveness of *Azospirillum* strains which are isolated from the wheat rhizosphere soil of saline stress environment under greenhouse conditions. In their experiment, they irrigated the wheat plants with different electrical conductivities of 0.7 dSm^{-1}, 4 dSm^{-1}, 8 dSm^{-1} and 12 dSm^{-1}. They reported that the plants inoculated with saline tolerant species of *Azospirillum* had higher nitrogen concentrations at all water salinity levels.

Ramyaanandan et al. (2013) selected five regions for paddy field rhizosphere soil collection and isolated 13 *Azospirillum* species. Among the 13 isolates, five were screened on the basis of sub-surface pellicle formation, size in micrometry, indole acetic acid production, and exopolysaccharide production. Out of five isolates, two *Azospirillum* isolates had shown more amounts of indole acetic acid and exopolysaccharide production, so that two isolates were selected for mass cultivation and seed dressing. With these two *Azospirillum* biofertilizers, pot culture experiments were conducted, and the plant growth parameters were evaluated under drought-prone conditions.

3.4 CHARACTERISTICS OF *AZOSPIRILLUM*

Azospirillum is a curve-shaped vibroid Gram-negative, measuring about 1–1.5 μm in diameter, motile because of possessing peritrichous flagella for swarming and a polar flagellum for swimming. The reserved cell inclusion energy is stored in the form of poly β-hydroxybutyrate (PHB) granules for preventing the cell during the lack of energy (Okon et al., 1976). Dobereiner et al. (1976) reported the *Azospirillum* sp. as an aerobic microorganism, but they are able to grow under microaerophilic conditions too that require oxygen at a low level. Surprisingly, it was observed that the *Azospirillum* sp. was highly effective in fixing atmospheric nitrogen under microaerophilic conditions.

Carbon source is the major nutrient requirement for the growth of bacteria and act as a backbone for the bacteria cell. Around 50% of the dry weight of bacteria is due to the carbon source. The salts present in the organic acids like acids like malate, succinate, lactate, and butyrate were preferably have been found to be an efficient and satisfactory carbon and energy sources for *Azospirillum* growth (Okon et al., 1976).

Loh (1982) reported that the growth of *Azospirillum brasilense* was good and luxuriant in the presence of succinate, but glucose does not showed any growth. In contrast, *Azospirillum lipoferum* could able to grow on both succinate and glucose. Both *Azospirillum brasilense* and *Azospirillum lipoferum* were differed in another way also. The *Azospirillum brasilense* could grow well and luxuriant on fructose, gluconate, galactose, and arabinose. Whereas, glucose, mannose, sorbose, and α-keto glutaric acid supported the growth of *Azospirillum lipoferum* to a greater extent. This shows that the nutritional requirements of *Azospirillum brasilense* and *Azospirillum lipoferum* were completely different (Del Gallo et al., 1984).

Chemotaxis is the process of movement of bacteria towards food (positive chemotaxis) and away from the toxic substance (negative chemotaxis). *Azospirillum* exhibit positive chemotaxis towards root exudates, organic acids, sugars, amino acids, and aromatic compounds (Zhulin and Armitage, 1993; Lopez de Victoria et al., 1994). Next to carbon source, the most needed nutrient requirement for microbial growth is nitrogen source. Ammonium or ammonia, nitrate, nitrite, amino acids, and molecular nitrogen can serve as nitrogen sources (Hartmann and Zimmer, 1994).

Azospirillum species can form the cyst under unfavorable conditions such as desiccation (extreme dryness) and nutrient limitation (lack of essential nutrients) (Sadasivan and Neyra, 1987). Tal et al. (1990) noticed that under conditions of stress and starvation, the accumulation of abundant poly-ß-hydroxybutyrate granules during encystations can serve as carbon and energy source for microorganisms. When compared to the other free living nitrogen fixing bacteria, *Azospirillum* is considered to be more efficient with nitrogenase properties comparatively better than other nitrogen fixing microorganisms (Okon, 1985). Phytohormone production and nitrogen fixation processes are the potential attributes for the plant growth-promoting character of *Azospirillum* (Steenhoudt and Vanderleyden, 2000).

3.5 *AZOSPIRILLUM*-PLANT INTERACTION

Nitrogen fixation is the key function of the *Azospirillum* sp. (Dobereiner and Day, 1976). Nitrogen fixation is the key function of the *Azospirillum* sp. The process of fixation of atmospheric nitrogen was carried out only under anaerobic conditions because the major enzyme required for nitrogen fixation is nitrogenase was sensitive to oxygen, and the process will be disturbed in the presence of oxygen (Steenhoudt and Vanderleyden, 2000). Further studies on the *Azospirillum* bioinoculant pinpointed the various beneficial effects on morphological and physiological changes in plants in the inoculated roots that would lead to an improvement in uptake multiple numbers of minerals and enhancement of water uptake (Okon and Kapulnik, 1986). Other physiological changes like getting resistance against the abiotic stresses also reported in the *Azospirillum* inoculated plants. Creus et al. (2005) inoculated the *Azospirillum* on the wheat seedlings and observed the development of osmotic stress. The growth efficiency of wheat with *Azospirillum* inoculated and non-inoculated seedlings was compared. It was found that the bioinoculant inoculated seeds has showed more fresh weight and yield of wheat when compared to non-inoculated seeds.

Plants exposed to salt stress have a possibility of suffering from water deficient. It was proved that inoculating with 10^8 cells of *Azospirillum brasilense* on root seedlings and thereafter exposed to mild and severe salt stress significantly reversed part of the negative effects. Casanovas et al. (2003) carried out the field experiments with the bioinoculant *Azospirilla* on *Sorghum bicolor* and *Zea mays*. Results have shown a significant increase in the growth, yield, and biochemical composition of crops along with better water and mineral uptake. Creus et al. (2004) reported that the *Azospirillum*-inoculated with the wheat seedlings were able to survive when exposed to the excess saline concentration up to 320 mM NaCl (sodium chloride) for an average of three days. The inoculation technology with *Azospirillum* sp. was extended to arid soil regions where the water scarcity was very high in order to protect the crops against drought conditions. Results of the experiments showed the significant increase in water content of plant, relative water content, water uptaking potential, apoplastic water fraction and lower cell wall modulus of elasticity values in *Azospirillum*-inoculated plants.

The beneficial effect of *Azospirillum* sp. in plants relies on the good colonization of roots and rhizosphere region. Root colonization is considered an important factor for creating the relationship of microorganisms with the plant, not only in infection caused by soil-borne phytopathogens (Negative effect) but also for the beneficial association with the plant growth supporting microorganisms (positive effect). During the positive effect, the first event in the colonization process is the adhesion of the bacteria to the plant roots. This *Azospirillum*-root interaction is a two-step process. The first step is adsorption mediated by the help of bacterial proteins. The second step is anchoring involved by the bacterial polysaccharides. The root colonization of *Azospirillum* sp. was highly depending on the active motility and chemotaxis towards the root exudates (Creus et al., 2004).

The distribution of *Azospirillum* isolates in the plant roots was studied by using the technique which analyzes the GFP-protein and tag bacteria. Liu et al. (2003) confirmed the colonizing pattern of *Azospirillum* sp. with plant roots. Some nitrogen-fixing strains of *Azospirillum lipoferum* and *Azospirillum brasilense* are established on the plant root surface, but other strains are not capable of colonizing the root interior surface in the apoplast and intercellular spaces. This ability means the lower vulnerability to harsh conditions that are imposed by the soil environment, which in turn supports the plant growth-promoting activities (Sturz and Nowak, 2000). The rhizobacteria which are establishing their relationship inside the plant roots are considered as endophytes. These microorganisms stimulate the plant growth by producing plant-growth-promoting hormones, enhance plant disease resistance against plant pathogens or improve the mobilization of nutrients in soil.

3.6 NITROGEN FIXATION BY *AZOSPIRILLUM*

Biological nitrogen fixation is a combination of microbiological and biochemical process in which atmospheric nitrogen is converted into simple ammonia that plants can use. Many workers reported the beneficial role of *Azospirillum* inoculation in enhancement of total nitrogen content, shoot, and grain yield of multiple plants (de Freitas, 2000; Saubidet et al., 2002; Kim et al., 2005; Felici et al., 2008; Bashan and DeBashan, 2010). In the beginning, the mode of action of *Azospirillum* sp. has not been well defined and not agreed up to 20 years of research. Although, Hartmann and Zimmer (1993) well defined the physiological properties of *Azospirillum*. Among them, the important characteristics are phytohormones production and its activities, nitrogen fixation ability, undefined signal molecules that interfere with plant metabolism, nitrite production and uptake of minerals.

The free-living *Azospirillum* bacterial strains are playing a vital role in the nitrogen cycle by fixing the atmospheric nitrogen efficiently, and they are associated with the rhizosphere region of the plant roots and participate in the nitrogen cycle (Heulin et al., 1987). The nitrogen balances in the plants are balanced through the nitrogen cycle. Among the various roles, the major role played by the *Azospirillum* species for improving the plant growth is nitrogen fixation through the nitrogen cycle. The acetylene reduction assay (ARA) is the method which was used to evaluate the incorporation of atmospheric nitrogen into the host plant by *Azospirillum* sp. (Van Berkum and Bohlool, 1980).

Tarrand et al. (1978) reported the efficiency of seed inoculation with *Azospirillum* in enhancing the seedling growth and vigor index in maize, sorghum, sunflower, bhendi, cotton, and rice. Kapulnik et al. (1985) reported an increase in the total nitrogen content in the shoots and grains of the plants inoculated with *Azospirillum* sp. According to Lima et al. (1987), 50% of the nitrogen content was supplied to the agricultural crops like Sugarcane, *Panicum maximum* and *Paspalum notatum* through *Azospirillum* sp. Mallik et al. (1987) showed that the grass derived its 30% to 60% of atmospheric nitrogen from *Azospirillum brasilense*. Kucey (1988) indicated that the inoculation of *Azospirillum* sp. in wheat and maize maximize the efficiency of nitrogen fixation in plants up to 18%.

Inoculation of *Azospirillum* increased the dry weight of root, shoot, and grain yield in finger millet (Rai, 1991), growth of jute (Bali and Mukerji, 1991), fruit production in pepper (Bumgardner and Mardon, 1992), increased germination in giant cactus (Puente and Bashan, 1993), increased number of seed per cob in maize (Fulchieri and Frioni, 1994) and increased seed germination and vigor index of chilies (Devi et al., 1995). Ignatov et al. (1995) proposed that the efficiency of nitrogen fixation was enhanced by *Azospirillum brasilense* in addition with wheat germ agglutinin. Bashan and Dubrovsky (1996) reported the nitrogen-fixing efficiency of *Azospirillum* isolates as 4.0 g N ha^{-1} day^{-1} in sorghum and 15 to 25 g N ha^{-1} day^{-1} in Corn.

Nitrogenase enzyme is an enzyme that plays a major role in the fixation of atmospheric nitrogen. Nitrogen fixation by an aerobic bacterium is a highly energy-requiring process that requires an efficient oxidative phosphorylation process for the generation of ATP (adenosine triphosphate), while the oxygen is toxic for the nitrogenase complex. The nitrogenase activity of single *Azospirillum* inoculum is comparatively less when compared to the dual *Azospirillum* inoculum and in other bacterial mixed cultures even if they originate from the entirely different ecosystem (Holguin and Bashan, 1996).

Ghosh and Puste (1997) concluded that the *Azospirillum lipoferum* inoculation in rice-wheat cropping sequence have the ability to increase the nitrogen content in soil between 35% and 36% and 7% to 9%. Christiansen-Weniger (1997) found that the nitrogenase activity induced by *Azospirillum* sp. significantly increased the tumor structure inhibiting efficiency in plants when compared to the untreated control plants. Unexpectedly, Yoneyama et al. (1997) reported that the contribution of *Azospirillum* sp. in biological nitrogen fixation and nitrogen uptake in different varieties of sugarcane. Marchal and Vanderleyden (2000) suggested that the inoculation of *Azospirillum* species might improve the nitrogen fixation efficiency and population of beneficial bacteria on roots of agricultural crops and consequently increased the production of plant growth-promoting substances.

In numerous greenhouse experiments and field experiments, inoculation of *Azospirillum* significantly reduces the required dose of nitrogen fertilizers in agricultural crops due to its replacement (Raja et al., 2006; Cassan et al., 2009). Oedjijono et al. (2014) observed the occurrence of *Azospirillum* spp. in iron-rich sands and rhizosphere soils. It was reported that the *Azospirillum* isolates are capable of fixing atmospheric nitrogen, producing indole acetic acid, and solubilizing inorganic phosphates under laboratory conditions. Most of *Azospirillum* isolates were capable of fixing atmospheric nitrogen ranging from 5.73–99.539 ppm of iron. Bassam et al. (2014) evaluated the efficiency of *Azospirillum* bioinoculant for its biological nitrogen fixation capability in the presence of molybdenum and found that the pre-inoculation with *Azospirillum* increases the yield and biochemical characters of soybean. Siddhartha et al. (2014) evaluated the association of *Cymbopogon winterianus* with the nitrogen-fixing bacteria, *Azospirillum*.

3.7 GROWTH HORMONE PRODUCTION BY *AZOSPIRILLUM*

Apart from nitrogen fixation, many *Azospirillum* isolates were proved to secrete other biologically significant growth hormones include indole acetic acid (IAA), indole lactic acid, indole-3-butyric acid (IBA), indole-3-ethanol, indole-3-methanol, unidentified indole compounds, gibberellins, abscisic acid (ABA) and cytokinins in lower quantities and auxin like phenylacetic acid (Bottini et al., 2004). *Azospirillum* uses the amino acid tryptophan as a precursor for IAA production. Since root exudates contain tryptophan, IAA production by *Azospirillum* will be of significant importance in the plant rhizosphere regions (Thuler et al., 2003).

Horemans et al. (1986) compared the efficiency of IAA production by *Azospirillum* sp. Comparatively, *Azospirillum brasilense* produce extremely high amount of IAA than *Azospirillum lipoferum*. Zakharova et al. (2000) performed the biochemical studies and high-pressure liquid chromatography (HPLC) analysis of *Azospirillum brasilense* and revealed high motive force towards tryptophan for IAA synthesis and this shows that indole act as the precursors for IAA in a tryptophan-independent pathway. The concentration of IAA was varying during different growth phases of the bacteria. The IAA concentration was low during the logarithmic or exponential growth phase and rapidly increased in the beginning of stationary phase. The significant increase of IAA was due to the presence of tryptophan in the culture medium (Ona et al., 2003; Tank and Saraf, 2003; Malhotra and Srivastava, 2009).

Phytohormones synthesized by *Azospirillum* indirectly influenced the uptake of minerals in the inoculated plants through (a) development of root hairs, (b) rate of respiration, (c) metabolism proliferation and (d) root proliferation (Molla et al., 2001; Radwan et al., 2002; Thuler et al., 2003). Naiman et al. (2009) inoculated the wheat with dual inoculants *Azospirillum brasilense* and *Pseudomonas fluorescens* strains and demonstrated that the development of roots and nitrogen-fixing efficiency was supported by the production of plant growth-promoting substances. The beneficial effect of *Azospirillum* sp. on plants has been initiated due to the production of gibberellins. The production of gibberellins has the ability to increase the root hair density in plants.

3.7.1 INDOLE ACETIC ACID (IAA)

In the culture medium supplemented with tryptophan, the *Azospirillum* sp. produced high quantities of indole acetic acid (IAA). El-Khawas and Adachi (1999) found that the addition of supernatants of filter-sterilized *Azospirillum* culture have the tendency to increase the root elongation, root surface area, root dry matter, and development of lateral roots and root hairs. The concentration of the supernatant should be optimum. In case, the concentration of the supernatant is high means it strongly inhibited the root elongation, lateral root development and caused nodule-like tumors on the roots. According to Dobbelaere et al. (1999), the addition of tryptophan enhanced the root morphology of plants by replacing *Azospirillum* cells with indole acetic acid.

Endophytic colonization of *Azospirillum* induces the formation of para-nodules in the roots of rice through the Auxins, 2,4-D, naphthalene acetic acid (NAA), and IAA enhanced polygalacturonase activity (Sekar et al., 1999). Combination of plant growth-promoting substances and auxin produced as a result of *Azospirillum* inoculation leads to morphological changes in the plant roots (Bashan and Holguin, 1997). Further evaluation by Dobbelaere et al. (1999) showed that the biosynthesis of auxin by *Azospirillum brasilense* showed an alteration in morphology of roots in wheat. It shows that the indole acetic acid synthesized by *Azospirillum brasilense* helps in the proliferation of wheat roots.

Zakharova et al. (2000) assessed the efficiency of *Azospirillum brasilense* for the production of indole acetic acid by using chemical methods and with High-Performance Liquid Chromatography (HPLC) of possible precursors like indole, anthranilic acid and tryptophan. It was revealed that the high motive force was must needed for the synthesis of tryptophan from chorismic acid and synthesis of indole acetic acid from tryptophan and unlikely this makes it anthranilic acid and indole that can act as a precursor for indole acetic acid through tryptophan-independent pathway. The regulation of indole acetic acid synthesis by *Azospirillum brasilense* was also supported by vitamins. Trace amount of vitamin B, especially the pyridoxine and nicotinic acid supported the *Azospirillum brasilense* for the indole acetic acid production.

Molla et al. (2001) reported that the cell-free supernatant of *Azospirillum brasilense* induced the highest number of roots and increased root length in soybean. Ona et al. (2003) conducted the research on indole acetic acid production by several strains of *Azospirillum* and confirmatory that the production of indole acetic acid was highly dependent on the culture media used and the availability of an amino acid tryptophan as a precursor. Among the multiple strains tested, *Azospirillum brasilense* produced the highest concentration of indole acetic acid among the strains tested. It was also observed that the pH of the culture medium has a significant effect on the level of indole acetic acid produced.

3.7.2 GIBBERELLINS

Next to indole acetic acid production, the beneficial effect of *Azospirillum* on plant was the production of gibberellins. The plant growth-promoting hormone gibberellins are responsible for increasing the root hair density. The gibberellin-producing strains *Azospirillum lipoferum* and *Azospirillum brasilense* produce gibberellin when cultured in the presence of glucosyl ester or glucoside of gibberellin A20 in the culture medium and hydrolyze the both conjugates. The findings of Piccoli et al. (1997) *in vitro* study supported the hypothesis that plant growth promotion was induced as a result of a combination of both gibberellin production and gibberellin-glucoside or glucosyl ester deconjugation by the *Azospirillum* species.

Lucangeli and Bottini (1997) explained the involvement of *Azospirillum* produced gibberellin in promoting the growth and yield of maize. Piccoli et al. (1999) showed the effect of oxygen concentration and water potential on growth of *Azospirillum lipoferum* and its ability of gibberellin A_3 production. It was observed that the increase in oxygen concentration and water potential was directly proportional to the production of gibberellin. Cassan et al. (2001) found that the bacterial enzyme 2-oxoglutarate-dependent dioxygenase produced by *Azospirillum brasilense* and *Azospirillum lipoferum* are responsible for the biosynthesis of plant growth-promoting gibberellin.

3.8 SIDEROPHORE PRODUCTION BY *AZOSPIRILLUM*

Siderophores (a Greek word which means "iron carrier") is defined as an iron-chelating, low molecular weight substance (500–1000 Daltons) with virtual specific ligand and the term Siderophore was first coined by Lankford (1973). Iron is the major component which supports the biosynthesis of siderophores, and they support the iron to the microbial cells. Siderophores play a major role in the uptake of iron by microbial cells. The PGPR secretes the iron-chelating compounds which are involved in sequestering of iron molecules in the root zone, making it unavailable to some rhizosphere microorganisms. Siderophores also bind to other metals such as Manganese, Chromium, Cadmium, and Lead (Birch and Bachofen, 1990) and act as a growth factor for enhancing plant growth and phytopathogenic suppressive agents for controlling plant diseases (Calvente et al., 2001).

Siderophores produced by root colonizing microbes may provide Fe to plant that can use the predominant siderophores types (Crowley and Wang, 2004; Lenin and Jayanthi, 2012). Many microorganisms have the ability to produce siderophores. Iron act as a co-factor for various enzymes such as peroxidase, aconitase, catalase, nitrogenase complex and ribonucleotide diphosphate reductase (Byers and Arceneaux, 1977). Suslow and Sehroth (1982) concluded that many microorganisms are able to produce the siderophore produced by the *Azospirillum* species helps in improving the iron nutrition and offers protection and resistance against the phytopathogens in plants.

3.9 ENHANCED MINERAL UPTAKE

Inoculation of *Azospirillum* biofertilizer enhanced the uptake of minerals like NO_3^-, NH_4^+, PO_4^{2-}, K^+, Rb^+ and Fe^{2+} (Sarig et al., 1988) and this kind of minerals uptake results in an increase in dry matter production and minerals accumulation in stem and leaves of plants. At the time of the plant reproductive period, these uptake minerals are transferred to the particles and spikes of plants and gives higher yield (Malhotra and Srivastava, 2009).

An increase in the efficiency of mineral uptake by plants inoculated with *Azospirillum* sp. has been observed as an important factor for the significant increase in the volume of root and not as a specific enhancement because of the normal iron uptake mechanism (Murthy and Ladha, 1988). Increased mineral uptake was observed in the inoculated roots, provided by enhancement in proton efflux activity of plant root particularly wheat inoculated with *Azospirillum* bioinoculant (Bashan et al., 1989; Bashan, 1995). This proton efflux activity is directly related to the balance of ions in plant roots. Conformity studies on the proton efflux phenomenon in wheat showed that *Azospirillum* inoculation enhanced the proton efflux and root elongation in wheat (Carillo et al., 2002; Creus et al., 2004).

3.10 CONCLUSION AND RECOMMENDATIONS

In the present-day situation, intensive agriculture needs the use of integrated nutrient management systems involving organic and inorganic sources of plant nutrients to sustain the yield of crop plants. Biofertilizers form an integral part of IPNS (integrated plant nutrition system) and organic farming, which constitutes the present as well as the future mandate of agriculture. *Azospirillum* spp. is of ubiquitous distribution globally in tropical, subtropical, and temperate climatic conditions and with a widespread of agricultural crops grown in different spectrum of soil types. *Azospirillum* spp. fixes atmospheric nitrogen and has been isolated from the rhizosphere of a variety of tropical and sub-tropical non-leguminous plants.

The occurrence of nitrogen-fixing *Azospirillum* species was observed in the roots of rice, sorghum, and maize. Various cereals have responded differently to field inoculation with this organism. Worldwide data accumulated over the past 20 years on field inoculation with *Azospirillum* has concluded that these bacteria are capable of increasing the yield of various agricultural crops in different soil and climatic regions.

Some of the most promising organisms, capable of colonizing roots in large numbers and exerting beneficial effects on plants belong to the genus *Azospirillum*. *Azospirillum* enhances the proper usage of fertilizers applied and enriching the soil with nitrogen that is fixed in association with the roots. The *Azospirillum* inoculant is efficiently used so far as carrier-based inoculant, and this is high adsorptive, easy to process, and non-toxic to *Azospirillum*. The use of a microbial inoculant is primarily concerned with crop productivity, not bacterial physiology nor ecology. Inoculant manufacturers must accept this reality and ensure that the probability of successful inoculation is maximized. Since crop response to inoculation, in most instances, is the result of appropriate strain selection and target organism population.

Azospirillum is considered a commercial inoculant for improving crop yields. The development of an effective seed or soil applied legume inoculant requires the integration of physical, chemical, and biological processes was stated by various researchers. Much research in this universe is done on *Azospirillum* strain selection and inoculation response. However, research conducted on inoculant production and formulation technologies is limited. A breakthrough is needed in inoculant technology to improve the shelf life and field efficacy of biofertilizers in India to make them commercially viable and acceptable to farmers.

KEYWORDS

- **acetylene reduction assay**
- **agricultural crops**
- ***Azospirillum* species**
- **enhanced mineral uptake**
- **gibberellins**
- **indole acetic acid**
- **induced systemic resistance**
- **nitrogen fixation**
- **nitrogenase enzyme**
- **phytohormones**
- **plant growth-promoting rhizobacteria**
- **plant rhizosphere**
- **poly-ß-hydroxybutyrate granules**
- **rhizosphere soil**
- **siderophores**

REFERENCES

Amer, H. A., Eid, M., Hegazi, N. A., & Monib, M., (1977). A survey on the occurrence of N_2 fixing *Spirillum* in Egypt. *Non-symbiotic N_2 Fixation Letters, 5*(7), 101, 102.

Baldani, J. I., Caruso, L., Baldani, V. L. D., Goi, S. R., & Dobereiner, J., (1997). Recent advances in BNF with non-legume plants. *Soil Biology and Biochemistry, 29*(9), 911–922.

Bali, M., & Mukerji, K. G., (1991). Interaction between VA mycorrhizal fungi and root microflora of jute. *Agricultural Management and Ecology, 24*(1), 396–401.

Bashan, Y., & de-Bashan, L. E., (2010). How the plant growth-promoting bacterium *Azospirillum* Promotes plant Growth-A critical Assessment. *Advances in Agronomy, 108*(5), 78–122.

Bashan, Y., & Holguin, G., (1995). *Azospirillum*-plant relationships: Environmental and physiological advances (1990–1996). *Canadian Journal of Microbiology, 43*(1), 103–121.

Bashan, Y., & Holguin, G., (1997). Nitrogen fixation and phytohormones production by *Azospirillum*. *Canadian Journal of Microbiology, 54*(7), 560–568.

Bashan, Y., & Levanony, H., (1989). Factors affecting adsorption of *Azospirillum brasilense* Cd to root hairs as compared with root surface of wheat. *Canadian Journal of Microbiology, 35*(2), 936–944.

Bashan, Y., & Levanony, H., (1990). Current status of *Azospirillum* inoculation technology-*Azospirillum* as a challenge for agriculture. *Canadian Journal of Microbiology, 36*(4), 591–608.

Bashan, Y., (1995). Short exposure to *Azospirillum brasilense* cd. inoculation enhanced proton efflux of intact wheat roots. *Canadian Journal of Microbiology, 36*(9), 419–425.

Bashan, Y., Holguin, G., & De Bashan, L. E., (2004). *Azospirillum*-plant relationships: Physiological, molecular and environmental advances (1997–2003). *Canadian Journal of Microbiology, 50*(3), 521–577.

Bassam, K. A. J., Halimi, M. S., Radziah, O., Sheikh, H. H., Hossain, K., & Saikat, H. B., (2014). Effect of *Azospirillum* in association with molybdenum on enhanced biological nitrogen fixation, growth, yield and yield contributing characters of soybean. *Journal of Food, Agriculture and Environment, 12*(2), 302–306.

Becking, J. H., (1963). Fixation of molecular nitrogen by an aerobic *Vibrio* or *Spirillum*. Antonie von leeweenhook. *Journal of Microbiology, 29*(3), 326–328.

Beijerinck, M. W., (1925). Zentrabl. bakteriol. parasitenkol, infekionskr. *Journal of Agricultural Advances, 263*(2), 353–359.

Bhattacharya, P., & Kumar, R., (2000). Liquid biofertilizer-Current knowledge and future prospect. *Paper Presented in National Seminar on Development and Use of Biofertilizers* (pp. 45–53). Biopesticides and organic manures, Kalyani, West Bengal, India.

Bilal, G., Rasul, J. A., Quershi, K. A., & Malik, A., (2014). Characterization of *Azospirillum* and related diazotrophs associated with roots of plant growing in saline soils. *World Journal of Microbiology and Biotechnology, 6*(4), 46–52.

Birch, L., & Bachofen, R., (1990). Complexing agents from microorganisms. *Experimental Biology, 46*(1), 827–834.

Bottini, R., Cassan, F., & Piccoli, P., (2004). Gibberellin production by bacteria and its involvement in plant growth promotion and yield increase. *Applied Microbiology and Biotechnology, 65*(2), 497–503.

Breed, R. S., Murray, E. G. D., & Smith, N. R., (1957). *Bergey's Manual of Determinative Bacteriology* (pp. 254–257). Williams and Wilkins, Baltimore, Maryland.

Bumgardner, C. J., & Mardon, D., (1992). Effects of *Azospirillum lipoferum* on dry matter accumulation and fruit production in greenhouse-grown bell pepper (*Capsicum annuum*) plants. *Transactions of the Kentucky Academy of Science (USA), 53*(9),101–108.

Byers, B. R., & Arceneaux, J. E. L., (1977). Microbial transport and utilization of iron. In: Weintberg, E.D., & Marceli, D., (eds.), *Microorganisms and Minerals*. New York.

Calvente, V., De Orellano, M. E., Sansone, G., Benuzzi, D., & Sanz De, T. M. I., (2001). Effect of nitrogen sources and pH on siderophore production by *Rhodotorula* strains and their application to biocontrol of phytopathogenic moulds. *Journal of Industrial Microbiology and Biotechnology, 26*(7), 226–229.

Carillo, G. A., Li, C. Y., & Baston, T., (2002). Increased acidification in the rhizosphere of cactus seedlings induced by *Azospirillum brasilense*. *Naturwissenschaften, 89*(2), 428–432.

Casanovas, E. M., Barassi, C. A., Andrade, F. H., & Sueldo, R. J., (2003). *Azospirillum* inoculated maize plant responses to irrigation restraints imposed during flowering. *Cereal Research Communications, 31*(3), 395–402.

Cassan, F., Bottini, R., Schneider, R., & Piccoli, P., (2001). *Azospirillum brasilense* and *Azospirillum lipoferum* hydrolyze conjugates of GA20 and metabolize the resultant aglycones to GA1 in seedlings of rice dwarf mutants. *Plant Physiology, 125*(5), 2053–2058.

Cassan, F., Maiale, S. O., Masciarelli, A., Vidal, V., Luna, V., & Ruiz, O., (2009). Cadaverine production by *Azospirillum brasilense* and its possible role in plant growth promotion and osmotic stress mitigation. *European Journal of Soil Biology, 45*(5), 12–19.

Chaudhury, S., & Sengupta, A., (1991). Association of nitrogen-fixing bacteria with leaves of *Avicennia officinalis* L. a tidal mangrove tree of Sundarban. *Indian Journal of Microbiology, 31*(7), 321–322.

Creus, C. M., Sueldo, R. J., & Barassi, C. A., (2004). Water relations and yield in *Azospirillum*-inoculated wheat exposed to drought in the field. *Canadian Journal of Botany, 82*(4), 273–281.

Creus, C., Graziano, M., Casanovas, E., Pereyra, M., Simontacchi, M., Puntarulo, S., Barassi, C., & Lamattina, L., (2005). Nitric oxide is involved in the *Azospirillum brasilense*-induced lateral root formation in tomato. *Planta, 221*(2), 297–303.

Crowley, D. E., & Wang, D., (2004). Utilization of microbial siderophores in iron acquisition by oat. *Plant Physiology, 87*(8), 680–685.

Day, J. M., & Dobereiner, J., (1976). Physiological aspects of nitrogen fixation by a *Spirillum* from *Digitaria* roots. *Soil Biology and Biochemistry, 8*(6), 45–50.

De Freitas, J. R., (2000). Yield and N assimilation of winter wheat (*Triticum aestivum* L., var Norstar) inoculated with rhizobacteria. *Pedobiologia, 44*(3), 97–104.

DeConinck, K., Horemans, S., Randombage, S., & Vlassak, K. (1988). Occurrence and survival of *Azospirillum* spp. in temperate regions. *Plant Soil, 110*(1988), 213–218.

Del, G. M., Martinez, D. G., Goebel, E. M., Burris, R. H., & Krieg, N. R., (1984). Carbohydrate metabolism in *Azospirillum* spp. In: Veeger, C., & Newton, W. B., (eds.), Advances in Nitrogen Fixation Research (p. 220). Martin Nijhoff. Dr. W. Junk Publishers, The Hague.

Devi, S. N., Vahab, M. A., & Mathew, S. K., (1995). Seedling vigor of chili as influenced by seed treatment with *Azospirillum*. *South Indian Horticulture, 43*(9), 54–56.

Dobbelaere, S., Croonenborghs, A., Thys, A., Van De, B. A., & Van, D. L. J., (1999). Phytostimulatory effect of *Azospirillum brasilense* wild type and mutant strains altered in IAA production on wheat. *Plant and Soil, 212*(7), 155–164.

Dobereiner, J., & Day, J. M., (1976). First international on nitrogen fixation, In: Newton, W. E., & Nyman, C. J., (eds.), *Proceedings of International Symposium on Nitrogen Fixation* (pp. 518–538). Washington, State University Press, Pullman, W.A.

Dubrovsky, J. G., Puente, M. E., & Bashan, Y., (1994). *Arabidopsis thaliana* as a model system for the study of the effect of inoculation by *Azospirillum brasilense* Sp 243 on root heir growth. *Soil Biology and Biochemistry, 26*(4), 1657–1664.

El-Khawas, H., & Adachi, K., (1999). Identification and quantification of auxins in culture media of *Azospirillum* and *Klebsiella* and their effect on rice roots. *Biology and Fertility of Soils, 28* (1), 377–381.

Fages, J., & Lux, J., (1991). *Azospirillum* and its effect on paddy cultivation. *Bioresource Technology, 13*(5), 1156–1162.

Felici, C., Vettori, L., Giraldi, E., Forino, L. M. C., Toffanin, A., Tagliasacchi, A. M., & Nuti, M., (2008). Single and co-inoculation of *Bacillus subtilis* and *Azospirillum brasilense* on *Lycopersicon esculentum*: Effects on plant growth and rhizosphere microbial community. *Applied Soil Ecology, 40*(2), 260–270.

Fulchieri, M., & Frioni, L., (1994). *Azospirillum* inoculation on maize (*Zea mays*): Effect of yield in a field experiment in Central Argentina. *Soil Biology and Biochemistry, 26*(4), 921–924.

Gamo, T., & Ahn, S. B., (1991). Growth promoting *Azospirillum* spp. isolated from the roots of several non-gramineous crops in Japan. *Soil Science and Plant Nutrition, 37*(7), 455–461.

George, M., (1990). *Azospirillum* for nitrogen fixation in coconut (a research note). *Philippines Journal of Coconut Studies, 15*(10), 1–3.

Ghosh, A., & Puste, A. M., (1997). Effect of irrigation and seed inoculation with *Azospirillum lipoferum* on growth and yield of wheat and residual total soil Nitrogen in rice-wheat cropping sequence. *Oryza, 34*(12), 77–79.

Hartmann, A., & Bashan, Y., (2009). Ecology and application of *Azospirillum* and other plant growth-promoting bacteria (PGPB). *European Journal of Soil Biology, 45*(9), 1–2.

Hartmann, A., & Zimmer, W., (1994). Physiology of *Azospirillum*. In: Okon, Y., (ed.), *Azospirillum-Plant Associations*. CRC Press, Florida, USA.

Heulin, T., Guckert, A., & Balandreau, J., (1987). Stimulation of root exudation of rice seedlings by *Azospirillum* strains: Carbon budget under gnotobiotic conditions. *Biological Fertilizers and Soils, 4*(1), 9–14.

Holguin, G., & Bashan, Y., (1996). Nitrogen-fixation by *Azospirillum brasilense* Cdis promoted when co-cultured with a mangrove rhizosphere bacterium (*Staphylococcus* sp). *Soil Biology and Biochemistry, 28*(4), 1651–1660.

Horeman, S., De coninck, K., Neuray, J., Hermans, R., & Vlassak, K., (1986). Production of plant growth substances by *Azospirillum* sp. and other rhizosphere bacteria. *Symbiosis, 5*(11), 341–346.

Idris, M., Memon, G., & Vinther, F. P., (1981). Occurrence of *Azospirillum* and *Azotobacter* and potential nitrogenase activity in Danish agriculture soils under continuous barley cultivation. *Acta Agricultural Science, 31*(2), 433–437.

Ignatov, V., Stadnik, G., Iosipenko, O., Selivanov, N., Iosipento, A., & Sergeeva, E., (1995). Interaction between partners in the association. Wheat *Azospirillum brasilense* sp. 245. *NATO Series, 37*(6), 271–278.

Kapulnik, Y., Okon, Y., & Henis, Y., (1985). Changes in root morphology of wheat caused by *Azospirillum* inoculation. *Canadian Journal of Microbiology, 31*(8), 881–887.

Kavitha, K., (2000). *Studies on Azospirillum Associated with Rice (Oryza sativa L.)* (p. 139). MSc (Ag.) Thesis, TNAU, Coimbatore.

Kim, C., Kecskes, M. L., Decker, R. J., Gilchrist, K., Newpeter, B., Kennedy, I. R., Kim, S., & Tongmin, R., (2005). Wheat root colonization and nitrogenase activity by *Azospirillum* isolates from crop plants in Korea. *Canadian Journal of Microbiology, 51*(11), 948–956.

Kirchhof, G., Schloter, M., Assmuss, B., & Hartmann, A., (1997). Molecular microbial ecology approaches applied to diazotrophs associated with non-legumes. *Soil Biology and Biochemistry, 29*(7), 853–862.

Krieg, N. R., (1976). Taxonomic studies of *Spirillum lipoferum*. In: Hollander, A., (ed.), *Genetic Engineering for Nitrogen Fixation* (pp. 462–472). Plenum Press, New York.

Kucey, R. M. N., (1988). Alternation of size of wheat root systems and nitrogen fixation by associative nitrogen-fixing bacteria measured under field conditions. *Canadian Journal of Microbiology, 34*(3), 735–739.

Lankford, C., (1973). Bacterial assimilation of iron. *CRC Critical Reviews of Microbiology, 2*(5), 273–331.

Lenin, G., & Jayanthi, M., (2012). Indole acetic acid, gibberellic acid and siderophore production by PGPR isolates from rhizospheric soils of *Catharanthus roseus*. *International Journal of Pharmaceutical and Biological Archives, 3*(4), 933–938.

Lima, E., Boddey, R. M., & Doberenier, J., (1987). Quantification of biological nitrogen fixation associated with sugarcane using a ^{15}N aided nitrogen balance. *Soil Biology and Biochemistry, 19*(7), 165–170.

Liu, Y., Chen, S. F., & Li, J., (2003). Colonization pattern of *Azospirillum brasilense* yu62 on maize roots. *Acta Botany, 45* (2), 748–752.

Loh, W. H., (1982). *Carbon Metabolism of Azospirillum and Associative Nitrogen Fixing Microorganism* (p. 147). Ph.D. Thesis, Ohio State University, Ohio.

Lopez De, V. G., Fielder, D. R., Zimmer, R. K., & Lovell, C. R., (1994). Motility behavior of *Azospirillum* species in response to aromatic compounds. *Canadian Journal of Microbiology, 41*(9), 705–711.

Lucangeli, C., & Bottini, R., (1997). Effects of *Azospirillum* spp. on endogenous gibberellin content and growth of maize (*Zea mays* L.) treated with uniconazole. *Symbiosis, 23*(3), 63–72.

Malhotra, M., & Srivastava, S., (2009). Stress-responsive indole-3-acetic acid biosynthesis by *Azospirillum brasilense* SM and its ability to modulate plant growth. *European Journal of Soil Biology, 45*(9), 73–80.

Mallik, K. A., Zafar, Y., Bilal, R., & Azam, F., (1987). Use of ^{15}N isotope dilution for quantification of N_2 fixation associated with roots of Kallar grass (*Leptochloa fusca*). *Biological Fertilizers and Soils, 4*(3), 103–108.

Marchal, K., & Vanderleyden, J., (2000). The oxygen paradox of dinitrogen-fixing bacteria. *Biological Fertilizers and Soils, 30*(1), 363–373.

Michiels, K., Vanderleyden, J., & Van, G. A., (1989). *Azospirillum* plant root associations. *Biological Fertilizers and Soils, 8*(2), 356–368.

Molla, A. H., Shamsuddin, Z. H., Halimi, M. S., Morziah, M., & Puteh, A. B., (2001). Potential for enhancement of root growth and nodulation of soybean co-inoculated with *Azospirillum* and *Bradyrhizobium* in laboratory systems. *Soil Biology and Biochemistry, 33*(5), 457–463.

Murthy, M. G., & Ladha, J. K., (1988). Influence of *Azospirillum* inoculation on the mineral uptake and growth of rice under hydroponic conditions. *Plant Soil, 108*(8), 281–285.

Naiman, A., Latronico, A., & Salamone, G., (2009). Inoculation of wheat with *Azospirillum brasilense* and *Pseudomonas fluorescens* impact on the production and culturable rhizosphere microflora. *European Journal of Soil Biology, 45*(2), 44–51.

Nath, D. J., Bhattacharjee, R. N., Devi, M. R., & Patgori, S. R., (1997). Widespread occurrence of *Azospirillum* in North Eastern Region of India. *Advances in Plant Science, 10*(5), 189–194.

Nayak, D. N., (1981). *Studies on Factors Influencing Nitrogen Fixation in Rice Soils* (p. 171). Ph.D., Thesis, Utkal University, Bhubaneswar.

Neyra, C. A., & Dobereiner, J., (1977). Nitrate reduction and nitrogenase activity in *Spirillum lipoferum*. *Canadian Journal of Microbiology, 23*(5), 306–310.

Nosko, P., Bliss, L. C., & Cook, F. D., (1994). The association of free-living nitrogen-fixing bacteria with the roots of High Arctic graminoids. *American Journal of Applied Research, 26*(4), 180–186.

Oedjijono, N., Endang, S. S., Sukarti, M., & Heru, A. D., (2014). Promising plant growth-promoting rhizobacteria of *Azospirillum* spp. isolated from iron sand soils, Purworejo coast, Central Java, Indonesia. *Advances in Applied Science Research, 5*(3), 302–308.

Okon, Y., & Kapulnik, Y., (1986). Development and function of *Azospirillum*-inoculated roots. *Plant and Soil, 90*(10), 3–16.

Okon, Y., (1985). *Azospirillum* is a potential inoculant for agriculture. *Trends in Biotechnology, 3*(5), 223–228.

Okon, Y., Albrecht, S. L., & Burris, R. H., (1976). Carbon and ammonia metabolism of *Spirillum lipoferum*. *Journal of Bacteriology, 128*(7), 592–597.

Ona, O., Smets, I., Gysegom, P., Bernaerts, K., Impe, J. V., Prinsen, E., & Vanderleyden, J., (2003). The effect of pH on indole-3-acetic acid (IAA) biosynthesis of *Azospirillum brasilense* sp7. *Symbiosis, 35*(5), 199–208.

Pereyra, M. A., Gonzalez, R. L., Creus, C. M., & Barassi, C. A., (2007). Root colonization vs. seedling growth, in two *Azospirillum*-inoculated wheat species. *Cereal Research Communications, 35*(7), 1621–1629.

Piccoli, P., Lucangeli, C. D., Schneider, G., & Bottini, R., (1997). Hydrolysis of (17,17-$2H_2$) gibberellin A20-glucoside and (17,17-$2H_2$) gibberellin A20-glucosyl ester by *Azospirillum lipoferum* cultured in a nitrogen-free biotin-based chemically-defined medium. *Plant Growth Regulation, 23*(7), 179–182.

Piccoli, P., Masciarelli, O., & Bottini, R., (1999) Gibberellin production by *Azospirillum lipoferum* cultured in chemically defined medium as affected by oxygen availability and water status. *Symbiosis, 27*(12), 135–145.

Pindi, G., (2012). Gibberellin production by bacteria and its involvement in plant growth promotion. *Applied Microbiology and Biotechnology, 65*(10), 497–503.

Puente, M. E., & Bashan, Y., (1993). Effect of inoculation with *Azospirillum brasilense* strains on the germination and seedling growth of the giant columnar card on cactus *(Pachycereus pringlei)*. *Symbiosis, 15*(6), 49–60.

Purushothaman, D., (2002). Biology of *Azospirillum* associated with cashew (*Anacardium occidentale*). In: *National Symposium and XII Southern Regional Conference on Microbial Inoculants*. Annamalai University, India.

Purusothaman, D., & Oblisami, G., (1985). Occurrence of *Azospirillum* in certain problem soils of Tamilnadu. *Symposium on Soil Biology*. HAU, Hissar.

Radwan, F. I., Innes, R. W., Kuempel, P. L., & Rolfe, B. C., (2002). Production of vitamins by *Azospirillum brasilense* in chemically-defined media. *Plant Soil, 153*(2), 97–101.

Rai, R., (1991). Studies on nitrogen fixation by antibiotic-resistant mutants by *Azospirillum brasilense* and their interaction with Cheena (*Panicummili aceum* L.) in calcareous soils. *Journal of Agricultural Science, 105*(4), 57–58.

Raja, P., Uma, S., Gopal, H., & Govindarajan, K., (2006). Impact of bio inoculants consortium on rice root exudates, biological nitrogen fixation and plant growth. *Journal of Biological Science, 6*(6), 815–823.

Ramyaanandan, A., Lakshmipriya, D., & Rajendran, P., (2013). Screening of *Azospirillum* for enhanced indole acetic acid production and exopolysaccharide for maximizing its survival rates in drought-prone paddy fields. *International Journal of Biology, Pharmacy and Allied Sciences, 2*(12), 2300–2311.

Rao, K. V. B., & Charyulu, P. B., (2003). Population dynamics and nitrogen fixation by diazotrophic bacteria in the rhizosphere of foxtail millet. *Indian Journal of Microbiology, 43*(9), 233–236.

Ravi, K. S., Ramanthan, G., Subha, N., Jeyaseeli, L., & Sukumaran, M., (2002). Qualification of halophilic *Azospirillum* from mangroves. *Indian Journal of Modern Science, 31*(7), 157–160.

Reinhold, B., Hurek, T., & Fendrik, I., (1987). Strain specific chemotaxis of *Azospirillum* sp. *Journal of Bacteriology, 162*(8), 190–195.

Sadasivan, L., & Neyra, C. A., (1987). Cyst production and brown pigment formation in aging cultures of *Azospirillum brasilense* ATCC 29145. *Journal of Bacteriology, 169*(3), 1670–1677.

Saha, A. K., Deshpande, M. V., & Kapadnis, B. P., (2001). Studies on survival of *Rhizobium* in the carriers at different temperatures using green fluorescent protein marker. *Current Science, 80*(5), 669–671.

Saleena, L. M., Rangarajan, S., & Nair, S., (2002). Diversity of *Azospirillum* strains isolated from rice plants grown in saline and non saline sites of coastal agricultural ecosystem. *Microbial Ecology, 44*(7), 271–277.

Sarig, S., Blum, A., & Okon, Y., (1988). Improvement of the water status and yield of field-grown grain sorghum (*Sorghum bicolor*) by inoculation with *Azospirillum brasilense*. *Journal of Agricultural Sciences, 110*(10), 271–277.

Saubidet, M. I., Fatta, N., & Barne, A. J., (2002). The effect of inoculation with *Azospirillum brasilense* on growth and nitrogen utilization by wheat plants. *Plant Soil, 245*(10), 215–222.

Sekar, C., Prasad, N. N., & Sundaram, M. D., (1999). Enhancement of polygalacturonase activity during auxin induced paranodulation and endorhizosphere colonization of *Azospirillum* in rice roots. *Indian Journal of Experimental Biology, 38*(11), 80–83.

Shanon, M. C., (2013). Effect of salinity on growth and accumulation of organic and inorganic ions in cultivated and wild tomato species. *Journal of American Society and Horticulture Science, 112*(20), 446–449.

Siddhartha, P. S., Adrita, G., Kalpataru, D. M., Animesh, G., Rumi, K., Himangshu, L., & Nabanita, H., (2014). Effect of 2,4-D treatment and *Azospirillum* inoculation on growth of *Cymbopogon winterianus*. *African Journal of Microbiology Research, 8*(9), 955–960.

Singh, K. P., (1992). Influence of stimulated water stress on free proline accumulation in wheat (*Triticum aestivum* L.). *Indian Journal of Plant Physiology, 29*(3), 319–321.

Somayeh, H. N., Mohammad, J., Farhad, R., & Verma, A., (2012). Yield and yield components of wheat as affected by salinity and inoculation with *Azospirillum* strains from saline or non-saline soil. *Journal of the Saudi Society of Agricultural Sciences, 11*(12), 113–121.

Steenhoudt, O., & Van, D. L. J., (2000). *Azospirillum* a free-living nitrogen-fixing bacterium closely associated with grasses: genetic, biochemical and ecological aspects. *FEMS Microbiology Reviews, 24*(10), 487–506.
Sturz, A. V., & Nowak, J., (2000). Endophytic communities of rhizobacteria and the strategies required to create yield enhancing associations with crops. *Applied Soil Ecology, 15*(2), 183–190.
Sumner, M. E., (1990). Crop responses to *Azospirillum* inoculation. *Advances in Soil Science, 12*(11), 53–123.
Suslow, T. V., & Schroth, M. N., (1982). Rhizobacteria of sugar beet: Effect of seed application and root colonization on yield. *Phytopathology, 72*(2), 199–206.
Tal, S., & Okon, Y., (1990). Purification and characterization of β-hydroxybutyrate dehydrogenase from *Azospirillum brasinlense*. *Journal of General Microbiology, 136*(10), 645–653.
Tank, N., & Saraf, M., (2003). Phosphate solubilization, exopolysaccharide production and indole acetic secretion by rhizobacteria isolated from *Trigonella foenum-graecum*. *Indian Journal of Microbiology, 43*(9), 37–40.
Tarrand, J. J., Krieg, N. R., & Dobereiner, J., (2015). A taxonomic study of the *Spirillum lipoferum* group with description of a new genus, *Azospirillum* and two species, *Azospirillum lipoferum* (Beijerinck) comb.nov. and *Azospirillum brasilense* nov. *Canadian Journal of Microbiology, 24*(12), 967–980.
Thuler, D. S., Floh, E. J. S., Handro, W., & Barbosa, H. R., (2003). Plant growth regulators and amino acids released by *Azospirillum* sp. in chemically defined media. *Letters in Applied Microbiology, 37*(9), 174–180.
Tilak, K. V., & Murthy, B. N., (1981). Occurrence of *Azospirillum* in association with the roots and stems of different cultivars of barley *(Hordeum vulgare)*. *Current Science, 50*(1), 496–498.
Tilak, K. V., (1997). Recent developments in *Azospirillum* plant associations. In: Reddy, S. M., Srivastava, H. P., Purohit, K., &. Ram, R. S., (eds.), *Microbial Biotechnology* (pp. 41–51). Scientific Publishers, New Delhi.
Usha, D. K., & Kanimozhi, K., (2011). Isolation and characterization of saline tolerant *Azospirillum* strains from paddy field of Thanjavur district. *Advances in Applied Science Research, 2*(3), 239–245.
Van, B. P., & Bohlool, B. B., (1980). Evaluation of nitrogen fixation by bacteria in association with roots of tropical grasses. *Microbiology Reviews, 44*(8), 491–517.
Webber, O. B., Baldani, V. L. D., Teiveira, K. R. S., Kirchoff, G., Baldani, J. I., & Dobereiner, J., (1999). Isolation and characterization of diazotrophic bacteria from banana and pineapple plants. *Plant and Soil, 210*(10), 103–113.
Weniger, C. C., (1997). Ammonium-excreting *Azospirillum brasilense* C3: *Gus A* inhabiting induced tumors along stem and roots of rice. *Soil Biology and Biochemistry, 29*(12), 943–950.
Yasmin, S., Rahman, B. M. A., Malik, K. A., & Hafeez, F. Y., (2004). Isolation characterization and beneficial effect of rice associated plant growth-promoting bacteria from Zanzibar soils. *Journal of Basic Microbiology, 44*(7), 241–252.
Yoneyama, T., Muraoka, T., Kim, T. H., Dacaney, E. V., & Nakanishi, Y., (1997). The natural ^{15}N abundance of sugar cane and neighboring plants in Brazil, the Philippines and Miyako (Japan). *Plant Soil, 189*(3), 239–244.

Yu, Z., & Mohn, W. W., (2001). Bacterial diversity and community structure in aerated lagoon by ribosomal intergenic spacer analysis and 16S ribosomal DNA sequencing. *Applied and Environmental Microbiology, 67*(5), 1565–1574.

Zakharova, E. A., Iosipenko, A. D., & Ignatov, V. V., (2000). Effect of water-soluble vitamins on the production of indole-3-acetic acid by *Azospirillum brasilense. Microbiology Research, 155*(7), 209–214.

Zhulin, I. B., & Armitage, J. P., (1993). Motility, chemokinesis and methylation-independent chemotaxis in *Azospirillum brasilense. Journal of Bacteriology, 175*(10), 952–958.

CHAPTER 4

APPLICATION OF PHOSPHATE SOLUBILIZING MICROORGANISMS FOR EFFECTIVE PRODUCTION OF NEXT-GENERATION BIOFERTILIZER: A PANACEA FOR SUSTAINABLE ORGANIC AGRICULTURE

CHARLES OLUWASEUN ADETUNJI,[1]
OSIKEMEKHA ANTHONY ANANI,[2] DEVARAJAN THANGADURAI,[3]
and SAHER ISLAM[4]

[1]*Microbiology, Biotechnology, and Nanotechnology Laboratory, Department of Microbiology Edo University Iyamho, Edo State, Nigeria*

[2]*Laboratory of Ecotoxicology and Forensic Biology, Department of Biological Science, Faculty of Science, Edo University Iyamoh, Edo State, Nigeria*

[3]*Department of Botany, Karnatak University, Dharwad, Karnataka, India*

[4]*Institute of Biochemistry and Biotechnology, Faculty of Biosciences, University of Veterinary and Animal Sciences, Lahore, Pakistan*

ABSTRACT

Phosphorus has been identified as one of the most numerous elements that are available in the earth's crust which exists in the form of both inorganic and organic forms. The presence of phosphorus is available in a very high concentration but only 0.1% of the entire phosphorus are made available to the plant which might be linked to their poor solubility and it could form a complex with other metallic elements in the soil such as Fe, Ca, Al to form ferrous phosphate, calcium phosphate, and aluminum phosphate which are unavailable to the plants. The application of synthetic fertilizer has several health and environmental hazards such as soil fertility depletion, pollution of the environment, and eutrophication has necessitated the need to search for an alternative approach that could help in feeding the ever-increasing population. The application of phosphate-solubilizing microorganisms has been identified as a sustainable, ecofriendly, green, biocompatible biotechnological tool that could lead to an increase in agricultural production and ecorestoration of heavily polluted soil. Therefore, this chapter intends to provide a holistic detail about the application of phosphorus solubilizing microorganisms and their interaction with the agricultural crop, their application as a plant growth promoter, and their utilization for the ecorestoration of heavily polluted soil. The modes of action utilized by these phosphate-solubilizing microorganisms were also highlighted. Special emphasis was also laid on the application of bacterial strain and fungal strains that have been reorganized as phosphorus solubilizing microorganisms while their relative performance when exploring under in situ conditions that could lead to enhanced phosphorus solubilization was also discussed.

4.1 INTRODUCTION

The sudden rise in the level of the global population has necessitated an increase in the request for enhanced food production and the high rates of anthropogenic problems majorly on the environment have resulted in several challenges in Agriculture yield (Adetunji et al., 2019a–d; Adetunji, 2019; Adetunji and Ugbenyen, 2019). These problems have led stimulates several reactions, and it is of global concern. Therefore, if rapid attention is not given to these highlighted challenges, it might lead to high rates of inefficiency in feeding the global population (Ladeiro, 2012). The present population of the globe has been approximating to be 7 billion but there is a greater tendency that it will rise drastically to 10 billion in the next 50 years. Hence, there is a need to identified agricultural methodologies and stratagems that could help to resolve the problem of feeding the ever-increasing populations (Glick, 2014). Some other specific examples include rapid loss in the production of agricultural produce, which might be linked to abiotic and biotic stress imposed on crops planted in the field.

Abiotic stress has been identified as a significant factor that mitigates against an increase in the production of agricultural production. Examples of such abiotic stress include environmental temperature, soil salinization, drought, soil pH, soil sodification, and rapid loss of soil nutrition. Among all them, soil sodification and soil salinization have been highlighted as the major pervasive soil degradation practices that affect the earth jeopardizing the prospective utilization of soils (Rengasamy, 2006; Ladeiro, 2012). Also, water and wind erosion is another abiotic factor that could cause a dangerous impact and led to a high rate of soil degradation (Ladeiro, 2012).

Biotic stress has been highlighted as another serious factor that could result in a drastic reduction of agricultural yield, which might lead to diseases and high infestation by agricultural pests. Typical examples of these microorganisms include parasites, bacteria, viruses, and fungi. These pathogenic microorganisms could result in a 30% reduction in annual agricultural increases (Fisher et al., 2012). Therefore, there is a need to search for a sustainable agricultural solution such as the utilization of safe and more effective pesticides, enhanced agricultural land management, enhanced application of chemical fertilizers, improved planting of transgenic crops, and more farm mechanization. Hence, all the aforementioned solutions could only resolve agricultural challenges for a limited period of time because of limited resources. This implies that there is a need to search for more effective solutions that could support sustainable organic agriculture. The application of eco-friendly biological solutions through the application of beneficial microorganisms has been highlighted as a striking technology hypothesized that could mitigate various challenges affection agricultural challenges.

Phosphorous has been recognized as one of the numerous metallic elements discovered in the earth's crust available in the soil in both organic and inorganic forms (Gyaneshwar et al., 2002). It is normally absorbed and utilized by plants in the form of the inorganic form (H_2PO_4-and HPO_{42}–) (Hinsinger, 2001). It plays a crucial function in the metabolic process which includes prevention of plant diseases, enhancement of crop qualities, energy transfer, fixing of nitrogen, improvement in the rate of photosynthesis, signal transduction, and all these features are related to phosphorous nutrition (Khan et al., 2014). Moreover, it has been discovered that phosphorus constitutes major parts of microorganisms and plays a vital role such as the movement and storage of energy, and constitutes a major part of the DNA (deoxyribonucleic acid) of all microorganisms.

Phosphorus plays a crucial role in the carbon metabolism, membrane formation and photosynthesis, elongation of plant roots, and it has been observed that the deficiency of phosphorus might lead to high instability in the root architecture (Williamson et al., 2001; Wu, 2005). The significant constituent of phosphorus takes in by the plant is build up ingrain in the form of phytic acid which later become immobile to plants and consequently affect the development of crops (Richardson, 1994).

The stratagem of improving the presence of a low quantity of phosphorus fertility production has been recognized as a problem most especially in acid soils because the high level of phosphorus that will be required might not be economically and environmentally viable in bioremediation of these soils (Hinsinger, 2001). The microbial population has been recognized as a major component of the soil-plant continuum where they play a crucial role in the development of plants, most especially in sustainable agriculture (Vassilev et al., 2006).

Phosphorus solubilizing microorganisms have been that they possess the capability to mineralize phosphorus and solubilize from organic and inorganic pools of total soil phosphorus and may be applied as inoculants to enhance P-availability to plants (Richardson, 1994, 2001).

Therefore, this chapter intends to provide a comprehensive review of the application of phosphorus solubilizing microorganism and their application in sustainable agriculture for adequate enhancement of agriculture. Special emphasis was laid on some of their role as biofertilizers, mineralization, and solubilization of phosphorus, and their methodologies used in the selection of these potential strains were also highlighted. The mechanism of action involved in the solubilization was also discussed in detail.

4.2 ISOLATION OF MICROBIAL STRAINS WITH PHOSPHORUS SOLUBILIZING CAPABILITY

Several beneficial microorganisms have been recognized for their capability to perform their beneficial role in the plant. Many of them are related to the plant rhizosphere, and they are commonly referred to as rhizobacteria and because they possess the capability to enhance plant growth, they are called plant growth-promoting rhizobacteria. Typical examples of such bacteria include *Serratia*, *Alcaligenes*, *Acinetobacter*, *Rhizobium*, *Arthrobacter*, *Pseudomonas*, *Azospirillum*, *Paenibacillus*, *Burkholderi*, *Bacillus*, *Flavobacterium*, *Erwinia*, and *Enterobacter*. They are utilized as a biofertilizer for the enhancement of crops. Phosphorus has been identified as the second mineral element to nitrogen that could prevent adequate development of crops. Phosphorus has been identified as an essential element necessary for the adequate development of plants and constitutes about 0.2% of plant dry weight. The plant needs phosphorus from the soil solution as phosphate anions. Conversely, it has been discovered that phosphate anions are exceedingly reactive and may be immobilized through precipitation with cations such as Al^{3+}, Ca^{2+}, Fe^{3+}, Mg^{2+} which varies depending on the types of the soil. This makes phosphorous not to be available to plant in this form. But several reports have been documenting the capability of bacterial species to solubilize insoluble inorganic phosphate compounds such as dicalcium phosphate, rock phosphate (RP), tricalcium phosphate (TCP), and hydroxyapatite.

Microbial solubilization of phosphorus from insoluble phosphates is an environmentally friendly and cost-effective approach in sustainable soil management. Introducing the indigenous microorganisms to soil requires a shorter adaptation period and causes fewer ecological distortions than exogenous microorganisms. This study was conducted to isolate and identify the indigenous fungi for phosphate solubilization in Mazandaran, Iran. A potent phosphate solubilizing fungus was isolated from an Iranian phosphate mine and selected for solubilization of RP. The identified fungus was characterized by calmodulin-based polymerase chain reaction (PCR) method as *Aspergillus tubingensis* SANRU (Sari Agricultural Sciences and Natural Resources University). The phosphate solubilization ability of the fungal strain was carried out in shake flask leaching experiments containing various concentrations of RP (1%, 2%, 4%, or 8% w/v). The maximum P solubilization rate of 347 mg/l was achieved at 1% of RP concentration on day 9. The regression analysis indicated that the P solubilized mainly through acidification. This study shows the possibility of using *A. tubingensis* SANRU for application in the management of P fertilization. It has been recognized that numerous proportions of phosphate solubilizing microorganisms (PSMs) is available in the rhizosphere, and they show more activity than any other sources (Vazquez et al., 2000).

Normally, 1 gram of fertile soil contains 101 to 1010 bacteria and their live weight may go beyond 2,000 kg ha^{-1}. It has been observed that the total number of soil bacterial varies in shapes which include a high proportion of PSM is concentrated in the rhizosphere, and they are metabolically more active than from other sources (Vazquez et al., 2000). Usually, one gram of fertile soil contains 101 to 1010 bacteria, and their live weight may exceed 2,000 kg ha^{-1}. Soil bacteria are in spiral (1–100 μm), cocci (sphere, 0.5 μm), and bacilli (rod, 0.5–0.3 μm).

However, bacilli have been documented as the most common of all of them but spirilli are very rare in natural environments (Baudoin et al., 2002). The population of the phosphorus solubilizing microorganism varies from the soil but depends majorly on the numerous soil features such as phosphorus content, physical, and chemical properties, organic matter, and cultural activities (Kim et al., 1998; Yahya and Azawi, 1998; Fallah, 2006).

4.3 PHOSPHATE SOLUBILIZING MICROORGANISMS (PSMS) AS BIOFERTILIZER ON THE FIELD, GREENHOUSE, POT ASSAY, AND LABORATORY TRIALS

The agriculture sector plays a pivotal role in food production. Food is a major necessity for human health and well-being. Day in, day out, crop productivity in agro-activities requires the use of inorganic fertilizer to boost its viability and consistent availability to the teeming hungry population of the world. However, this chemical (fertilizers), especially the widely used form; N (Nitrogen), P (Phosphorus), and K (Potassium) in different ratios, have been used indiscriminately to advance crop output.

Phosphorus, one of the major ingredients used as inorganic fertilizer is very important in plant development and growth. However, just a small percentage of it is directly absorbed by the root system of plants, while the rest is washed away by erosion to nearby water bodies. This might serve as a precursor of eutrophication if the quantity used or applied during planting and persistent usage is very high. More so, some of the residues will combine with specific trace metals (Aluminum, Iron, and Calcium) that the plants cannot absorb directly because they are in an insoluble state or form. The role of PSMs in the sustainable utilization and regulation of insoluble phosphate chelate with specific trace metals in the agro-fertilizer application is very important in plummeting the encumbrance faced by farmers in improving the health, ecological, and productivity status of food crops. It has been revealed that Phosphate-solubilizing fungi possess the potential to improve the availability of phosphorus in the soil, most especially in the soil where there are P-limiting regions. In this regard, Zhang et al. (2018) isolated two strains AspN-JX16 and TalA-JX04 from the rhizosphere soil of moso bamboo (*Phyllostachys edulis*). These strains isolated have been affirmed to be P-limiting regions in China. The strains were identified as *Aspergillus neoniger* and *Talaromyces aurantiacus* based on morphologic and phylogenetic characteristics. They were cultured in a submerged medium containing potato dextrose liquid medium with six different initial pH values which vary from 6.5 to 1.5 to assess acid resistance. The phosphate-solubilizing fungi were cultured in Pikovskaya's (PVK) liquid media with numerous pH which entails $C_6H_6Ca_6O_{24}P_6$, $Ca_3(PO_4)_2$, $AlPO_4$, $FePO_4$, and $CaHPO_4$ to assess their P-solubilizing potential. The results obtained show that there was no significant difference in the biomass of all the fungi cultivated in the media with numerous initial pH showing that these fungi could be cultivated when subjected to acid stress. The P-solubilizing potential of TalA-JX04 had the maximum value followed by $CaHPO_4$, followed by $Ca_3(PO4)_2$, $FePO_4$, $C_6H_6Ca_6O_{24}P_6$, and $AlPO_4$ among all the six types of initial pH treatment while the P-solubilizing capacity of AspN-JX16 was more significant when compared to that of TalA-JX04. It was also observed that the fungi enhance the dissolution of phosphorous potential, which indicates that it had a negative correlation to the pH of the fermentation broth. This indicates that the dissolution of Phosphorus sources by producing organic acids. Our result indicated that AspN-JX16 and TalAJX04 could survive in an acidic environment and these two fungi have the capability to liberate soluble phosphorus by breaking the P-breaking down compound, most especially that contain aluminum, phytate, calcium, iron phosphates. The two had the capacity to be utilized in the mass production of environment-friendly biofertilizers most especially in the subtropical bamboo ecosystem.

Phosphorus has been highlighted as one of the most crucial macronutrients that are important for plant development and growth. It has been observed that inorganic phosphorus (P), which constitutes 70% of the total P content in soils could exist inform of iron-complexed, calcium, and aluminum which are not made available for plant utilization. P_2O_5 has been applied as a fertilizer to augment the nutrient for crop development as a consequence of this mineral phosphorus. The application of phosphate-solubilizing microorganisms has been applied to decrease the incorporation of mineral phosphorus to agricultural soils. Sharon et al. (2016) discovered that some bacteria could solubilize phosphate present in the soil. It was discovered that *Pantoea* sp. could solubilize TCP ($Ca_3(PO_4)_2$) which was determined at the rate of 956 mg L^{-1}. The phosphorus solubilizing producing bacterial strain could generate different types of gluconic, propionic acids, organic acids, formic, and acetic. The experiment performed in the greenhouse with a tomato plant that has been inoculated with *Pantoea* sp. showed more enhanced biomass and phosphorus when compared to the uninoculated.

Walpola et al. (2013) isolated a phosphate-solubilizing bacterium and evaluated their influence on the development of mung bean (*Vigna radiata* [L.] R. Wilczek) plants. Their study showed that only two strains possess that capability to solubilize phosphorous which were *Burkholderia anthina* (PSB-2) and *Pantoea agglomerans* (PSB-1). The highest phosphorus solubilizing value of 720.75 μg mL^{-1} was obtained from the cultures that were co-inoculated with *B. anthina* and *P. agglomerans*. Their result also indicated that some level of correlation exists between the phosphorus solubilization, pH, the concentration of soluble phosphorus, and titratable acidity (TA). The two tested strains indicate an increase in the shoot and root dry matter, phosphorus uptake of mung bean plants, while the addition of TCP with PSB inoculation enhanced the development of the tested plant. The incorporation of TCP and phosphorus solubilizing bacteria strains enhanced the development and uptake of phosphorus mung bean plants. Their study indicated that they could be applied as a sustainable alternative that could reduce the challenges of phosphorus in agricultural soils.

The application of phosphate solubilizing bacteria (PSB) for the production of biofertilizer has enhanced the absorption of phosphorus adsorption in plants and enhanced the yields in numerous crop species. Baliah et al. (2016) evaluate the capability of different PSB that dwell in the rhizosphere of the soil from eggplant, okra, tomato, chili, and cotton. The result indicated that the population dynamics of PSB were detected in higher content in the rhizospheric soil of tomato followed by brinjal. It was observed that 6 strains out of the 10 evaluated strains could solubilize phosphate. The isolated strains include *Pseudomonas putida*, *P. fluorescens*, and *Bacillus megaterium*. Their study showed that the selected strains could utilize amino acid, vitamin, carbon, and nitrogen as their sources of the nutrient.

The phosphate-solubilizing capability of the microbial community that dwells in the rhizosphere of mangroves plants was carried out with culture media. Besides the insoluble type tribasic calcium phosphate was incubated with the root of white [*Laguncularia racemosa* (L.) Gaertn.] and black (*Avicennia germinans* L.) mangrove. Vazquez et al. (2000) demonstrated the phosphate-solubilizing effectiveness of some bacterial strains isolated from the rhizosphere of the mangroves. Examples of the isolated strain include *Chryseomonas luteola*, *Enterobacter aerogenes*, *Bacillus amyloliquefaciens*, *Bacillus licheniformis*, *Pseudomonas stutzeri*, *Kluyvera cryocrescens*, *Bacillus atrophaeus*, *Enterobacter asburiae*, *Paenibacillus macerans*, *Enterobacter taylorae*, and *Xanthobacter agilis*. Their study showed that *Aspergillus niger* was isolated from the rhizosphere of black mangroves. The phosphate-solubilizing activity of all the isolates was detected using a solid medium having tribasic calcium phosphate which serves as a source of phosphorus. Their study showed that the Spectrophotometric quantification of the solubilized phosphate indicated that the *A. niger* and all the isolated bacterial strains had an enhanced solubilization activity when tested in submerged fermentation. Hence, it was observed that *V. proteolyticus* was the best solubilizing strain among all the isolated bacteria. The gas chromatographic evaluation showed the presence of 11 numerous unidentified non-volatiles and volatiles organic acids. The highest isolated compounds include isobutyric, acetic acids, lactic, and isovaleric. It was observed that all the bacterial species generated more than one organic acid while the *A. niger* generates only succinic acid. Their study showed that the generation of organic acid by these mangrove rhizosphere microorganisms represent the modes of action these microorganisms applied in the solubilization of insoluble calcium phosphate.

Inorganic phosphate-solubilizing bacteria have been highlighted as a significant component of microbial populations that occur in the lake sediments. Moreover, the liberated phosphorous becomes a significant source of phosphorous for eutrophic algae. Li et al. (2019) isolated and screened inorganic phosphate-solubilizing bacteria mainly from the sediments of Sancha Lake using the National Botanical Research Institute's phosphate (NBRIP) plates. The molecular characterization of these strains was later performed using the 16S rDNA method. The TCP-solubilizing capability of the Inorganic phosphate-solubilizing strains was eased utilizing PVK liquid medium and NBRIP-bromophenol blue (BPB). The capability of the Inorganic phosphate-solubilizing to liberate phosphorous from the sediments was carried out by mimicking the lake environment. The authors isolated and screened a total of 43 Inorganic phosphate-solubilizing strains from the sediments of Sancha Lake that could be categorized into 10 genera, eight families, and three phyla. Their study indicated that SWSI1734 and SWSI1728 which could be classified as *Bacillus* and a novel strain SWSI1719 belong to the family Micromonosporaceae. It was discovered that the inorganic phosphate-solubilizing were very diverse, while it was discovered that *Paenibacillus* and *Bacillus* were the two dominant strains. It was also observed that 30 of the 43 out of the inorganic phosphate-solubilizing bacteria showed a clear zone of inhibition on the plates when tested in the liquid culture experiment, and it was affirmed that all the strains possessed the capacity to soften TCP. The phosphate-solubilizing capability of the strains was varied significantly and the strain SWSI1725 of the *Bacillus* genus indicated the strongest capability with enhanced phosphate-solubilizing content of 103.57 mg/L. The sterilized systems indicated significantly enhanced breaking down of phosphorus hydrochloride (HCl-P) and liberation from the sediments after inoculation of inorganic phosphate-solubilizing strains, but there was no significant influence for phosphonium hydroxide (NaOH-P). Their study indicated that Inorganic phosphate-solubilizing available in the sediments of Sancha Lake possesses rich diversity and the capability to liberate phosphorus in sediments.

Nutrient management has been identified as one of the most crucial factors required for the successful planting of plants. Biofertilizers can influence the quality and quantity of the crop. The low phosphate solubility has been recognized as one of the most crucial factors that prevent adequate growth of plant growth in the soil. Numerous microorganisms can enhance the solubility of phosphate, but little has been documented regarding the magnitude of their phosphorus solubility.

The local population of phosphate-solubilizing bacteria and fungi were studied in numerous rhizospheres soil samples collected from banana plant and its effect on spinach plant (*Amaranthus cruentus* L.) was studied by Reena et al. (2013). Their result indicated that *Aspergillus* species (234.12 mm) was the largest phosphate solubilizers together with (160.82 mm) followed by *Pseudomonas aeruginosa* (126.11), *Penicillium* sp. (99.02), and *Micrococcus* sp. (89.4). The best optimization condition that showed that the maximum condition for fungi and bacteria were the temperature of 37 and 28°C.

Trichoderma sp. has been identified as a known biological control agent against numerous phytopathogens. Given the aforementioned, Kapri et al. (2010) evaluates the potentials of 14 strains of *Trichoderma* sp. isolated from forest tree rhizospheres of oak, pinus, guava, and bamboo using *Trichoderma* selective medium. These isolates were evaluated for their P-solubilizing capability utilizing National Botanical Research Institute Phosphate (NBRIP) broth having TCP as the sole Phosphorus source in comparison to the standard culture of *T. harzianum*. The result obtained shows that the cultures were discovered to possess the capability to solubilize TCP but with different potentials. It was discovered that isolate DRT-1 exhibited the highest level of soluble phosphate with 404.07 $\mu g.ml^{-1}$ followed by the 386.42 $\mu g.ml^{-1}$ obtained for the standard culture of *T. harzianum* after 96 hours of incubation of 30+10C. The alkaline phosphatases and extra-cellular acid produced from fungus were induced in the presence of insoluble phosphorus source while isolate DRT-1 had the highest extra-cellular alkaline phosphatase activity value of 14.50 $U.ml^{-1}$ followed by 13.41 $U.ml^{-1}$ recorded by the standard culture at 72 hours. It was observed that the culture revealed a much small acid phosphatase activity. It was observed that *Trichoderma* sp. inoculation enhanced the chickpea (*Cicer arietinum*) growth parameters, most especially in the phosphorus-deficient soil having only bound phosphate when performed under the greenhouse. The shoot weight was enhanced by 33% and 23% by the isolate DRT-1 when amended with 200 and 100 mg TCP kg^{-1} soil, respectively after 60 d of sowing. Their study indicates the capability of *Trichoderma* sp. for the solubilization of fixed phosphates available in the soil which could improve the level of soil fertility and plant growth.

Numerous microorganisms have been established to possess the capability to change the insoluble forms of phosphorus to an accessible soluble form, thereby contributing to the rate of plant nutrition as plant growth-promoting microorganisms (PGPM). Oliveira et al. (2009) isolated, screen, and assessed the phosphate solubilization potential of microorganisms available in the rhizosphere of maize for the management of soil microbial communities and to obtain potent microbial inoculants. The authors isolated 45 isolates from 371 colonies derived from the rhizosphere of the soil of maize cultivated in an oxisol of the Cerrado Biome having phosphorus deficiency. These microorganisms were selected based on their capability to solubilize organic and inorganic phosphate sources using a modified PVK liquid medium culture containing TCP ($Ca_3(PO_4)_2$), sodium phytate (phytic acid), aluminum phosphate ($AlPO_4$), and soybean lecithin. These isolated were characterized using internal transcribed spacer (ITS) rDNA for fungi and 16S ribosomal DNA (rDNA) for bacteria using the data obtained from the nucleotide sequence of these microorganisms. Their result indicated that bacteria generated the highest level of solubilization in medium having TCP. The most efficient strain was identified as *Burkholderia* sp. B5 and *Bacillus* sp. B17 with 58.5 and 67% total P ($Ca_3(PO_4)_2$) after 10 days after 10 days and were obtained from the rhizosphere of phosphorus efficient L3 maize genotype when performed under phosphorus stress. It was documented that the fungal population showed the most effective solubilizing phosphorous sources of lecithin, aluminum, and phytate. Moreover, more diversity of phosphorus solubilizing microorganisms was observed in the rhizosphere of phosphorous efficient maize genotypes which shows that the phosphorous effectiveness in these cultivars could be related to the capability to increase the microbial relationship of phosphorous solubilizing microorganisms.

Beneficial microorganisms that dwell around the rhizosphere of most plants have been shown to have high solubilization potentials to insoluble phosphate compounds in crop nutrition (Chen et al., 2006; Malboobi et al., 2009). These potentials are based on the ability of the microorganisms to utilize phosphate in soil via soluble radical forms (carboxyl and hydroxyl) concerning their low molecular mass, thence converting them into usable and absorbable forms. PSMs have been used widely as one of the green biotechnologies in agriculture as a potential green biofertilizer that could be applied to replace the most widely NPK based inorganic fertilizers. Because of this, about 30–40 kg of P_2O_5 per ha and 50–60% PSMs are needed to boost the quality of crops in farm operation or productivity. Apart from other strains that dwells in the rhizosphere, other strains of microorganisms like P13 (*Pseudomonas putida*), P5 (*Pantoea agglomerans*), and P7 (*Microbacterium laevaniformans*) have been identified as a potential PSMs (Baas et al., 2016). PSMs have been shown to have inducing potentials by immobilizing certain heavy metals (HMs) in the soil such as Pb and Fe, and remediating polluted soil with a low amount of phosphate that cannot cause eutrophication (Park et al., 2011; Romano et al., 2017).

Mondal et al. (2017) evaluated the use of PSMs as biofertilizer in *Zea mays* field at West Bengal in India. The authors stated that phosphorus plays a significant role in plant growth. That is because of its poor absorption via the soil to the root system of plants, a more efficient and green application of PSMs is highly needed. That PSMs aid the rapid increase and uptake of phosphate by the plant roots. The authors tested and screened different strains of microorganisms with Halo-zone and pH background checks. The results of their biological controlled study showed that PSMs can improve the soil quality, elicit plant growth, and developments when applied to the maize fields. The authors recommend PSMs as first-class biofertilizer because they have a greater potential to increase plant productivity when used, their eco-friendly nature, and their economic benefits.

Anand et al. (2016) did an extensive review on the potentials of PSMs as substitute biofertilizer in farmland soil. The author stressed the need for phosphorus as an active ingredient in the ecosystem because of the essential role it plays in the physiology of plants' growth and development. However, it is known as one of the least mobile earth macronutrients as compared to its likes. The utilization of PSMs to facilitate the exchange and the possible availability of phosphorus to plants via chelate mineralization is highly recommended. The mode of action is enhanced by the pH, enzyme secretion from the microbes of the soil, and through the mineralization of certain organic acids. In conclusion, they suggested the non-application of inorganic phosphates openly to the soil because of the potential environmental complications, it portends and recommends the utilization of biofertilizers (PSMs) as a substitute because it is cheap and eco-friendly.

Din et al. (2019) tested the efficacy of phosphorus solubilizing *Aspergillus niger* on *Lagenaria siceraria* (bottle gourd) and *Abelmoschus esculentus* (Okra) as biofertilizers on agricultural field trials. The Vanadomoybdate technique was used in quantity the solvable phosphorous and the Heinonen technique was used to evaluate the concentration of the soil enzymes (phytase and phosphatase) in the media. The results of the laboratory study indicated that *Aspergillus* strain gave out specific catalytic chemicals (phytase and phosphatase) of amount (133 and 170UI), respectively, after 48 hours of fermentation, which has the potential to solubilize the hard soil phosphate and make it accessible or absorbable for plant uptake. Further field trials using the same plant species, indicated that the phosphate solubilizing microbe (*Aspergillus* sp.) was able to elicit significant growth and development in several parts of the studied plants when compared with their control. Also, plants that were inoculated together with a nitrogen-fixing microorganism (*Azotobacter*) and phosphate solvable microorganism (*Aspergillus* sp.) had well developed and vigorous recital as compared to those that were inoculated along with separate biofertilizer only. In conclusion, the authors recommend *Aspergillus* sp. as a first-class eco-friendly, non-toxic, and economical biofertilizers for agricultural practices.

Adnan et al. (2017) tested and evaluated the antagonistic impact of PSB on the bioavailability of phosphate in highly alkaline and calcification soils. The authors examined different media (inoculated and control); SSP (single super phosphorus) (SSP), RP, PM (poultry manure), and FYM (farmyard manure) with different soil, sea-green (lime) concentrations; 4.78, 10, 15, and 20%, respectively in the highly alkalized soils for a period of 56 days. The results of their study indicated that PSB increased the phosphate contents in the PM and FYM more than the SSP and RP. The addition of the sea-green to the agricultural soils sharply decreased the bioavailability of the soil phosphate, but this was contrarily adverse and buffered by the inoculation of the PSB when added to the media. In conclusion, they discovered that there were antagonistic impacts on the soil calcium and phosphate levels. They recommend that PSB with organic manure utilization will aid in improving the agriculture soil nutrients when used synergistically in agricultural purposes.

Kaur and Reddy (2014) tested and evaluated two PSBs (*Pantoeo* sp. and *Pseudomonas* sp.) isolates on agricultural fields to ascertain their solubility potentials on RP to elicit fertility substances in soil (nitrogen fixation) and crop productivity (siderophores and indole-acetic acid; IAA). The results of their study indicated that both microbes efficiently solubilized and utilized the RP and free important amount of the phosphate (> 271 $\mu g ml^{-1}$) in the laboratory medium. The application of the laboratory trial in the field study, which lasted for 2 years on wheat and maize plants with and without the RP, yielded a positive correlation and caused the growth, developments, and increased enzymatic activities of the plants and soil fertility as related to the organic carbon contents and RP fertilization associated with their controls. The authors finally suggested that *Pantoea* sp. and *Pseudomonas* sp. alongside RP, play a significant part in refining crop efficiency in natural farming.

Wu et al. (2019) tested and evaluated the impact of two natural PSBs JX285 (*B. aryabhattai*) and HN038 (*P. auricularis*) on the physiological and biological uptake of nutrients in Tea-oil seedlings. The results of their study indicated a sharp significant preferment of the growth of Tea-oil plants after inoculation three times, stimulation of the nitrogen and phosphorus concentrations in the leaf of the plant, and elevation of the nitrogen, phosphate, and potassium concentrations in the examined field soils. The authors, in conclusion, recommended PSBs as an alternative biofertilizer because it aids in the boosting of the productivity of Tea-oil plants, aid in the decrease of soil pollution and improves nutrient stabilization in soil.

Walpola and Yoon (2013) isolated and tested the impact of PSBs on the growth and development of Mung bean. Out of the 31 of the isolated strains, 2 strains [PSB-1 (*Pantoea* sp.) and PSB-2 (*Burkholderia* sp.) were identified to be more efficient strains and were utilized for advanced studies. The results of their study revealed high phosphate solubilization of 720.75 μgmL^{-1} obtained from a combined inoculated culture of *Burkholderia anthina* and *Pantoea agglomerans*. Besides, a positive relationship was observed between the solvable phosphate and the pH of the medium, and the TA with the solvable phosphate. The effective strains were later tested in greenhouse conditions and were observed to improve the growth and development of all the parts of the Mung bean plants. The authors suggested the utilization of PSBs as a favorable substitute to abate the phosphorus issues in cultivated soils.

Suleman et al. (2018) isolated and characterized PSBs to improve the calcium and phosphate uptake in wheat plants. Around 15 PSBs were used in this study and were isolated from rhizospheric wheat soils in two regions of Pakistan. The isolates were later identified with the aid of a light microscope and a 16S rRNA protein sequence machine correspondingly. Among the identified isolates, two species or strains [MS16 (*Pseudomonas* sp.) and MS32 (*Enterobacter* sp.)] that have the potential to efficiently solubilize phosphate were resolved numerically (136±280 μgmL^{-1}) and found to both have the index of solubility of 3.2±5.8. This laboratory *in vitro* of the efficient strains yielded several acids [22.5±11.8 (gibberellic acid), SI 2.8±3.3 (solubilized zinc compounds), 1-Aminocyclopropane-1-carboxylic acid deaminase, and 25.6±28.1 μgmL^{-1} (IAA)]. The results of the *in vitro* study revealed that the PSBs treated with wheat sprouts developed effectively with an 11% increment as related to the control. There were positive effects in terms of growth when the strain *Pseudomonas* sp. was used in the grain pot assay of 38.5% augmentation and the field trials of 17±18% augmentation as related to their control, respectively. The results of the acids produced by the effective strains revealed that the PSBs gluconic acid-producing strains (*Pseudomonas* sp.) could be an economical and eco-friendly contender for advanced plant development and phosphate uptake when used to cultivate wheat plants.

Saeed et al. (2015) did comparative work on the impact of chemical and biofertilizers on the development, biology, and chemical structures of *Cucumis sativus* (cucumber). The biological control experiment was done in four duplicates; T1: Control, T2: Biofertilizer, T3: Chemical, and T4: Blend treatment (biofertilizer and half-chemical). The results of their study revealed the momentous variance between the chemical and biofertilizer when applied to the cucumber. They had a distinct impact and improved the production and development characters of the study plants. The results of the relationship between the fruit output and sum fruit mass per crop showed a strong positive regression of 0.89 which yielded about 50% of the fruit output difference. The authors recommended the use of the biofertilizer s because it has improved the crop yield and biological contents significantly.

Atia (2018) isolated and tested the potentials of rhizospheric PSBs in promoting the characters, growth, and development of Wheat (*Triticum aestivum*). Around 30 isolates were obtained and derived from the rhizosphere and 10 efficient solubilizing strains (WumS-3, WumS-4, WumS-5, WumS-11, WumS-12, WumS-21, WumS-24, WumS-25, WumS-26, and WumS-28). The results from their study revealed the high solubilizing ability of the effective designated strains were 4–7 μg/ml index of solubilization and in the agar dish and 30–246 μg/ml in liquid bouillabaisse respectively. The best possible environmental conditions for the phosphorus solubilization in the *in vitro* state were at pH 7 (neutral) and temperature 35°C, glucose as a good carbon source, and ammonium nitrate as a good nitrogen source. The findings of the study also revealed that the selected strain was able to produce plant hormones (IAA), siderophore, HCN, and $NH_{3.}$ The PSBs showed significant improvement in the seed growth (50–80%), shoot dimension, and shoot elongation (10–90%) at $P<0.05$ to their control in the laboratory. While in normal (natural) settings, there was a significant difference at $P<0.05$ in the seed growth (40–80%) shoot elongation (5–34.8%) and shooting dimension (5–96%) were noticed also. The authors concluded that the phosphate solubilizing rhizospheric bacteria (PSRB) elicited the plant growth and development rate of the plant. They recommended the utilization of the effective strains as a perfect substitute for inorganic fertilizer for the cultivation of the wheat plant.

Wu et al. (2005) evaluated and tested the efficacy of the sustainable use of biofertilizer obtained from microbes as a potential candidate and substitute to inorganic fertilizer for soil productivity and crop output in agriculture. The authors utilized four biofertilizers comprising of strains of arbuscular mycorrhizal fungus, nitrogen-fixing bacteria, PSB, and potassium solubilizing bacteria on the soil characteristics and the development of maize plants. The result of their study showed that the utilization of the mycorrhizal fungus and the three species of bacteria meaningfully amplified the development of the maize plant. The microorganisms not only boosted the nutritional values of the plant but also elicited the soil characteristics. In contrast, of these fungi beneficial amplification, they also caused or provoked inhibiting impacts on the PSMs or bacteria, which resulted in the inability for them to fix the soil nutrients, instead of forming a higher association of colonization during the greenhouse trials. They recommended multiple applications of different biofertilizers in agricultural practices, but with the exception of the ones that have inhibiting potentials.

Vikram and Hamzehzarghani (2008) assessed and tested the effect of PSBs on the nodule-growth and development features of Greengram (*Vigna radiata*). Around 16 strains of PSBs were used in the biological control experiment (greenhouse settings) to improve the nodule-growth and development features of *Vigna radiata*. Triplicates of 18 treatments were used. The results of their study indicated that the *Vigna radiata* seeds (PSBV-13 and PSBV-14) had the highest protuberance amount in terms of biomass; wet, dry, and total matter of the root and shoot system after 45 days of planting. Most of the PSBs (PSBV-4, PSBV-9, PSBV-12, PSBV-13, PSBV-14, and PSBV-15), strained used in the study also tested positive in stimulating the nodule-growth and developmental parameters of the plant. In conclusion, the authors recommended that the competent PSBs could be verified for their effectiveness in field settings before endorsing them for marketable use.

Din et al. (2019) evaluated the impact and utilization of N-fixing bacteria (*Azotobacter*)-SR-4 and PSBs (*Aspergillus niger*)-as biofertilizer on *Lagenaria siceraria* (bottle gourd) and *Abelmoschus esculentus* (okra). The method of Kjeldahl was used to determine the biological field control strains, the Vanadomoybdate method was used to quantify the PSBs and the Heinonen method was used to evaluate the amount of the enzymes (phosphatase and phytase) in the media. The results of the field-controlled experiment revealed that the N-fixer bacteria were able to fix 35.08 mg of N/g of C after 3 days (72 h) fermentation. In the same way, the PSBs strain was able to produce biological enzymes of 133 and 170 UI (phosphatase and phytase) respectively, after 2 days (48 h) which can break and make soluble rock phosphorus, and make it accessible to the crops. The findings of their study showed that the tested biofertilizers were able to elicit the growth and developmental patterns of the plants by increasing their shoot, root, leaves, and fruits when related to their control. In conclusion, the authors recommend the inoculation of the seeds of the tested plants as well as related species with the tested biofertilizer because they may substitute expensive and non-eco-friendly noxious inorganic fertilizers.

Fenta and Assefa (2017) isolated and characterized PSBs in Rhizosphere tomato soil, and evaluated their effect under Green House settings. The authors examined 11 PSBs (PSB1, PSB2, PSB4, PSB5, PBS6, PSB7, PSB8, PSB9, PSB10, and PSB11) of which PSB1, PSB2, PSB4, PSB5, and PSB7 were designated out of the 11 strains and were identified to be in the genus *Pseudomonas*. Several media were used in culturing the PSBs [TCP, RP, and BP (bone phosphate)] and were found to be effective after 5, 10, and 20 days incubation periods. The results of their study revealed that all the PSBs species inoculated in the TCP solubilized effectively at $P<0.05$ as against the control. Higher phosphate solubilization (7.64 mg/50 ml) was observed on day 5, caused by the strain PSB1 while strains PSB4 and PSB5 had 4.79 mg/50 ml on the same day. On day 10, the highest phosphate solubilizing efficiency was recorded in strains PSB1 and PSB2; 8.19 mg/50 ml and 8.10 mg/50 ml, respectively. The highest phosphate solubilizing efficiency of 11.77 mg/50 ml and 11.33 mg/50 ml in PSB2 and PSB7, respectively were recorded on day 20. While the least phosphate solubilizing efficiency (5.44 mg/50 ml) was recorded in the strain PSB5. The results of the RP revealed no significant difference at $P>0.05$ as against the control at day 5. On day 10, the highest phosphate solubilizing efficiency (5.02 mg/50 ml) was recorded in the isolate PSB5 while the lowest (3.46 mg/g) was recorded from strain PSB4. The highest phosphate solubilizing efficiency (7.928 mg/50 ml) was recorded in the isolate PSB7 on day 20 while the lowest (4.025 mg/50 ml) efficiency was in PSB4. The results of the BP showed that the highest phosphate solubilizing efficiency (3.020 mg/50 ml) was recorded in strain PSB7 as related to the control. The highest phosphate solubilizing efficiency (7.37 mg/50 ml and 7.025 mg/50 ml) values on day 10 were noticed in isolated PSB5 and PSB7, respectively while the least (5.47 mg/50 ml) was noticed in isolated PSB2. The highest phosphate solubilizing efficiency (11.09 mg/ 50 ml) recorded on day 20 was noticed in the isolate PSB7 while the least (9.06 mg/50 ml) was noticed in strain PSB4. The findings of their biological controlled greenhouse experiment revealed rapid root and shoot elongation of the plant. The biomass contents of the treated plant showed positive growth exempting PSB4 with TCP while PSB2, PSB4, and PSB7 revealed strong maximum uptake of phosphate when compared with the positive control. In conclusion, the authors recommend the use of PSMs (Rhizosphere strains) as a first choice in substituting the health and environmental hazards posed by the inorganic fertilizer because of their non-toxic nature and eco-safety.

Karpagam and Nagalakshmi (2014) isolated and characterized PSMs for agricultural utilization. The sum of 37 PSMs associations was isolated in agar medium (PVK) comprising unsolvable TCP sourced from farm soil. Colonies showing Halo zone were classified as phosphate solubilization. Around 38 of the PSMs were used in this study in which eight effective phosphate solubilizing strains were utilized for further study having PSI (phosphate solubilizing index) of 1.13–3.00. The results of the study showed that the 8 effective isolates had three maximum agar phosphate solubilization PS1 (0.37 mgL^{-1}), PS2 (0.30 mgL^{-1}), and PS6 (0.28 mgL^{-1}) in the bouillabaisse media. The findings of the biological controlled experiment showed that the 3 potent strains were from the genus *Pseudomonas*, *Bacillus*, and *Rhizobium* correspondingly. This fact was based on their structural differences, biology, and chemical characterizations. In conclusion, the potency isolates were chosen as first-class candidates (biofertilizer) over the toxic inorganic fertilizer because they were able to elicit growth in plants when used for agricultural purposes.

Sharon et al. (2016) isolated and evaluated the efficiency of PSBs on tomato plant growth. The results of the biological pot experiment showed that the most efficient microorganism, *Pantoea* sp. in Pot 1, was able to solubilize TCP ($Ca_3(PO_4)^2$) at the proportion of 956 mg L^{-1}. It was able to manufacture a variety of natural acidic; propionic, acetic formic, and gluconic. The results of the greenhouse experiment revealed that *Pantoea* sp. In pot 1 also solubilize and assimilate the soil phosphate, which is used to manufacture additional biomass as compared with the control but in contrast with other tomatoes not inoculated with *Pantoea* sp.

Bahadir et al. (2018) isolated and utilized phosphorus solubilizing bacteria species in promoting the growth of plants. Around 440 bacteria strains (*Bacillus* sp.) from diverse sources were selected qualitatively and quantitatively for their effectiveness in solubilizing phosphorus and producing organic acid. The results of the biological controlled experiments revealed the mean range of the phosphorus solubilization to be 6.9–95.5 μgmL^{-1} for the bacteria strain. While the mean range of the organic acid in considerable amounts was 70.70–619.20 μgmL^{-1}. The results of the relationships between the parameters in the media (phosphate, pH, and total organic acid contents) were positive. Six best phosphate solubilizing strains were further used in *in vitro* seed pot trial germination, tested for high IAA (indole-3-acetic acid) production and molecular identification. The findings of the controlled biological experiment showed that the strains were able to improve the plumule and radicle parts of the tested plants. On this ground, the authors recommend that the bacteria strains have the potential to elicit growth and development in crops and should be considered as a first-class biofertilizer.

4.4 ENVIRONMENTAL UTILIZATION OF PSMs FOR THE CLEANING OF POLLUTED ENVIRONMENT

Tirry et al. (2018) isolated and screened 27 resistant microorganisms that have the potential to stimulate the growth of the plant and abate heavy metal impacts. The microorganisms were confirmed for Cr (VI) diminution, phosphate solubilizing efficiency, production of IAA; IAA, siderophores, Cr, Zn, Cu, Ni, Pb, and Co reduction. The results of their study reveal that the microorganism strain was able to repel the metal contents in the media. The results of the phosphate solubilizing efficiency of the strains showed that about 37.14% of the strains have the ability of solubilization of phosphorus, about 28.57% of the spores (siderophores) were produced and all the strains were able to produce effective growth-stimulating hormones (IAA). The results of a 16S rDNA sequence, identified an isolate to be *Cellulosimicrobium* sp. A further Pot essay analysis done in a greenhouse setting indicated that the isolate was able to elicit growth and development of alfalfa plant as related to the control and regulated the uptake of the examined metals via the soil to the plant. In conclusion, the authors recommend *Cellulosimicrobium* sp. for the bio-remediation as well as for the promotion of plant growth because it has importance in the controlling of ecological contaminants in the soil.

Ahemad (2015) did a review on the role of PSBs in remediation of soil contaminants. The author stressed the importance of toxic metals in the soil. Those toxic metals reduce the efficiency of plant crops. That PSBs application to agricultural soil aid in the reduction of the setbacks encounters by crops and ensure bioavailability of the proper nutrient uptake by plants.

Lin et al. (2018) evaluated and tested the role of PSBs in improving copper polluted soil for the efficacy and growth of *Wedelia trilobata*. The results of their biological controlled experiment indicated that showed that the removal efficiency of Cu from soil was facilitated by the impact of the PSBs and it also spurred the growth and development of thc tested plant. The positive effect on the soil and the plant led to the Cu assimilation by the root system, translocation, and excellent results.

Min et al. (2017) evaluated a review of the remediation of soil contaminated by HMs using PSBs. The authors stressed the role of PSB; biofertilizers in remediating soil polluted with toxic metals. That PBSs can decontaminate HMs by improving the plant vigor and resistance capability. Besides, the authors also state that PSBs also improve the plants against numerous maladies and fortify the persistence effectiveness in the phyto-remediating toxic concentration of HMs. This chapter provides an overview of PSMs research status and summarizes the remediation effects and mechanisms of heavy metal contaminated soils by PSMs. The disadvantages of PSMs for remediation, heavy metal contaminated soils are also analyzed, and the future research orientation is pointed out accordingly.

Chen et al. (2019) tested and evaluated the application of basic bio-char obtained from rice husk and sludge with PSBs in the remediation of lead. The results of their study indicated that the bio-chars rice husk and sludge were able to remediate the Pb^{2+} efficiently of about 18.61 and 53.89% correspondingly. But when PSBs were introduced, the percentage remediating efficiency increased for both bio-chars rice husk and sludge for 24.11 and 60.85%, individually. Besides, the PSBs were able to improve significantly the creation of steady pyromorphite on superficial of sludge bio-char consequent of the uniformly spread of phosphate and controlled pH released by the microbes. Furthermore, minor elements lesser than <0.074 mm on the surface, indicated the production of high pyro-morphite induced by the microbes on the rice and sludge bio-chars. Nonetheless, the sludge biochar showed higher bio-sorption potentials which spur the microbes to provide an opposite platform to decontaminate the toxic metal. The authors, therefore, recommend the integration of PSBs and conglomerates of bio-char as the first-class candidate for the decontamination of toxic HMs in agricultural soils.

Ren et al. (2019) isolated and tested the effect of growth-promoting microorganisms in decontaminating high concentrations of Cu in agronomic soils. The authors used metal impervious growth-promoting microorganisms [J62 (*B. cepacia*), Y1-3-9 (*P. thivervalensis*), and JYC17 (*M. oxydans*)] in the toxic soil decontamination and enhancement of the growth of *B. napus*. The biological control experiment lasted for 50 days. The results of their study revealed that the Rape plant was able to absorb Cu when inoculated with *B. cepacia*, *P. thivervalensis*, and *M. oxydans* and increase of biomass (113.38, 66.26, and 67.91%) correspondingly. More so, there was evidence of an increase in a translocation rate/factor (0.85) when *B. cepacia* was inoculated after the 50 days trials, which later elicited the remediation potentials of the Rape plant to Cu bioavailability to Cu in the soil. The results also showed that *B. cepacia* and *P. thivervalensis* also impacted the soil Cu bioavailability and the water-Cu-water solubility potentials by 10.13 and 41.77% correspondingly, as related to the control. The results of the antioxidant actions in the plant leaves indicate that the plant was impacted positively by the tested microbes via increased concentrations of the antioxidant non-enzymatic materials; glutathione and ascorbic acid by 9.89–17.67% and 40.24–91.22% individually. These activities brought down the oxidative strain brought about by the toxic metal and the concentration of the peroxidase and thiobarbituric acid-reactive substances. A further analysis using the PCR-denaturing gradient gel electrophoresis indicated that the DGGE bands were dominated by the microbe's rhizosphere and endosphere, which serve as important components in the culturing media. The result of an advanced correlation showed a positive relationship between bacterial communities; dependent and impendent with the root rhizosphere and endosphere. The authors concluded that the microbes were able to decontaminate the soil and recommend the strains as perfect bioremediation organisms.

4.5 MECHANISMS OF PHOSPHORUS SOLUBILIZATION MICROORGANISMS

Some bacterial species possess the capability to solubilize and perform mineralization capacity for inorganic and organic phosphorus (Hilda and Fraga, 2000; Khiari and Parent, 2005). Phosphorus solubilizing activity is evaluated by the capability of the microorganisms to release metabolites which include organic acid using their carboxyl and hydroxyl groups chelate the cation joined to the phosphate, which is later converted to soluble forms (Sagoe et al., 1998). The process of Phosphate solubilization occurs mainly by numerous microbial processes and modes of action entailing proton extrusion and organic acid production (Surange, 1995). These beneficial microorganisms could recycle the insoluble organic and inorganic soil phosphates mainly by bacteria and fungi (Banik and Dey, 1982). Whitelaw (2000) stated that phosphorus solubilization is performed by numerous numbers of a large number of saprophytic fungi and bacteria which could solubilize soluble soil phosphates through chelation-mediated modes of action.

The inorganic phosphorus is solubilized by the effect of inorganic and organic acids produced by phosphorus solubilizing bacteria in which carboxyl and hydroxyl groups of acids chelate cations such as Ca, Al, and Fe and reduce the pH available in the basic soil (Stevenson, 2005). The phosphorus solubilizing bacteria possess the capability to dissolve the soil. Through the generation of low molecular weight organic acids which consists of keto gluconic acids and gluconic and together with the reduction in the pH of the rhizosphere (Deubel et al., 2000).

The reduction in the pH of the rhizosphere could be through the following process such as gaseous (O_2/CO_2) exchanges, biotical production of the proton, and bicarbonate release (anion/cation balance). The rate of phosphorus solubilization could be linked to the level of pH of the medium. The liberation of the root exudates such as organic ligands could change the concentration of phosphorus available in the soil solution (Hinsinger, 2001). The organic acid generated by the phosphorus solubilizing bacteria could solubilize the phosphorus by reducing the pH, chelation of cations, and conflicting with phosphate for adsorption locations available in the soil (Nahas, 1996). Some inorganic acids such as hydrochloric acid could also solubilize phosphate, but they have a lower efficacy when compared to organic acid when applied at the same pH (Kim et al., 1997). Moreover, phosphate solubilization is stimulated by the induction of phosphate starvation (Gyaneshwar et al., 1999).

4.6 CONCLUSION AND FUTURE RECOMMENDATION

This chapter has provided comprehensive information on the application of phosphorus solubilizing microorganisms and their wide application for the increment of agricultural crop production and their utilization for the ecorestoration of heavily polluted soil. The modes of action explored by these beneficial strains were highlighted. There is a need to utilize the application of genetic engineering for the beneficial microorganisms for the production of an improved strain that could increase the rise in the production of absorbable phosphorus by the plant. There is a need to explore the utilization of agricultural wastes for the mass production of effective strain that could lead to an increase in the production of phosphorus.

KEYWORDS

- **biofertilizer**
- **ecorestoration**
- **modes of action**
- **phosphorus solubilizing microorganism**
- **plant growth-promoting microorganisms**
- **sustainable agriculture**
- **tricalcium phosphate**

REFERENCES

Adetunji, C. O., & Ugbenyen, A. M., (2019). Mechanism of action of nanopesticide derived from microorganism for the alleviation of abiotic and biotic stress affecting crop productivity. In: *Nanotechnology for Agriculture: Crop Production and Protection*. Springer, Singapore, https://doi.org/10.1007/978-981-32-9374-8_7.

Adetunji, C. O., (2019). Environmental impact and ecotoxicological influence of biofabricated and inorganic nanoparticle on soil activity. In: Panpatte, D., & Jhala, Y., (eds.), *Nanotechnology for Agriculture*. Springer, Singapore. https://doi.org/10.1007/978-981-32-9370-0_12.

Adetunji, C. O., Egbuna, C., Tijjani, H., Adom, D., Khalil, L., Al-Ani, T., & Partick-Iwuanuyanwu, K. C., (2019a). Pesticides, home made preparation of natural biopesticides and application. In: *Natural Remedy for Pest. Diseases and Weed Control* (pp. 1–12), Elsevier.

Adetunji, C. O., Kumar, D., Raina, M., Arogundade, O., & Sarin, N. M., (2019b). Endophytic microorganisms as biological control agents for plant pathogens: A panacea for sustainable agriculture. In: Varma, A., Tripathi, S., & Prasad, R., (eds.), *Plant Biotic Interactions*. https://doi.org/10.1007/978-3-030-26657-8_1.

Adetunji, C. O., Oloke, J. K., Bello, O. M., Pradeep, M., & Jolly, R. S., (2019d). Isolation, structural elucidation and bioherbicidal activity of an eco-friendly bioactive 2-(hydroxymethyl) phenol, from *Pseudomonas aeruginosa* (C1501) and its ecotoxicological evaluation on soil. *Environ. Technol. Innov., 13*, 304–317.

Adetunji, C. O., Panpatte, D. G., Bello, O. M., & Adekoya, M. A., (2019). Application of nanoengineered metabolites from beneficial and eco-friendly microorganisms as biological control agents for plant pests and pathogens. In: Panpatte, D., & Jhala, Y., (eds.), *Nanotechnology for Agriculture: Crop Production and Protection*. Springer, Singapore, https://doi.org/10.1007/978-981-32-9374-8_13.

Adnan, M., Shah, Z., Fahad, S., Arif, M., Alam, M., Khan, I. A., Mian, I. A., Basir, A., et al., (2017). Phosphate-solubilizing bacteria nullify the antagonistic effect of soil calcification on bioavailability of phosphorus in alkaline soils. *Scientific Reports, 7*, 16131. https://doi.org/10.1038/s41598-017-16537-5.

Shumaila. B., & Atia, I., (2018). Phosphate solubilizing rhizobacteria as alternative of chemical fertilizer for growth and yield of *Triticum aestivum* (var. Galaxy 2013). *Saudi J. Biol. Sci.* https://doi.org/10.1016/j.sjbs.2018.05.024.

Baas, P., Bell, C., Mancini, L. M., Lee, M. N., Conant, R. T., et al., (2016). Phosphorus mobilizing consortium Mammoth P™ enhances plant growth. *Peer. J., 4*, e2121. doi: 10.7717/peerj.2121.

Bahadir, P. S., Liaqat, F., & Eltem, R., (2018). Plant growth-promoting properties of phosphate solubilizing *Bacillus* species isolated from the Aegean Region of Turkey. *Turk. J. Bot., 42*, 183–196. https://doi.org/10.3906/bot-1706-51.

Baliah, N. T., Pandiarajan, G., & Kumar, B. M., (2016). Isolation, identification and characterization of phosphate solubilizing bacteria from different crop soils of Srivilliputtur Taluk, Virudhunagar District, Tamil Nadu. *Trop. Ecol., 57*(3), 465–474.

Banik, S., & Dey, B. K., (1982). Available phosphate content of an alluvial soil as influenced by inoculation of some isolated phosphate solubilizing microorganisms. *Plant Soil, 69*, 353–364.

Baudoin, E., Benizri, E., & Guckert, A., (2002). Impact of growth stages on bacterial community structure along maize roots by metabolic and genetic fingerprinting. *Appl. Soil Ecol., 19*, 135–145.

Chen, H., Zhang, J., Tang, T., Su, M., Tian, D., Zhang, L., Li, Z., & Hu, S., (2019). Enhanced Pb immobilization via the combination of biochar and phosphate solubilizing. *Environ. Int., 127*, 395–401.

Chen, Y. P., Rekha, P. D., Arun, A. B., Shen, F. T., Lai, W. A., & Young, C. C., (2006). Phosphate solubilizing bacteria from subtropical soil and their tricalcium phosphate solubilizing abilities. *Appl. Soil Ecol., 34*(1), 33–41. https://doi.org/10.1016/j.apsoil.2005.12.002.

Deubel, A., Gransee, & Merbach, W., (2000). Transformation of organic rhizodeposits by rhizoplane bacteria and its influence on the availability of tertiary calcium phosphate. *J. Plant Nutr. Soil Sci., 163*, 387–392.

Din, M., Nelofer, R., Salman, M., Abdullaha, Faisal, H. K., Khana, A., Ahmada, M., Jalila, F., Ud, D., & Khand, M., (2019). Production of nitrogen-fixing Azotobacter (SR-4) and phosphorus solubilizing *Aspergillus niger* and their evaluation on *Lagenaria siceraria* and *Abelmoschus esculentus*. *Biotech. Rep., 20*, e00323.

Fallah, A., (2006). Abundance and distribution of phosphate solubilizing bacteria and fungi in some soil samples from north of Iran. In: *18th World Congress of Soil Science*. Philadelphia, Pennsylvania, USA.

Fenta, L., & Assefa, F., (2017). Isolation and characterization of phosphate solubilizing bacteria from tomato (*Solanum* L.) rhizosphere and their effect on growth and phosphorus uptake of the host plant under greenhouse experiment. *Int. J. Adv. Res.*, 1–49.

Fisher, M. C., Henk, D. A., Briggs, C. J., Brownstein, J. S., Madoff, L. C., McCraw, S. L., & Gurr, S. J., (2012). Emerging fungal threats to animal, plant and ecosystem health. *Nature, 484*, 186–194. https://doi.org/10.1038/nature10947.

Glick, B. R., (2014). Bacteria with ACC deaminase can promote plant growth and help to feed the world. *Microbiol. Res., 169*, 30–39. https://doi.org/10.1016/j.micres.2013.09.009.

Gyaneshwar, P., Parekh, L. J., Archana, G., Podle, P. S., Collins, M. D., Hutson, R. A., & Naresh, K. G., (1999). Involvement of a phosphate starvation inducible glucose dehydrogenase in soil phosphate solubilization by *Enterobacter asburiae*. *FEMS Microbiol. Lett., 171*, 223–229.

Hilda, R., & Fraga, R., (2000). Phosphate solubilizing bacteria and their role in plant growth promotion. *Biotech. Adv., 17*, 319–359.

Hinsinger, P., (2001). Bioavailability of soil inorganic P in the rhizosphere as affected by root-induced chemical changes: A review. *Plant Soil, 237*, 173–195.

Jamshidi, R., Jalili, B., Bahmanyar, M. A., & Salek-Gilani, S., (2016). Isolation and identification of a phosphate solubilizing fungus from soil of a phosphate mine in Chaluse, Iran. *Mycology, 7*(3), 134–142. https://doi.org/10.1080/21501203.2016.1221863.

Kapri, A., & Tewari, L., (2010). Phosphate solubilization potential and phosphatase activity of rhizospheric *Trichoderma* spp. *Braz. J. Microbiol., 41*(3), 787–795. https://doi.org/10.1590/S1517-83822010005000031.

Karpagam, T., & Nagalakshmi, P. K., (2014). Isolation and characterization of phosphate solubilizing microbes from agricultural soil. *Int. J. Curr. Microbiol. App. Sci., 3*(3), 601–614.

Kaur, G., & Reddy, M. S., (2014). Role of phosphate-solubilizing bacteria in improving the soil fertility and crop productivity in organic farming. *Archives of Agronomy and Soil Science, 60*, 4.

Khan, M. S., Zaidi, A., & Wani, P. A., (2007). Role of phosphate-solubilizing microorganisms in sustainable agriculture: A review. *Agron. Sustain. Dev., 27*, 29–43.

Khiari, L., & Parent, L. E., (2005). Phosphorus transformations in acid light-textured soils treated with dry swine manure. *Can. J. Soil Sci., 85*, 75–87.

Kim, K. Y. D., Jordan, D., & McDonald, G. A., (1997). Solubilization of hydroxyapatite by *Enterobacter agglomerans* and cloned *Escherichia coli* in culture medium. *Biol. Fert. Soils., 24*, 347–352.

Kim, K. Y., Jordan, D., & McDonald, G. A., (1998). Effect of phosphate-solubilizing bacteria and vesicular-arbuscular mycorrhizae on tomato growth and soil microbial activity. *Biol. Fert. Soils, 26*, 79–87.

Kumar, A., Baby, K., & Mallick, M. A., (2016). Phosphate solubilizing microbes: An effective and alternative approach ass biofertilizers. *Int. J. Phar. Pharm. Sci., 8*(2), 37–40.

Ladeiro, B., (2012). Saline agriculture in the 21st century: Using salt contaminated resources to cope food requirements. *J. Bot.* https://doi.org/10.1155/2012/310705.

Li, Y., Zhang, J., Zhang, J., Xu, W., & Mou, Z., (2019). Characteristics of inorganic phosphate-solubilizing bacteria from the sediments of a eutrophic lake. *Int. J. Environ. Res. Public Health., 16*, 2141. https://doi.org/10.3390/ijerph16122141.

Lin, M., Jin, M., Xu, K., He, L., & Cheng, D., (2018). Phosphate-solubilizing bacteria improve the phytoremediation efficiency of *Wedelia trilobata* for Cu-contaminated soil. *Int. J. Phytoremed.*, 813–822. https://doi.org/10.1080/15226514.2018.1438351.

Malboobi, M. A., Owlia, P., Behbahani, M., Sarokhani, E., Moradi, S., Yakhchali, B., Deljou, A., & Heravi, K. M., (2009). Solubilization of organic and inorganic phosphates by three highly efficient soil bacterial isolates. *World J. Microbiol. Biotech., 25*(8), 1471–1477. https://doi.org/10.1007/s11274-009-0037-z.

Min, L., Zedong, T., & Mingyang, S., (2018). Research advances in heavy metal contaminated soil remediation by phosphate solubilizing microorganisms. *Acta Ecologica Sinica, 38*(10) https://doi.org/10.5846/stxb201703100406.

Munees, A., (2015). Phosphate-solubilizing bacteria-assisted phytoremediation of metalliferous soils: A review. *3 Biotech, 5*, 111–121. https://doi.org/10.1007/s13205-014-0206-0.

Nahas, E., (1996). Factors determining rock phosphate solubilization by microorganisms isolated from soil. *World J. Microb. Biotechnol., 12*, 18–23.

Oliveira, C. A., Alves, V. M. C., Marriel, I. E., Gomes, E. A., Scotti, M. R., Carneiro, N. P., Guimaraes, C. T., et al., (2009). Phosphate solubilizing microorganisms isolated from rhizosphere of maize cultivated in an oxisol of the Brazilian Cerrado biome. *Soil Biol Biochem., 41*, 1782–1787.

Park, J. H., Bolan, N., Megharaj, M., & Naidu, R., (2011). Isolation of phosphate solubilizing bacteria and their potential for lead immobilization in soil. *J. Haz. Mat., 185*(2), 829–836.

Reena, T., Dhanya, H., Deepthi, M. S., & Pravitha, D., (2013). Isolation of phosphate solubilizing bacteria and fungi from rhizospheres soil from banana plants and its effect on the growth of *Amaranthus cruentus* L. *IOSR J. Pharm. Biol. Sci., 5*(3), 6–11.

Ren, X. M., Shi-Jun, G., Wei, T., Chen, Y., Han, H., Chen, E., Li, B. L., et al., (2019). Effects of plant growth-promoting bacteria (PGPB) inoculation on the growth, antioxidant activity, Cu uptake, and bacterial community structure of rape (*Brassica napus* L.) grown in Cu-contaminated agricultural soil. *Front. Microbiol., 10*, 1455.

Rengasamy, P., (2006). World salinization with emphasis on Australia. *J Exp. Bot., 57*, 1017–1023. https://doi.org/10.1093/jxb/erj108.

Richardson, A. E., (1994). Soil microorganisms and phosphorus availability. In: Pankhurst, C. E., Doulse, B. M., Gupta, V. V. S. R., & Grace, P. R., (eds.), *Soil Biota Management in Sustainable Farming System* (pp. 50–62). CSIRO, Australia.

Richardson, A. E., (2001). Prospects for using soil microorganisms to improve the acquisition of phosphorus by plants. *Austr. J. Plant Physiol., 28*, 897–906.

Romano, S., Bondarev, V., Kölling, M., Dittmar, T., & Schulz-Vogt, H. N., (2017). Phosphate limitation triggers the dissolution of precipitated iron by the marine bacterium *Pseudovibrio* sp. FO-BEG1. *Frontiers in Microbiology, 8*, 364. doi: 10.3389/fmicb.2017.00364.

Saeed, K. S., Ahmed, S. A., Hassan, I. A., & Ahmed, P. H. (2015). Effect of biofertilizer and chemical fertilizer on growth and yield in Cucumber (*Cucumis sativus*) in greenhouse condition. *Pak. J. Biol. Sci. 18*(3), 129–134.

Sagoe, C. I., Ando, T., Kouno, K., & Nagaoka, T., (1998). Relative importance of protons and solution calcium concentration in phosphate rock dissolution by organic acids. *Soil Sci. Plant Nutr., 44*, 617–625.

Sandhimita, M., Dutta, S., Banerjee, A., Banerjee, S., Datta, R., Roy, P., Podder, A., et al., (2017). Production and application of phosphate solubilizing bacteria as biofertilizer: Field trial at maize field, Uchalan, Burdwan District, West Bengal. *Int. J. Env. Agric. Res., 3*, 1.

Sharon, J. A., Hathwaik, L. T., Glenn, G. M., Imam, S. H., & Lee, C. C., (2016). Isolation of efficient phosphate solubilizing bacteria capable of enhancing tomato plant growth. *J. Soil Sci. Plant Nutr., 16*(2), 525–536.

Stevenson, F. J., (2005). *Cycles of Soil: Carbon, Nitrogen, Phosphorus, Sulfur, Micronutrients*. John Wiley and Sons, New York.

Suleman, M., Yasmin, S., Rasul, M., Yahya, M., Atta, B. M., & Mirza, M. S., (2018). Phosphate solubilizing bacteria with glucose dehydrogenase gene for phosphorus uptake and beneficial effects on wheat. *PLoS One, 13*(9), e0204408. https://doi.org/10.1371/journal.pone.0204408.

Surange, S., Wollum, A. G., Kumar, N., & Nautiyal, C. S., (1995). Characterization of *Rhizobium* from root nodules of leguminous trees growing in alkaline soils. *Can. J. Microbiol., 43*, 891–894.

Tirry, N., Joutey, T., Sayel, H., Kouchou, A., Bahafid, W., Asri, M., & El Ghachtouli, N., (2018). Screening of plant growth-promoting traits in heavy metals resistant bacteria: Prospects in phytoremediation. *J. Genet. Eng. Biotech., 16*, 613–619.

Vassilev, N., Vassileva, M., & Nikolaeva, I., (2006). Simultaneous P-solubilizing and biocontrol activity of microorganisms: potentials and future trends. *Appl. Microbiol. Biotechnol., 71*, 137–144.

Vazquez, P., Holguin, G., Lopez-Cortes, A., & Bashan, Y., (2000). Phosphate-solubilizing microorganisms associated with the rhizosphere of mangroves in a semiarid coastal lagoon. *Biol. Fertil. Soils., 30*, 460–468.

Vikram, A., & Hamzehzarghani, H., (2008). Effect of phosphate solubilizing bacteria on nodulation and growth parameters of green gram (*Vigna radiata* L. Wilczek). *Res. J. Microbiol., 3*, 62–72. https://doi.org/10.3923/jm.2008.62.72.

Walpola, B. C., & Yoon, M. H., (2013). Phosphate solubilizing bacteria: Assessment of their effect on growth promotion and phosphorous uptake of mung bean (*Vigna radiata* [L.] R. Wilczek). *Chilean J. Agr. Res., 73*(3). https://doi.org/10.4067/S0718-58392013000300010.

Whitelaw, M. A., (2000). Growth promotion of plants inoculated with phosphate solubilizing fungi. *Adv. Agron., 69*, 99–151.

Williamson, L. C., Ribrioux, S. P. C. P., Fitter, A. H., & Leyser, H. M. O., (2001). Phosphate availability regulates root system architecture in *Arabidopsis*. *Plant Physiol., 126*, 875–882.

Wu, F., Li, J., Chen, Y., Zhang, L., Zhang, Y., Wang, S., Shi, X., Li, L., & Liang, J., (2019). Effects of phosphate solubilizing bacteria on the growth, photosynthesis, and nutrient uptake of *Camellia oleifera* Abel. *Forests, 10*, 348. https://doi.org/10.3390/f10040348.

Wu, H., (2005). Identification and characterization of a novel biotin synthesis gene in *Saccharomyces cerevisiae*. *Appl. Environ. Microbiol., 71*(11), 6845–6855.

Wu, S. C., Cao, Z. H., Lib, Z. G., Cheung, K. C., & Wonga, M. H., (2005). Effects of biofertilizer containing N-fixer, P and K solubilizers and AM fungi on maize growth: A greenhouse trial. *Geoderma, 125*, 155–166.

Yahya, A., & Azawi, S. K. A., (1998). Occurrence of phosphate solubilizing bacteria in some Iranian soils. *Plant Soil, 117*, 135–141.

Zhang, Y., Chen, F. S., Wu, X. Q., Luan, F. G., Zhang, L. P., Fang, X. M., Wan, S. Z., et al., (2018). Isolation and characterization of two phosphate-solubilizing fungi from rhizosphere soil of moso bamboo and their functional capacities when exposed to different phosphorus sources and pH environments. *PLoS One, 13*(7), e0199625. https://doi.org/10.1371/journal.pone.0199625.

CHAPTER 5

RECENT TRENDS IN THE UTILIZATION OF ENDOPHYTIC MICROORGANISMS AND OTHER BIOPESTICIDAL TECHNOLOGY FOR THE MANAGEMENT OF AGRICULTURAL PESTS, INSECTS, AND DISEASES

CHARLES OLUWASEUN ADETUNJI,[1]
OSIKEMEKHA ANTHONY ANANI,[2] SAHER ISLAM,[3] and
DEVARAJAN THANGADURAI[4]

[1]*Microbiology, Biotechnology, and Nanotechnology Laboratory, Department of Microbiology Edo State University Uzairue, Auchi, Nigeria*

[2]*Laboratory of Ecotoxicology and Forensic Biology, Department of Biological Science, Faculty of Science, Edo State University Uzairue, Auchi, Nigeria*

[3]*Institute of Biochemistry and Biotechnology, Faculty of Biosciences, University of Veterinary and Animal Sciences, Lahore, Pakistan*

[4]*Department of Botany, Karnatak University, Dharwad, Karnataka, India*

ABSTRACT

The high rate of over-dependence on the utilization of synthetic chemicals for crop production has resulted in numerous injurious effects such as the outbreak of secondary pests, pest resurgence, high level of pest resistance which led to the over-application of these synthetic pesticides to achieve the expected control outcome. The application of biopesticides derived from a beneficial microorganism such as endophytic microorganisms has been highlighted as a biotechnological technique that could help in the mitigation of all the highlighted challenges. This might be linked to the fact that microbial pesticides possess some special and unique features such as relatively low toxicity, broad-spectrum against pests and pathogens, insects, high level of compatibility with smallholder farming, easy production. Therefore, this chapter discussed the general overview of the application of endophytic microorganisms and other recent biopesticidal approaches for the management of pests and pathogens affecting agricultural productivity. The modes of action exploited by the endophytic microorganism and other advances in biopesticidal technology such as nanotechnology, metagenomics, and encapsulations were also highlighted. The merits and demerits of biopesticides over synthetic pesticides were also discussed. Necessary suggestions and future recommendations were also provided.

5.1 INTRODUCTION

Pesticides are utilized for protecting crops from the adverse effect of harmful agricultural pest and pathogens which normally affect the increase in agricultural production and normally leads to economic losses. The application of synthetic pesticides has been observed to show several adverse effects on natural resources and nature with hazardous effects on the environment and human health (Sinha, 2012; Adetunji et al., 2019a–d; Adetunji, 2019; Adetunji and Ugbenyen, 2019). Typical examples of such synthetic pesticides include pyrethroids, DDT (dichlorodiphenyltrichloroethane), methyl bromide, organophosphates which have all been highlighted to exhibit numerous health challenges, environmental, development of high resistance on targeted pests as well as an increase in the public awareness about the adverse effect of synthetic pesticides to discover substitutes for crop protection. Therefore, there is a need to search for an alternative and sustainable which supports eco-friendly approaches for the management of these agricultural pests and pathogens. The application of biopesticides has been recognized as a sustainable, eco-friendly, the economical approach that could mitigate against all the highlighted challenges associated with synthetic pesticides (Khater, 2012; Fareed et al., 2013).

The world health organization has estimated the level of mortality as high as 20,000 globally every year besides the adverse effect of pesticides which includes accumulation as food residues, carcinogenicity, neural disorders, longer degradation periods, and high and acute residual toxicity.

The application of biopesticides has been identified as a biotechnological tool with several merits when compared to synthetic pesticides, which include enhancement, water solubility, complete biodegradability, and eco-friendly (Thakore, 2006). The biopesticides derived from microorganism has been recognized as an alternative path because of their high safety to the ecosystem and human and their non-target effect to both individual and organism whenever the biopesticides are applied in individual applications and within integrated pest management (IPM) (Gašić and Tanović, 2013).

The application of endophytic microorganisms has been highlighted to be rising most especially in crop production. This might be linked to the role in the stimulation of plant growth, the supply of nitrogen to plants, stress alleviation, biological control of pests and pathogens (Beltran-Garcia et al., 2014b; Santoyo et al., 2016; Maksimov et al., 2018; White et al., 2018). The increase in crops by endophytic microorganisms could be linked to the production of plant growth regulation. 1-aminocyclopropane-1-carboxylic acid (ACC) deaminase activity fixing of nitrogen, and solubilization of phosphorus. Furthermore, diazotrophic endophytes have been recognized as versatile microbes because they possess the capability to give nutrients to the plant even in the presence of nodules which could be called 'associative nitrogen fixation' (Carvalho et al., 2014).

Therefore, this chapter intends to discuss detailed information on the application of endophytic microorganisms. The modes of action utilized by these beneficial microorganisms were also highlighted. Some examples of recent advances in biopesticides were also highlighted, such as nanotechnology, metagenomics, encapsulations were also highlighted. Future direction and recommendation of the application of biopesticides were elucidated.

5.2 AGRICULTURAL SIGNIFICANCE OF ENDOPHYTIC MICROORGANISMS AS A POTENTIAL BIOTECHNOLOGICAL TOOL

Endophytic microorganisms have been recognized as a biotechnological tool that could be used in the management of several agricultural challenges mitigating against the increase in agricultural production, most especially increase in food production. Some of these endophytic microorganisms have the potential to protect the plant against pests, diseases, and insects, which have been recognized as potential enemies of crops (Azevedo et al., 2000). The biotechnological potential of endophytic microorganisms has been identified in several sectors ranging from industries, agriculture, food, pharmaceutical but most especially in the agrochemical industries for the mass production of agrochemicals as well as genetic vectors (Souza et al., 2004). These endophytic microorganisms have been identified to possess the capability to generate several biologically active compounds, antibiotics, toxins, which play a crucial role in the maintenance of their host plant by offering higher resistance to several biotic and abiotic stress conditions, generation of phytohormones, and altering physiological features (Azevedo et al., 2000). It has been highlighted that these beneficial endophytic micro-organisms possess the capability to induce a systemic resistance against several pests and pathogens, which constitutes the major basis for their utilization as a biological control agent for agricultural pests and pathogens. They could alter the cell structure of the walls (deposition of lignin) as well as physiological and biological alteration which could lead to the generation of proteins and chemical substances which play a crucial role in the defense of their host. Moreover, the significance of plant growth-promoting bacterial could be linked to their capability to stimuli generated in the plant itself by the action of various metabolites such as ethylene, salicylic acid, and jasmonic acid. The compound could serve as elicitors that could induce a high level of resistance to external pathogens or pests. This process could lead to resistance induction which could make the plant generate the enhanced generation of proteinase-inhibitor compounds as well as plants inhibitors such as aspartate, serine, and cysteine. The plant may stimulate the build-up of secondary metabolites such as phenylpropanoids, synthesis of siderophores and phytoalexins as well as generation of pathogenesis-related proteins (PR-protein). Phytoalexins has been recognized as a microbial feature which build-up around the infection site while Phenylpropanoids could catalyze the development of trans-cinnamic acid-precursor of numerous compounds used in plant defense. PR-protein, most especially the peroxidase of phenols, is involved in the lignification of plant cells and cell walls which play a crucial role during their defense against pests and pathogens. This could also trigger hypersensitive response and development of papilla which could prevent the invasion of pathogenic microorganisms and establishment in their host. Therefore, the interpretation of the ecological roles of these beneficial microorganisms as an effective biotechnological tool could lead to sustainable organic agricultures most especially as a protective agent and enhancement of plant growth through the prevention of pests and pathogens (Peixoto et al., 2002). Most of this growth-promoting could be evaluated using molecular techniques which permit the quantification of genes involved in the anticipated characteristics while biochemical test could also help in the establishment of these microorganisms which have several potential such as phosphate solubilization, nitrogen fixation, production of indole acetic acid (Kuklinsky-Sobral et al., 2004), production of siderophores (Compant et al., 2005), generation of high resistance, biological control of diseases and pests (Ramamoorthy et al., 2001). Endophytic microorganisms are one of the most promising microbes that intend to boost agricultural output, defend flora from herbivores by generating ancillary metabolites with unpleasant and/or noxious taste to the plant-eating animals and display pesticidal features examples are *I. diterpenes*, *Nodulisporium* sp. and *B. daphnoides*. AR37 and AR1 strains have been recognized or rated to have potentials in combating plant-insects invasion. *P. indica* and *E. coenophiala* are the most widely utilize endophytes that colonize the root system of plants, which assist in the resilience, development, and growth in plants and aid in the resistance to abiotic and biotic environmental stresses. *P. eupatoriis* aids in the inhibition of the phytopathogen, *P. infestans*. Some species of endophytes also increase the uptake of soil phosphate and nutrients in plants.

5.3 MANAGEMENT OF AGRICULTURAL PEST AND INSECTS USING ENDOPHYTES

Entomopathogenic fungi are universally utilized as inundate sprays to prevent plants from the attached of agricultural pests. They possess the capability to establish themselves as fungal endophytes, which consequently give several benefits to including protection against numerous agricultural pests. Given the aforementioned, Bamisile et al. (2019) evaluated one strain of *Isaria fumosorosea* and two fungal strains of *Beauveria bassiana* and assessed their pathogenicity test against adults of Asian citrus psyllid (*Diaphorina citri*). The result obtained shows that the tested biological control strains cause a 50% reduction in the survival rate of *D. citri* adults after exposure for 5 days. Moreover, the capability of the 3 fungal strains to endophytically colonize Citrus limon was also evaluated while the influence of the endophytic fungi to stimulate systemic establishment on 3 sequential generations of *D. citri* feeding on colonized plants was assessed. The Citrus seedlings were also inoculated at 4 months post-planting with these fungal strains using foliar spraying. It was observed that *B. bassiana* effectively colonized the seedlings while BB Fafu-13, which was one of the *B. bassiana* colonization was maintained for 12 weeks in the colonized seedlings while strain BB Fafu-16 was obtained back from the plant after 8 weeks post-inoculation. Moreover, it was observed that *Isaria fumosorosea* (IF Fafu-1) did not show any colonization on the plant while the two strains of *B. bassiana* stimulates enhance the development of plant height and level fabrication in endophytically colonized seedlings. Moreover, the endophytic *B. bassiana* resulted in 10–15% *D. citri* adult death after exposure for 7 days while it was observed that the Female *D. citri* that fed on the plants inoculated with *B. bassiana* laid smaller eggs in comparison to the plant without endophytes inoculation. Also, a decrease in adult emergence was observed on the plant that was treated with *B. bassiana*. Their study showed that *B. bassiana* could be used as endophytic fungi for the biological control of agricultural pests that normally affect citrus plants.

Banana (*Musa* spp.) has been recognized as a significant crop globally but is normally affected by black Sigatoka disease which could be linked to the fungus *Pseudocercospora fijiensis* that affects fruit production. Macedo-Raygoza et al. (2019) evaluated the significance of *Klebsiella pneumonia* and *Enterobacter cloacae* as endophytic microorganisms with biological potential against *P. fijiensis*. The role of these endophytic strains for the support of plant nutrients in the soil that has a small amount of N-transfer. These strains were identified using 16S rRNA sequence evaluation and MALDI-TOF (matrix-assisted laser desorption/ionization-time of flight) mass spectrometry (MS). The result obtained indicated that the endophytic bacteria possess the capability to solubilize phosphate, fix nitrogen, produce indole acetic acid, negative for 1-aminocyclopropane 1-carboxylic acid deaminase, and demonstrated a biological control activity against *P. fijiensis*. To determine the influence of the endophytes on the plant growth, a non-endophyte using *E. coli* strain and two endophytic plant-bacteria were inoculated for 60 days as heat-killed cells (HKC) and active cells (AC) into plant microcosms deprived of nutrients and likened to a water-only treatment and a mineral nutrients solution (MMN) treatment. The result obtained shows that the plant treated with the bacterial endophytes enhances all the growth parameters and mitigates enhanced senescence which was detected in MMN treatments which serve as the control. There was the mortality of the plant after 20 days of applying irrigation water while the plant treated with MMN enhanced the plants to produce new leaves with almost 30% weight loss during this period. Moreover, it was observed that the plant inoculated endophytic bacteria demonstrated an enhanced growth while those plants treated with *E. cloacae* AC treated plants exhibited an increase in biomass when compared to the *E. cloacae* HKC. Moreover, the plant treated with *E. cloacae* AC demonstrated the presence of intracellular bacteria within root cells, which indicated that an established symbiosis exists between the endophytic bacterial and the host plant. The authors also validated the movement of organic N from bacteria into the plants using three bacteria that were cultivated with $Na_{15}NO_3$ and NH_4Cl as a source of nitrogen source. The rate of movement of the nutrient from the endophytic bacteria into the plant was evaluated using pheophytin isotopomer abundance. The relative abundance of the isotopomers obtained at 874.57, 876.57, 872.57, 875.57, and 873.57 showed that the plant absorbed 15N atoms directly from bacterial cells utilizing them as a source of nitrogen to enhance the plant growth in the soil deprived of nutrients. Their study showed that *E. cloacae* could be utilized for the improvement of plant growth and increasing the health of banana crops.

Azevedo et al. (2000) did a review on the current developments in the control of agricultural insects of tropical plants using endophytic microbes. The authors addressed the mechanism of action and the role of endophytic microbes in stimulating eco-physiological balance against pest and insect invasion, rebound, and its influence in boosting plant ability to cope with environmental stress concerning how endophytic microbes regulate the invasion of plant pest and insect is based on their ability to secrete noxious substances that can influence the biological genome of the plant to respond to the pest-insect attack as well as to adapt to variation of environmental changes. The authors also stressed the importance of entomo-pathogenic endophytic fungi for the control of insects and pests. In conclusion, they proposed the utilization of endophytic bacteria and fungi micro-biota sourced from several citrus trees and medicinal plants as a future driver for the control of insects and pests' invasion and rebound in agricultural land.

Kumar et al. (2008) did a review on the management of pests and diseases using endophytic fungi. The authors stated that endophytic microbes reside in the inside of vascular tissues of good physical shape plants. That the microbes have healthy potentials are yet to be tapped. They recounted that various wild and domesticated plants have been shown to have endophytic metabolites such as isocoumarin, sesquiterpenes, indole, alkaloids pyrrolizidine, and guanidine derivatives which have been shown to have special antimicrobial and pesticidal against plant pests and diseases causing organisms. In conclusion, the authors proposed new ways of engineering endophytes for sustainable agriculture.

Clifton et al. (2018) tested and evaluated the impact of the endophytic-entomopathogenic strain of fungi on *Glycine max* to *Aphis glycines* and the proof of identity of *Metarhizium brunneum* and *Beauveria bassiana* strains from agro fields. The authors stated that land-dwelling plants tend to harbor endophytes on their root system, especially strains of fungi, which may cause vicissitudes in plant chemical make-up as well as affect the relationship with other plant-eating animals. The results from their study showed that the population of *Glycine max* was increased with the inoculation of *Metarhizium brunneum* strain. However, there was no significant impact of *Beauveria bassiana* on *Aphis glycines*. An endophytic relationship was established in the stem system of the examined plant, indicating *Beauveria bassiana* to be more than *Metarhizium brunneum*. The authors also stated *Metarhizium brunneum* was recovered from the plant stem only, while *Beauveria bassiana* was recovered from the leaves and stem, respectively. The study also showed that some entomo-pathogenic endophytes in *Glycine max* might harm the plant as related to their growth, development, and susceptibility to *Aphis glycines*. To mitigate this, the authors did genetic phylogenetic sequence analysis on the strain *Metarhizium* and identify *M. robertsii* as the causative agent. The finding of this study revealed a close wide range of what was obtained in previous works. In conclusion, the authors suggested that dynamic results are important in understanding what sustainable environmental approaches can be utilized in the mitigation of pest-insect invasion and rebound.

Ben et al. (2016) did a mini-review on the sustainable utilization of endophytic fungi for good crop yield and output. The authors stated that endophytic fungi play a vital role in the ecosystem habitation of plants, boost plant protection and performance against pests, diseases, and ecological stressors. They can do these because of encrypted second level organic unstable metabolites. More so, endophytic fungi are also used to correct post-harvest problems faced in agricultural practices, especially in cereals preservation and management. The authors contended that agricultural practices such as uncontrolled chemical application, seed treatments, and plant breeding, are one of the reasons endophytic fungal loss their vegetative spores, which in turn reduce their effectiveness and diversity in their primary symbiotic role in the ecosystem. In conclusion, they proposed sustainable approaches to mitigate these negative effects in contemporary agro practices.

Li et al. (2015) isolated, tested, and evaluated the interaction of some endophytic plant bacteria on some plant pests. Around 84 endophytic bacteria were isolated from six congregations of plants designated for required characteristics (cultivars) of attractive plantain lilies (*hostas*). From the isolates, five strains were identified after morphological characterization using rDNA-16S sequence scrutiny. The results from the DNA scrutiny revealed that the isolates 'Blu-v2' was identified from a 'blue Umbrella' cultivar as *Bacillus amyloliquefaciens* and exhibited strong anti-fungal qualities and the ability to prevent feeding by *Spodoptera fruigiperda* (Fall armyworms). The result of the lipo-peptides of Blu-v2 did not display any noxiousness to the larvae of *Spodoptera fruigiperda*. A further test of strain (Blu-v2) in a greenhouse setting inoculated into small sprouts of hosta leaves of infected bacteria and non-affected bacteria, in a 7-day trial period, revealed a significant reduction in the weights of the Fall armyworms larvae nourished with infected bacteria (E+) as compared to non-affected bacteria (E−) plants. The results of the death rate of Fall armyworms larvae nourished with non-affected bacteria leaves was lesser (3.33%) as compared to that of Fall armyworms larvae fed infected bacteria leaves (30%). In conclusion, the authors recommend that endophytic bacteria of strain Blu-v2 has probable worth as a bio-control mediator to decrease harm from fungal causative agents or pathogens as well as insect-pest plantain lilies cultivars.

Harman and Uphoff (2019) did a review on the ecological benefits derived from soil-root endophytic symbiotic microorganisms. The authors opined those plants should not be considered as an object onto themselves, nonetheless as the noticeable fragment of flora-microbe multiplexes which are best known as genome or "holobiomes." That some microbes, when given the chance to dwell in the rooted flora will take the advantage of becoming root-rot symbiont. In such association, the microorganisms spur the development, growth, and yield of the roots and shoots system of the plant via absorption, fixation, and uptake of nutrients, enhancement of photosynthetic reactions, protection against ecological stress (abiotic and biotic), and resistance against pest invasion and rebounds. An extensive phylogenetical literature was done on various improved holobiomes on *Piriformospora indica*, *Trichoderma*, Rhizobiaceae, and Arbuscular mycorrhizae. These species exhibit various styles of inhabiting flora parts especially the root system. However, they all possess useful to localize and systemic variations in the flora genome; reactive oxygen species (ROS), which are biochemical reactive species encompassing oxygen which are alike. The authors recounted that ROS have been shown to regulate overexcitation of photosynthetic pigment and reduce ecological pressures on flora. They also stated that ROS controls the cellular environment of floral biochemical compositions as well as the gene expression. Enhanced plant holobiomes (EPHs) have the ability to intensify photosynthesis, development, and growth of plants alongside the organic matter of the soil. The synergy of the EPHs and the soil organic matter (SOM) have been shown to improve the CO_2 quality in the atmosphere, reduction of greenhouse gases (GHGs), and aid in the stabilization of climatic factors in the ecosystem. In conclusion, the author proposes the development of EPHs as an enterprise that has the prospect to improve farm crops against pest and insect attack as well as reduces global warming episodes.

Gonzalez et al. (2016) did a review on the novel prospects for the incorporation of microbes into natural pest regulatory flora greenhouse settings. The authors stated the need to use natural enemies of pests to combat pest infestation in an ever-multifaceted ecosystem. They also stated that these natural pest enemies are too expensive to purchase, rarely available, and not really effective enough. Based on these, the use of natural control microbes ('microbials'), a novel development, are utilized. The authors opined in their review, that, there is the possible need to integrate natural enemies of the pest with microbials to explore better future ways of controlling plant pests and climatic management. They also explore different ways of the utilization of microbes, for example, endophytes in the bio-control of insect-microbes associations. In conclusion, the authors recommend the integrated optimization of insect-microbials as natural enemies to arthropod pests in greenhouse agro cultivation.

Kitherian (2017) did a review on the control of insect pests using bio-nano and nano-particles. Food security has been on the forefront in a call for sustainable development worldwide. The call for safe food by food scientists has triggered diverse several green novel technologies ways of securing food for the present and future generations. The authors stressed the need for sustainable agriculture to mitigate the problems facing food such as pesticide accumulation, pest, and insect outbreaks, stunted growth in plants, pre, and post-harvest storage. The authors, in conclusion, propose the utilization of bio-nano particles for the control of pest-insect in agricultural farms, because, it is cheap, eco-friendly, and green.

Bong and Sikorowski (1991) discovered that *Pseudomonas maltophila* possess the capability to induce changes in larval development and a decrease in the development of *Helicoverpa zea*. Thuler et al. (2006) affirmed that strains of EN5 of *Alcaligenes piechaudii* and EN4 of *Kluyvera ascorbata* could decrease the viability of *Plutella xylostella* in about 80 to 50% which could be applied for intensive field trial for the management of these insects. Moreover, Praça (2012) discovered that *B. thuringiensis* strains S1905 and S2122 could induce 100% of mortality in caterpillars in the 3° instar of *Plutella xylostella* in assessment implemented 48 h after caterpillars had been wide-open to discriminating bioassays, whereas S2124 triggered 58.33% of death after the similar period and 98.33% of death after 96.

Moreover, the CL_{50} values vary from 2.33 to 4.84 ng/mL possesses a similar result when toxicity was carried out in comparison to the control. Viana et al. (2009) discovered that 12 out of the 58 strains of *B. thuringiensis* evaluated against caterpillars of *Plutella xylostella* possessed that capability to induce 100% mortality within 24–48 h: 3A.140, T3A.259, T08.024, E1, E26, 2.7L, 1.7L, E22, 22.7L, 49.19A, and E2. Also, Castelo Branco et al. (2003) reported 100% of death for larvae of *Plutella xylostella,* most especially in the 2° instar when *B. thuringiensis* was applied.

Castelo et al. (2003) observed 100% mortality for larvae of *Plutella xylostella* in the 2° instar with the application of *B. thuringiensis*. Macedo et al. (2012) also reported that some local strains of *B. thuringiensis* (S1301, S1264, S602) could inhibit 75% of mortality after dilution of 50 times. Moreover, it was discovered that among all the tested strains S602 and S1264 showed 75% mortality after dilution of 50 times to *D. saccharalis*. Melatti et al. (2010) also discovered that *B. thuringiensis* could prevent the development of cotton aphid (*Aphis gossypii*). It was later discovered that *Bacillus kurstaki* S1168 and *Bacillus aizawai* S1576, S29, S616, S40, exhibited the highest mortality rate against *Aphis gossypii* which led to more than 50% mortality rate. Amongst the evaluated strains, S1168 and S29 were the most active in the selective bioassay, producing deaths of 73 and 76% against *A. gossypii*, respectively.

In a similar study, Polanczyky et al. (2003) evaluated the biological potential of 58 subspecies of *Bacillus thuringiensis* against *S. frugiperda*. The result obtained indicated that *Bacillus thuringiensis morrisoni* triggered 80% of death in caterpillars. In a similar vein, Berlitz et al. (2003) evaluate the effect of 24 isolates of *Bacillus thuringiensis* derived from rice-growing regions of Rio Grande do Sul, Brazil for the biological control of *S. frigiperda*. The result obtained reveals the greatest death rates between 31.6% and 100% with only 5 isolates.

Also, Campanini et al. (2012) evaluated the biological control potential of *B. thuringiensis* against *Sphenophorus levis* and *Spodoptera frigiperda*. It was later discovered that IB8.2 and IB17.3 were the most effective in the biological control of caterpillars of *S. frugiperda* while it was discovered that IB26.2 showed more than 75% mortality against the larvae *S. levis*.

Lucho (2004) discovered that the CL_{50} of Cry proteins of *Bacillus thuringiensis aizawai* destroy the caterpillars of *S. frugiperda* in the 3° instar. The outcomes showed that CL_{50} of 2.22, 0.41, and 0.18 μg/mL for 2, 3, and 4 days after the treatment of the biological control agent. It was discovered that a very high level of inhibition was demonstrated by the following proteins such as Cry1D, Cry1Aa, Cry1C, and Cry1Ab respectively from *Bacillus thuringiensis aizawai* against *Spodoptera frugiperda*. Furthermore, Knaak et al. (2007) established that proteins Cry1Ac and Cry1Ab produced by *Bacillus thuringiensis kurstaki* HD73 and *Bacillus thuringiensis aizawai*

407 showed a CL_{50} 9.29 and 1.79 μg/cm^2 against caterpillars in the 1° instar. Azambuja and Fiuza (2003) also affirmed that two natural isolates *Bacillus thuringiensis* isolated from rice-growing regions of Rio Grande do Sul, Brazil could inhibit the caterpillars of *Anticarsia gemmatalis*.

Pinto et al. (2003) utilized six isolates of *Bacillus thuringiensis* that possess genes from the class *cry*3 or *cry*7 that could be stimulated and regulate the activity of class *cry*3 or *cry*7 that could synthesize the production of proteins insecticidal to coleopterous. The authors' evaluated the influence of these proteins against the *Oryzophagus oryzae*. The result revealed that 2 induced corrected killing of 100%, 3 between 59% and 67%, and 1 around 50%. Steffens et al. (2001) recorded 53.41% mortality of larvae of *O. oryzae* from isolates of *B. thuringiensis* having *cry*3 genes with inhibitory activity against coleopterous. The high rate of inhibition might be linked to the presence of the two genes such as *cry*7 and *cry*3 genes, which categorize proteins explicit to coleopterous.

Castilhos-Fortes et al. (2002) evaluated 57 strains of these bacteria and discovered that seven showed an enhanced activity. These strains had the following inhibitory such as *B. thuringiensis kurstaki* (Btk) with 28.57%, *B. thuringiensis roskildiensis* (Btr) and *B. thuringiensis sooncheon* (Bts) with 100%, *B. thuringiensis colmeri* (Btc) with 42.85%, *B. thuringiensis huazhongensis* (Bth) with 57.1%, *B. thuringiensis yunnanensis* (Bty) with 71.4% and *B. thuringiensis brasiliensis* (Btb) with 52.3% mortality after 7th day after treatment application. The author later utilized isolate *B. thuringiensis roskildiensis* and *B. thuringiensis sooncheon* which induced 100% of death in the pre-selective assays. Broderick et al. (2000) evaluated the synergetic effect of zwittermycin A of *Bacillus cereus* when combined with *B. thuringiensis* against lepidopterous *Lymantria dispar* (L.). The result obtained showed that the combinatory effect of zwittermycin A of *Bacillus cereus* when combined with *B. thuringiensis* resulted in 35% in the mortality of the lepidopterous *Lymantria dispar*. Ma et al. (2008) tested the effect of *B. bassiana* and Cry 1Ac protein of *B. thuringiensis* and evaluated their effectiveness against *Ostrinia furnacalis*. The result obtained showed that the induced high mortality against the tested larvae of *Ostrinia furnacalis*.

5.4 MANAGEMENT OF ROOT BORNE DISEASES AFFECTING CROPS USING ENDOPHYTES

Dutta et al. (2014) did a review on the utilization of endophytes as an apparatus for phyto-protection. The authors recounted that bacterial or fungi microbes that are endophytes are symptom-less organisms that are cosmopolitan in all living flora species parts. These symbiotic plant-microbes can inhabit the inner tissues of the plant and thus improving their performance. Numerous fungal-bacteria endophytic organisms secrete second-level growth hormones or metabolites, for example, gibberellin and auxin in the plant host they inhabit. Most of these secondary or second level metabolites elements or substances found in the plants have antibacterial and antifungal qualities that can powerfully inhibit the growing of additional microbes as well as disease-causing organisms in crops. The authors, in conclusion, recommend the utilization of bacterial-fungi endophytes in the mitigation of plant diseases.

Khare et al. (2018) did a review on the connections between different flora and endophytic organisms and their prospects. The authors stated the benefits derived from endophytic microbes such as defense retort against diseases causing organisms. An interesting aspect of this is the possession of metabolites that aid in the protection of the plant against pathogens. The authors were able to explore several areas endophytic microbes can be utilized in agriculture as well as in human health protection from plant pathogens.

Gond et al. (2010) did a review on the role endophytic fungal play in plant protection. The authors recounted that every flora inhabits endophytes in their tissues without endangering or resulting in any disease therein. However, they harbor latent natural or biological products that serve as warfare against disease-causing organisms. Examples of such biological properties are immunosuppressive substances, insecticide, nematicide, antioxidant, antiviral, antimycotic, antibiotic, and anticancer. Some of the natural products derived from endophytic microbes are taxol, 3-hydroxypropionic acid, loliterm B, cryptocandin, jesterone, oocydin, ambuic acid, noxious alkaloids, and cytochalasines. The authors, in conclusion, propose microbial endophytes as first-class engineering industries.

Muthukumar et al. (2017) did a review on the role of endophytes in the control of plant disease. The authors stated that plant roots are dominated with endophytes in the following ranks fungi > bacteria > virus. That the present techniques used in the control of flora disease are the application of agricultural chemicals. This method has been known for its noxious ecological and human health concerns. However, a novel natural method using biological optimize endophytic microbes in the biological control of several flora pathogens. They can combat diseases using a series of associations like trophobiotic, commensalism, cannibalistic, and mutualistic in enhancing plant-microbes resistance. The authors concluded that endophytes are the first choice in phyto-protection of plants against pre- and post-harvest preservations.

Rabiey et al. (2019) did a review on the bio-control of tree pathogens using endophytic microorganisms. Tress disease has been managed with agrochemicals which have been shown to have residual effects in the ecosystem. Nonetheless, several pressures on the concern ecological and health risks associated with their usage have necessitated the proposition of a green bio-endophytic anti-disease agent for the management of tree diseases. Its usage has shown to be environmentally friendly, cheap, and increase the plant's adaptability to future disease rebounds. The authors stressed when these endophytic microorganisms are deploying in the soil as inoculants, they stimulate the environment of the plants (phyto-sphere). Examples of such organisms are the phylloplane microbes (fungi, yeasts, and bacteria). These microorganisms have the potential to bio-control plant pathogens and overcome several challenges faced by the plants. The utilization and optimization of these groups of microbes are competent and more effective in phyto-disease control because of the genome found in their chromosomes. However, the authors stated that the selective latent advantageous and utilizable pattern of most endophytic organisms has not been fully understood. In conclusion, the authors were able to explain how the endophytes can upset and interact with each other in the bio-control of diseases.

Waqas et al. (2015) did a tested and evaluated how endophytic microbes-fungi promote growth in plants as well as mitigate root-rot disease. The authors opined that crops that are resistant to disease have a genetic trait that can be used for an agro purpose. Endophytes, especially fungi have growth stimulating enzymes-gibberellins, which have little or no role in the prevention of plant disease. The authors looked at the role of two strains of fungi (*P. citrinum* and *A. terreus*) in stimulating growth and bio-control of diseases in *Helianthus annuus* as well as the ability to control signaling hormone as regards to plant defense against *Sclerotium rolfsii*, in a biological controlled experiment for three days interval; in a 12-day trial period (3, 6, and 12 days). The results of their study indicated that chlorophyll content, photosynthesis, transpiration, diameter, length, and biomass of the shoot system were elicited by the inoculated fungi as compared to their control within the trial periods. The findings of their study revealed other negative influences caused by *Sclerotium rolfsii*. The authors also noticed that *P. citrinum* had a more positive influence on the sunflower than *A. terreus* to *Helianthus annuus*. In conclusion, the endophytic fungi were able to restructure the flora growth during the pathogenic invasion and respond swiftly in the defense of the flora immune system against the pathogen. They propose the utilization of both strains of fungi for the sustainable management of crops because of their eco-friendly nature and economical usage as against the widely used agro fungicide(s).

5.5 MANAGEMENT OF ENVIRONMENTAL POLLUTION BY HEAVY METAL USING ENDOPHYTES

Ikram et al. (2018) tested and evaluated the potential effects of indole-3-acetic acid (IAA) generating endophytic fungi in the remediation of heavy metal in wheat crops cultivated in polluted soils. The authors stressed the need of utilizing endophytic microorganisms for the remediation of heavy metals (HMs) in seriously impacted agricultural soils. Those cheap and effective ways are needed to make different strains of endophytes available for efficient plant growth. The authors applied IAA *P. roqueforti* strain as well as a mixture of it with waste effluents, to remediate heavy metal in wheat polluted soils. The results of the study revealed that the mixtures (*P. roqueforti* and wastes effluents), elicited the nutrients level, growth of the roots and reduced drastically the HMs in the soils of shoots and roots of the plants. In contrast, the wheat plants not inoculated under HMs stress were underdeveloped with signs of chlorosis. The findings from their study revealed that the endophyte used was able to establish a closed nonparasitic association with the plant host and was a useful tool for phytoremediation and bio-stabilization of soil HMs in and agroecosystem.

Domka et al. (2019) did a review on the role of endophytic fungi in the adaptability of HMs noxiousness in soil cultivated plants. The authors recounted the environmental problems caused by HMs in the ecosystem. That micro-remediation is an effective substitute for a sustainable environment as against other conventional methods. Microbes such as endophytic fungi have been shown to have high tolerance limits to HMs toxicity and other soil stressors. In conclusion, the authors proposed endophytic fungi as a first-class candidate for effective remediation of HMs toxicity in agricultural soils.

Rizvi et al. (2017) tested and evaluated the stress mitigation of HMs and the bio-toxic influence on *Triticum aestivum* (wheat plant), using effective endophyte, *Pseudomonas aeruginosa.* The authors opined that speedy development and uncontrolled HMs release the ecosystem have warranted serious concerns both locally and globally. The biological controlled experiment involved the bio sequencing of strains of rhizosphere using 16S rDNA. The results of the analysis indicated *Pseudomonas aeruginosa* as the main isolates. The strain was later subjected to some HMs (Cr, Cd, and Cu) of various concentrations. The results of the biological controlled experiments revealed that *P. aeruginosa* strain was able to endure a high level of Cr, Cd, and Cu of concentrations 1000 $\mu g\ ml^{-1}$, 1000 $\mu g\ ml^{-1}$, and 1400 $\mu g\ ml^{-1}$ respectively. The results of the phytotoxic impacts on the *Triticum aestivum,* amplified with increases in the levels of Cr, Cd, and Cu. The strain inoculated *Triticum aestivum* had better yields and growth under the studied HMs stress. The dry mass (biomass) of the roots inoculated flora root was improved by 48, 28, and 44% at the rate of 204 mg/kg (Cr), 36 mg/kg (Cd), and 2007 mg/kg (Cu) correspondingly. The findings of this study showed that *P. aeruginosa* strain had the potential to mitigate HMs in polluted soils as well as stimulate growth and development in *Triticum aestivum* plant.

5.6 THE APPLICATION OF NANOAGROPARTICLES AS A REPLACEMENT FOR PESTICIDES

Several pieces of research around the world have raised numerous concerns on the detrimental effects of synthetic pesticides. Some typical examples include neonicotinoids, clothianidin, and thiamethoxam have been highlighted to be very toxic to terrestrial and aquatic microorganisms, and a high level of pollution has been recorded in the ecosystem (Krupke et al., 2012). Numerous organophosphorus pesticides (OPPs) have been reported to lead to a high level of toxicity and they possess the capability to build up in the body which results in a high level of biomagnification in higher tropical organisms, which might lead to disruption in the food chain and ecosystem (Gill and Garg, 2014). Some OPPs possess the capability to enter into the water bodies and give rise to a high level of pollution in drinking water (Baker et al., 2013c). Moreover, some commonly applied OPPs the herbicide glyphosate was discovered to have a diverse effect against human beings (Guyton et al., 2015). The OPPs possess the capability to enter into the food system of a human being, most especially their diet, and their presence has been identified in the breast milk. Some adverse effect associated with these synthetic pesticides includes diabetes, cancer, infertility, neurological effects, respiratory diseases, genetic disorders, Parkinson disorder, and fetal diseases (Hu et al., 2015).

Several drastic measures have been put in place for the reduction of the high level of pollutants using conventional techniques such as transportation of pollutants to offsite dumping yards, vitrification, the addition of reactants, excavation, excavation (Singh et al., 2016). However, most of these processes involved several expensive techniques for the management of these pollutions most especially in developing countries where there is a lack of economic development and scientific advancement (Kang, 2014).

Therefore, the application of biopesticides has been identified as sustainable from beneficial microorganisms for the production of agrochemical-based products that are environmentally friendly in nature. Furthermore, the introduction of nanotechnology for the synthesis of nanobiopesticides has been identified as a sustainable biotechnological tool that could help to overcome all the highlighted challenges associated with synthetic pesticides. One of the merits of eco-friendly based biopesticides is that they are unable to meet up with the market demand and deprived performance, which has instigated the scientific community to invent new alternative strategies. Hence, the utilization of beneficial microorganisms such as endophytic microorganisms for the synthesis of nanobiopesticides will go a long way in mitigating all the negative challenges associated with synthetic pesticides.

Teodoro et al. (2010) stated the insecticidal action of nanostructured aluminum against *Rhyzopertha dominica* and *Sitophilus oryzae*, which are a major pest of stored food. The result obtained showed that the nano-aluminum exhibited a high death rate when compared to the commercial insecticidal dust used as control. Liu et al. (2006) also evaluated the effect of silica nanoparticles as a vehicle for effective delivery of pesticides such as validamycin for improvement of activity and decrease the level of toxicity in comparison to free validamycin. The study established that the action of nanoparticles and their release rate depend mainly on the temperature and pH of the dissolution medium. Also, the action of silver-zinc and silver combined nanoparticles was evaluated against *Aphis nerii* Boyer de Fonscolombe which led to mortality LC_{50} at 539.46 and 424.67 mg/mL respectively (Rouhani et al., 2012).

5.7 DIFFERENT TYPES OF NANOPARTICLES UTILIZED IN THE AGRICULTURAL SYSTEM

Nanoagroparticles have been identified as a new paradigm that could enhance and led to an increase in agricultural production and that could mitigate against several challenges affecting the increase in agricultural production. There is numerous class of nanoparticles used in agriculture and their action depend on the mechanism of action. Examples of nanoparticles include aluminum, silver, zinc, copper, silica, gold, multiwalled carbon nanotubes (Sabir et al., 2014). In sustainable agriculture, their application depends on features of the nanoparticles such as control release of the component, solubility, non-self-decomposition, specificity most especially, the nanoagroparticles normally utilized as conjugated or tagged with carrier molecules that could lead to the development of liposome-based products, polymerization, immobilized association, emulsion, and hydrogel. Some other features of designed nanoagroparticles include highly sensitive with negligible usage and maximum action, cost-effective, biocompatible that could enhance and lead to an increase in agricultural crop productivity.

5.8 APPLICATION OF NANOAGROPARTICLES AS POTENT ANTIMICROBIAL AGENTS AGAINST PHYTOPATHOGENS

In the agriculture sector, effective management of microbial infestation is an important factor, and pathogenic microorganisms can cause a deleterious effect on crop productivity (Bhardwaj et al., 2014). In recent years, the emergence of drug-resistant microorganisms has posed a risk to mankind (Baker and Satish, 2015). The inappropriate and over usage of microbicidal chemicals agents, especially fungicides and bactericides, have resulted in the development of the application of beneficial microorganisms as an active antimicrobial agent in the preparation of nanoagroparticles against phytopathogens. In the agricultural sector, the application of nanoagroparticles has been recognized to play a crucial role in the management of microbial infestation and most especially for the treatment of pathogenic microorganisms which cause deleterious effect on crop productivity (Bhardwaj et al., 2014). The increase in the high level of resistance from the constant application of synthetic pesticides has led to the development of a high resistance which results in the pathogens invading crops (Baker and Satish, 2015). Therefore, the application of nanoagroparticles has been recognized as sustainable and effective eco-friendly nanobiopesticides with a novel mechanism of action that could reduce the high level of resistance caused by highly resistant pathogenic strains. The sizes of these nanoagroparticles enable them to destroy the cell wall of different pathogenic microorganisms and cause a high level of destabilization on the cell membrane, which eventually leads to an eruption of the cellular contents and which eventually causes the interaction of the nanoagroparticles. The crucial cellular contents of the pathogenic microorganisms lead to disruption in the metabolism (Baker and Satish, 2012, 2015; Baker et al., 2015).

5.9 APPLICATION OF NANOAGROPARTICLES AS POTENTIAL FUNGICIDES

Fungal diseases have been highlighted to be responsible for almost 70% of the crop diseases in comparison with pathogens which are responsible for high economic losses on agricultural food commodities (Agrios, 2005). Several crops such as grapevine, rice, groundnut, wheat, barley have b discovered to have a high susceptibility to the attack of fungal pathogens (Dhekney et al., 2007). The application of fungicides has been utilized for the management of this fungal pathogen, but their application is endowed with several environmental and health hazards that affect the ecosystem and also non-targeted microorganisms (Patel et al., 2014).

Therefore, the application of nanoagroparticles derived from a beneficial microorganism such as endophytes could be a sustainable biotechnological tool for the management of agricultural fungal pathogens (Marziye et al., 2014). The utilization of silver nanoparticles has been described as an effective nanobiopesticides with enhanced activity against rice blast disease triggered by *Magnaporthe grisea*. The utilization of silver nanoparticles has been recorded to prevent the utilization of synthetic fungicides such as azoxystrobin and isoprothiolane.

Ouda (2014) discovered that 15 mg/L of copper nanoparticles could suppress the effect of *Botrytis cinerea* (BC) and *Alternaria alternata*. Also, Wani and Shah (2012) discovered that magnesium and zinc oxide (ZnO) exhibited an enhanced inhibitory effect against the spores of several pathogenic fungal such as *Mucor plumbeus*, *A. alternata*, *Rhizopus stolonifer*, *Fusarium oxysporum* silver nanoparticles fabricated from Al-Othman et al. (2014), also validated the effect of *Aspergillus terreus* (KC462061) against aflatoxin producing isolates of *Aspergillus flavus*. Parizi et al. (2014) established that magnesium oxide nanoparticles could be used against *F. oxysporum* f. sp. *lycopersici* which have been highlighted to be responsible for the wilting in the tomato plant.

Ramy and Ahmed (2013) evaluated the influence of ZnO nanoparticles against *F. oxysporum* and *Penicillium expansum* in an *in vitro* antifungal assay. The result obtained indicates that the antifungal action exhibited by the nanoparticles is concentration-dependent. The 12 mg/L ZnO nanoparticle's highest inhibitory effect was recorded against the mycelia growth was discovered to be 100 and 77% against *P. expansum* and *F. oxysporum*, respectively. Furthermore, these nanoparticles also reduced the level of mycotoxins production. The scanning electron microscopy was used to establish that the nanoparticles from the ZnO-induced several malformations on the treated pathogens. Ouda (2014) also compared the antifungal activities of copper and silver nanoparticles against *B. cinerea* and *A. alternata*. The result obtained showed that 15 mg/L concentration of silver nanoparticles demonstrated a very high inhibitory effect against fungal hyphae growth. The microscopic evaluation also affirmed that the nanoparticles exhibited a high level of damage against the conidia and the hyphae. Also, high damage was observed on the lipid content, sugar, n-acetyl glucosamine when treated with silver nanoparticles.

5.10 APPLICATION OF NANOAGROPARTICLES AS POTENTIAL NANOBACTERICIDES

Several scientists have validated the effect of nanoparticles with enhanced antimicrobial activity against several pathogenic bacteria (Baker et al., 2015). Chowdhury et al. (2014) tested the antibacterial of silver nanoparticles against multidrug-resistant (LBA4404 MDR) *Agrobacterium tumefaciens* and the normal strain (LBA4404). Their study showed that spherical silver nanoparticles with 5–40 nm exhibited the highest antifungal effect against these tested bacterial pathogens.

5.11 APPLICATION OF BIONANO-HYBRID AGROPARTICLES AGAINST PHYTOPATHOGENS

Bio-nano hybrid agroparticles has been identified as a complex nanosystem which absolutely depends on bioconjugation chemistry. Bioconjugation gives significant interest in the biology of insightful activities. In this regard, biomolecules are utilized together with molecules or biomolecules with nanoparticles have been identified as significant Biotechnological tools. The biosynthesis of the bioconjugation approached highlighted the biochemical properties and physicochemical features of the nanomaterials as well as the biological materials. The relationship between biomolecules and nanoparticles depends on the functional moieties and electrostatics forces on the nanoparticles, which might lead to the development of functionalized nanoparticles in a reversible manner (Bagwea et al., 2003). The functional moieties can develop a hydrophobic relationship and covalent bonding (Baker and Satish, 2012, 2015). The utilization of active metabolites and biologically active molecules from endophytic microorganisms could be used for the development of bio-nano-hybrid conjugates which could prevent all the adverse effects of pesticides resistance developed by agricultural pathogen whenever synthetic pesticides are applied for the management of agricultural pests and diseases.

5.12 APPLICATION OF ENCAPSULATION FOR ENHANCEMENT OF BIOPESTICIDE ACTIVITY

Encapsulation has been identified as techniques used for the preservation of the active ingredients present in the formulation in close contact. The delivery of encapsulated viral particles has been recognized as a preferred delivery system to prevent the loss of the activity when exposed to solar radiation and protect the viral insecticidal activity. Nucleopolyhedrovirus of *Heliotis* and *Bacillus thuringiensis* have been successfully encapsulated in starch granules. Some encapsulation materials such as maize starch, gelatin, calcium alginate, and chitin have been used to maintain the viability of active ingredients. Moreover, *Spodoptera frugiperda* Nucleopolyhedrovirus (SfNPV) has been successfully encapsulated using Eudragit-S100 microparticles. It was discovered that the resulting particles showed more resistance to UV-inactivation rather than using Nucleopolyhedrovirus alone (Ravishankar and Venkatesha, 2010; Haase et al., 2015).

5.13 APPLICATION OF METAGENOMIC TECHNIQUES FOR ENHANCEMENT OF BIOPESTICIDE ACTIVITY

Metagenomic study involves identification, determination, characterization, and functional evaluation of novel strains, most especially from environmental samples, entails the isolation of DNA from the soil and detection, isolation, and screening of oppressive compounds from microorganisms. The utilization of next-generation, sequencing, new genomic techniques offer a novel and economical fast technology that could depict microbial diversity without any need to utilize cultural techniques (Hirsch et al., 2012). The application of metagenomics identification has been recognized in the identification of beneficial microorganisms, most especially the endophytic microorganisms, and other related microbes, and their interaction with agricultural pests. This has been recognized to be visible because metagenomics could be used to establish that some endophytic fungi possess the capability to exhibit toxic reactions for plant pathogens through the production of secondary metabolites and inhibition of their growth (Porras-Alfaro and Bayman, 2011).

5.14 CONCLUSION AND FUTURE RECOMMENDATION

This chapter has provided a comprehensive report on the application of endophytic microorganisms as well as their application for the biological control of pests, insects, and pathogens affecting increases in agricultural production. Moreover, the modes of action utilized by these beneficial endophytic microorganisms were also highlighted. Therefore, there will be a need to know the diversity of these endophytic, their function, and frequency to have proper knowledge that will expand the Biotechnological application of these beneficial microorganisms, most especially their application in the production of eco-friendly agrochemicals. This will also serve as eco-friendly biotechnology approaches for the management of a heavily contaminated environment. There is a need to explore the biotechnological potential of endophytic microorganism that dwell in a marine region and the detection as well as structural elucidation of diverse and unique biological active metabolites with enhanced activity for the management of pests and diseases. This will also enhance a resourceful, cost-effective, and ecological substitute for the resolution of damages instigated by pest insects in agriculture.

The regulatory authorities also need numerous knowledge on microbial pesticides an identification, evaluation of their adverse effect on the environment and humans as well as affirming their efficacy against pests. There is also a need to define and harmonize regulatory techniques for microbial pesticides in different regions of the world. The process of regional harmonization in facilitating regulations is one most significant barrier to the great implementation of microbial pesticides. There is a need to develop several areas of microbial pesticides adaptation and development. These include enhancing the product range targeting the major plant protection challenges, development of resilient microbes, production of quantity and quality product in the market, guaranteeing satisfactory intellectual property rights for examining, generate a healthy supervisory atmosphere, regulatory coordination across the region, and inspiring public-private partnerships to fascinate more small-medium entrepreneurship into the microbial pesticide industry.

KEYWORDS

- **endophytes**
- **heavy metals**
- **indole-3-acetic acid**
- **mineral nutrients solution**
- **modes of action**
- **nanotechnology**
- **organophosphorus pesticides**
- **pathogens**

REFERENCES

Adetunji, C. O., & Ugbenyen, A. M., (2019). Mechanism of action of nanopesticide derived from microorganism for the alleviation of abiotic and biotic stress affecting crop productivity. In: *Nanotechnology for Agriculture: Crop Production and Protection*. https://doi.org/10.1007/978-981-32-9374-8_7.

Adetunji, C. O., (2019). Environmental impact and ecotoxicological influence of biofabricated and inorganic nanoparticle on soil activity. In: Panpatte, D., & Jhala, Y., (eds.), *Nanotechnology for Agriculture*. Springer, Singapore. https://doi.org/10.1007/978-981-32-9370-0_12.

Adetunji, C. O., Egbuna, C., Tijjani, H., Adom, D., Khalil, L., Al-Ani, T., & Partick-Iwuanuyanwu, K. C., (2019a). Pesticides, home made preparation of natural biopesticides and application. In: *Natural Remedy for Pest, Diseases and Weed Control* (pp. 1–12) Elsevier.

Adetunji, C. O., Kumar, D., Raina, M., Arogundade, O., & Sarin, N. M., (2019b). Endophytic microorganisms as biological control agents for plant pathogens: A panacea for sustainable agriculture. In: Varma, A., Tripathi, S., & Prasad, R., (eds.), *Plant Biotic Interactions*. https://doi.org/10.1007/978-3-030-26657-8_1.

Adetunji, C. O., Oloke, J. K., Bello, O. M., Pradeep, M., & Jolly, R. S., (2019d). Isolation, structural elucidation and bioherbicidal activity of an eco-friendly bioactive 2-(hydroxymethyl) phenol, from *Pseudomonas aeruginosa* (C1501) and its ecotoxicological evaluation on soil. *Environ. Technol. Inno., 13*(2019), 304–317.

Adetunji, C. O., Panpatte, D. G., Bello, O. M., & Adekoya, M. A., (2019c). Application of nanoengineered metabolites from beneficial and eco-friendly microorganisms as biological control agents for plant pests and pathogens. In: Panpatte, D., & Jhala, Y., (eds.), *Nanotechnology for Agriculture: Crop Production and Protection*. Springer, Singapore, https://doi.org/10.1007/978-981-32-9374-8_13.

Al-Othman, M. R., El-Aziz, A. R. M. A., Mahmoud, M. A., Eifan, S. A., El-Shikh, M. S., & Majrashi, M., (2014). Application of silver nanoparticles as antifungal and antiaflatoxin B1 produced by *Aspergillus flavus*. *Dig. J. Nanomater. Biostruct., 9*, 151–157.

Azambuja, K., & Fiuza, L. M., (2003). Isolation of *Bacillus thuringiensis* pathogenic to *Anticarsia gemmatalis* (Lepidoptera, Noctuidae). In: *Shows of Scientific Initiation of Unisinos*. São Leopoldo.

Azevedo, J. L., Maccheroni, J. W., Pereira, J. O., & Araujo, W. L., (2000). Endophytic microorganisms: A review on insect control and recent advances on tropical plants. *Elect. J. Biotech., 3*(1), 40–65.

Bagwea, R. P., Xiaojun, Z., & Weihong, T., (2003). Bioconjugated luminescent nanoparticles for biological applications. *J. Dispers. Sci. Technol., 24*, 453–464.

Baker, S., & Satish, S., (2012). Endophytes: Towards a vision in synthesis of nanoparticles for future therapeutic agents. *Int. J. Bio-Inorg. Hybrid Nanomater., 1*, 67–77.

Baker, S., & Satish, S., (2015). Biosynthesis of gold nanoparticles by *Pseudomonas veronii* AS41G inhabiting *Annona squamosa* L. *Spectrochim. Acta Mol. Biomol. Spectrosc., 150*, 691–695.

Baker, S., Kumar, K. M., Santosh, P., Rakshith, D., & Satish, S., (2015). Extracellular synthesis of silver nanoparticles by novel *Pseudomonas veronii* AS41G inhabiting *Annona squamosa* L. and their bactericidal activity. *Spectrochim. Acta Mol. Biomol. Spectrosc., 136*, 1434–1440.

Baker, S., Vinayaka, A. C., Manonmani, H. K., & Thakur, M. S., (2013c). Development of dipstick-based immuno-chemiluminescence techniques for the rapid detection of dichlorodiphenyltrichloroethane. *Luminescence, 27*, 524–529.

Bamisile, B. S., Dash, C. K., Akutse, K. S., Qasim, M., Aguila, L. C. R., Wang, F., Keppanan, R., & Wang, L., (2019). Endophytic *Beauveria bassiana* in foliar-treated citrus lemon plants acting as a growth suppressor to three successive generations of *Diaphorina citri* Kuwayama (Hemiptera: Liviidae). *Insects, 10*, 176. https://doi.org/10.3390/insects 10060176.

Beltrán-García, M. J., (2019). *Enterobacter cloacae*, an endophyte that establishes a nutrient-transfer symbiosis with banana plants and protects against the black sigatoka pathogen. *Front. Microbiol., 10*, 804. https://doi.org/10.3389/fmicb.2019.00804.

Beltran-Garcia, M. J., White, J. F., Padro, F. M., Prieto, K. R., Yamaguchi, L. F., Torres, M. S., Kato, M. J., et al., (2014b). Nitrogen acquisition in *Agave tequilana* from degradation of endophytic bacteria. *Sci. Rep., 4*, 6938. https://doi.org/10.1038/srep06938.

Ben J. J. L., John R. C., & Linda, J. J., (2016). Fungal endophytes for sustainable crop production. *FEMS Microbiology Ecology, 92*(12), fiw194. https://doi.org/10.1093/femsec/fiw194.

Berlitz, D. L., Azambuja, D. L., Azambuja, A. O., Antonio, A. C., Oliveira, J. V., & Fiuza, L. M., (2003). New insulated and Cry Proteins from *Bacillus thuringiensis* applied in the control of caterpillars of *Spodoptera frugiperda*. In: *III Brazilian Congress of Irrigated Rice and XXV Meeting of the Irrigated Rice Culture*. Camboriú. Analls. Camboriú.

Bhardwaj, D., Ansari, M. W., Sahoo, R. K., & Tuteja, N., (2014). Biofertilizers function as key player in sustainable agriculture by improving soil fertility, plant tolerance and crop productivity. *Microb. Cell Fact., 13*, 66.

Bong, C. F. J., & Sikorowski, P. P., (1991). Effects of cytoplasmic polyhedrosis virus and bacterial contamination on growth and development of the corn earworm, *Helicoverpa zea*. *J. Invert. Pathol., 57*(3), 406–412.

Broderick, N. A., Goodman, R. M., Raffa, K. F., & Handelsman, J., (2000). Synergy between zwittermicin A and *Bacillus thuringiensis* subsp. *kurstaki* against gypsy moth (Lepidoptera: Lymantriidae). *Biol. Contr., 29*(1), 101–107.

Campanini, E. B., Davolos, C. C., Alves, E. C. C., & Lemos, M. V. F., (2012). Characterization of new isolates of *Bacillus thuringiensis* for control of important insect pests of agriculture. *Bragantia, 71*(3), 362–369.

Carvalho, T. L. G., Balsemao-Pires, E., Saraiva, R. M., Ferreira, P. C. G., & Hemerly, A. S., (2014). Nitrogen signaling in plant interactions with associative and endophytic diazotrophic bacteria. *J. Exp. Bot., 65*, 5631–5642. https://doi.org/10.1093/jxb/eru319.

Castelo, B. M., França, F. H., Pontes, L. A., & Amaral, P. S. T., (2003). Evaluation of the susceptibility to insecticides in the moth populations in moth-of-cruciferous some areas of Brazil. *Hort. Braz., 21*(3), 549–552.

Castilhos-Fortes, R., Matsumura, A. T., Diehl, E., & Fiuza, L. M., (2002). Susceptibility of *Nasutitermes ehrhardti* (Isoptera: Termitidae) to *Bacillus thuringiensis* subspecies. *Braz. J. Microb., 33*(3), 219–222.

Chowdhury, S., Basu, A., & Kundu, S., (2014). Green synthesis of protein capped silver nanoparticles from phytopathogenic fungus *Macrophomina phaseolina* (Tassi) goid with antimicrobial properties against multidrug-resistant bacteria. *Nanoscale Res. Lett., 9*, 365.

Clifton, E. H., Jaronski, S. T., Coates, B. S., Hodgson, E. W., & Gassmann, A. J., (2018). Effects of endophytic entomopathogenic fungi on soybean aphid and identification of *Metarhizium* isolates from agricultural fields. *PLoS One, 13*(3), e0194815. https://doi.org/10.1371/journal. pone.0194815.

Compant, S., Duffy, B., Nowak, J., Clement, C., & Barkai, E. A., (2005). Mini-review: Use of plant growth-promoting bacteria for biocontrol of plant diseases: Principles, mechanisms of action, and future prospects. *Appl. Environ. Microb., 71*(9), 4951–4959.

Dhekney, S., Li, A., Anaman, M., Dutt, M., Tattersall, J., & Gray, D., (2007). Genetic transformation of embryogenic cultures and recovery of transgenic plants in *Vitis vinifera, Vitis rotundifolia*, and *Vitis* hybrids. *Acta Hortic., 738*, 743–748.

Domka, A. M., Rozpaądek, P., & Turnau, K., (2019). Are fungal endophytes merely mycorrhizal copycats? the role of fungal endophytes in the adaptation of plants to metal toxicity. *Front. Microbiol.* https://doi.org/10.3389/fmicb.2019.00371.

Dutta, D., Puzari, K. C., & Gogoi, R., (2014). Endophytes: Exploitation as a tool in plant protection. *Biol. Technol., 57*, 5. http://dx.doi.org/10.1590/S1516-8913201402043.

Fareed, M., Pathak, M. K., Bihari, V., Kamal, R., Srivastava, A. K., & Kesavachandran, C. N., (2013). Adverse respiratory health and hematological alterations among agricultural workers occupationally exposed to organophosphate pesticides: A cross sectional study in North India. *PLoS One, 8*(7).

Gašić, S., & Tanović, B., (2013). Biopesticide formulations, possibility of application and future trends. *Pesticidii Fitomedicina, 28*(2), 97–102.

Gill, H. K., & Garg, H., (2014). In: Soloneski, S., (ed.), *Pesticides: Environmental Impacts and Management Strategies, Pesticides-Toxic Aspects*. InTech, London. http://dx.doi.org/10.5772/57399.

Gond, S., Verma, V. C., Mishra, A., Kumar, A., & Kharwar, R. N., (2010). Role of fungal endophytes in plant protection. In: Arya, A., & Perrello, A., (eds.), *Management of Fungal Plant Pathogens*. CAB International, UK.

Gonzalez, F., Tkaczuk, C., Dinu, M. M., Fiedler, Ż., Vidal, S., Zchori-Fein, E., & Messelink, G. J., (2016). New opportunities for the integration of microorganisms into biological pest control systems in greenhouse crops. *J. Pest Sci., 89*, 295–311. https://doi.org/10.1007/s10340-016-0751-x.

Guyton, K. Z., Loomis, D., Grosse, Y., El Ghissassi, F., Benbrahim-Tallaa, L., Guha, N., Scoccianti, C., et al., (2015). Carcinogenicity of tetrachlorvinphos, parathion, malathion, diazinon, and glyphosate. *Lancet Oncol., 16*, 490–491.

Haase, S., Sciocco-Cap, A., & Romanowski, V., (2015). Baculovirus insecticides in Latin America: Historical overview, current status and future perspectives. *Viruses, 7*, 2230–2267.

Harman, G. E., & Uphoff, N., (2019). Symbiotic root-endophytic soil microbes improve crop productivity and provide environmental benefits. *Scientifica.* https://doi.org/10.1155/2019/9106395.

Hirsch, J., Strohmeier, S., Pfannkuchen, M., & Reineke, A., (2012). Assessment of bacterial endosymbiont diversity in *Otiorhynchus* spp. (Coleoptera: Curculionidae) larvae using a multitag 454 pyrosequencing approach. *BMC Microbiology, 12*(1), S6. http://dx.doi.org/10.1371/journal.pone.0029268.

Hu, R., Huang, X., Huang, J., Li, Y., Zhang, C., Yin, Y., Chen, Z., et al., (2015). Long and short term health effect of pesticides exposure: A cohort study from China. *PLoS One., 10.*

Ikram, M., Ali, N., Jan, G., Jan, F. G., Rahman, I. U., Iqbal, A., & Hamayun, M., (2018). IAA producing fungal endophyte *Penicillium roqueforti* Thom. enhances stress tolerance and nutrients uptake in wheat plants grown on heavy metal contaminated soils. *PLoS One, 13*(11), e0208150. https://doi.org/10.1371/journal.pone.0208150.

Kang, J. W., (2014). Removing environmental organic pollutants with bioremediation and phytoremediation. *Biotechnol. Lett., 36*, 1129–1139.

Khater, H. F., (2012). Prospects of botanical biopesticides in insect pest management. *Pharmacologia, 3*(12), 641–656.

Kitherian, S., (2017). Nano and bio-nanoparticles for insect control. *Res. J. Nanosci. Nanotech., 7*(1), 1–9.

Knaak, N., Rohr, A. A., & Fiuza, L. M., (2007). *In vitro* effect of *Bacillus thuringiensis* strains and cry proteins in phytopathogenic fungi of paddy rice-field. *Braz. J. Microb., 38*(3), 1–7.

Krupke, C. H., Hunt, G. J., Eitzer, B. D., Andino, G., & Given, K., (2012). Multiple routes of pesticide exposure for honey bees living near agricultural fields. *PLoS One*, e29268. http://dx.doi.org/10.1371/journal.pone.0029268.

Liu, F., Wen, L., Li, Z., Yu, W., Sun, H., & Chen, J., (2006). Porous hollow silica nanoparticles as controlled delivery system for water-soluble pesticide. *Mater. Res. Bull., 41*, 2268–2275.

Lucho, A. P. R., (2004). Management *Spodoptera frugiperda* (J. E. Smith 1797) (Lepidoptera: Noctuidae) in irrigated rice. São Leopoldo: Dissertation (Master). *Diversity and Wildlife Management* (p. 73) Universidade do Vale do Rios dos Sinos, Sao Leopoldo, UNISINOS.

Ma, X. M., Liu, X. X., Ning, X., Zhang, B., Guan, X. M., Tan, Y. F., & Zhang, Q. W., (2008). Effects of *Bacillus thuringiensis* toxin Cry 1Ac and *Beauveria bassiana* on Asiatic corn borer (Lepidoptera: Crambidae). *J. Invert. Pathol., 99*(2),123–128.

Macedo, C. L., Martins, E. S., Macedo, L. L. P., Santos, A. C., Praça, L. B., Góis, L. A. B., & Monnerat, R. G., (2012). Selection and characterization of *Bacillus thuringiensis* strains effective against *Diatraea saccharalis* (Lepidoptera: Crambidae). *Pesq. Agrop. Braz., 47*(12), 1759–1765.

Macedo-Raygoza, G. M., Valdez-Salas, B., Prado, F. M., Prieto, K. R., Yamaguchi, L. F., Kato, M. J., Canto-Canché, B. B., et al., (2019). *Enterobacter cloacae*, an endophyte that establishes a nutrient-transfer symbiosis with banana plants and protects against the black sigatoka pathogen. *Front. Microbiol.* https://doi.org/10.3389/fmicb.2019.00804.

Maksimov, I. V., Maksimova, T. I., Sarvarova, E. R., & Blagova, D. K., (2018). Endophytic bacteria as effective agents of new-generation biopesticides. *Appl. Biochem., 54*, 128–140. https://doi.org/10.1134/S0003683818020072.

Marziye, A. P., Yazdan, M., Roostaei, A., Khani, M., Negahdari, M., & Rahimi, G., (2014). Evaluation of the antifungal effect of magnesium oxide nanoparticles on *Fusarium oxysporum* f. sp. *lycopersici*, pathogenic agent of tomato. *Eur. J. Exp. Biol., 4*(3), 151–156.

Melatti, V. M., Praça, L. B., Martins, E. S., Sujii, E., Berry, C., & Monnerat, R. G., (2010). Selection of *Bacillus thuringiensis* strains toxic against cotton aphid, *Aphis gossypii* glover (Hemiptera: Aphididae). *Bioassay, 5*(2), 1–4.

Muthukumar, A., Regunathan, U., & Ramasamy, N., (2017). Role of bacterial endophytes in plant disease control. In: *Endophytes: Crop Productivity and Protection*. https://doi.org/10.1007/978-3-319-66544-3_7.

Ouda, S. M., (2014). Antifungal activity of silver and copper nanoparticles on two plant pathogens, *Alternaria alternata* and *Botrytis cinerea*. *Res. J. Microbiol., 9*, 34–42.

Parizi, M. A., Moradpour, Y., Roostaei, A., Khani, M., Negahdari, M., & Rahimi, G., (2014). Evaluation of the antifungal effect of magnesium oxide nanoparticles on *Fusarium oxysporum* f. sp. *lycopersici*, pathogenic agent of tomato. *Eur. J. Exp. Biol., 4*, 151–156.

Patel, N., Desai, P., Patel, N., Jha, A., & Gautam, H. K., (2014). Agro-nanotechnology for plant fungal disease management: A review. *Int. J. Curr. Microbiol. Appl. Sci., 3*, 71–84.

Peixoto, N. P. A. S., Azevedo, J. L., & Araujo, W. L., (2002). Endophytic microorganisms. *Biotechnol. Sci. Dev., 29*, 62–77.

Polanczyk, R. A., Silva, R. F. P., & Fiuza, L. M., (2003). Screening of *Bacillus thuringiensis* isolates to *Spodoptera frugiperda* (J.E Smith) (Lepidoptera: Noctuidae). *Arq. Inst. Biol., 70*(1), 69–72.

Porras-Alfaro, A., & Bayman, P., (2011). Hidden fungi, emergent properties: Endophytes and microbiomes. *Phytopathology, 49*(1), 291.

Praça, L. B., (2012). *Interactions between Bacillus thuringiensis Strains and Hybrids of Cabbage for the Control of Plutella xylostella and Plant Growth Promotion* (p. 141). Brasília: UNB, 2012. Thesis (Doctoral)-Faculty of Agronomy and Veterinary Medicine, University of Brasília, Brasília.

Rabiey, M., Hailey, L. E., Roy, S. R., Mahira, K. G., Al-Zadjali, A. S., Barrett, G. A., & Jackson, R. W., (2019). Endophytes vs tree pathogens and pests: Can they be used as biological control agents to improve tree health? *Euro. J. Plant Pathol.*, 1–19.

Ramamoorthy, V., Viswanathan, R., Raguchander, T., Prakasan, V., & Samiyappan, R., (2001). Induction of systemic resistance by plant growth-promoting rhizobacteria in crop plants against pests and diseases. *Crop Protect., 20*(1), 1–11.

Ramy, S. Y., & Ahmed, O. F., (2013). *In vitro* study of the antifungal efficacy of zinc oxide nanoparticles against *Fusarium oxysporum* and *Penicillium expansum*. *Afr. J. Microbiol. Res., 7*, 1917–1923.

Ravishankar, B., & Venkatesha, M., (2010). Effectiveness of SlNPV of *Spodoptera litura* (Fab.) (Lepidoptera: Noctuidae) on different host plants. *J. Biopest., 3*(1), 168–171.

Rizvi, A., & Khan, M. S., (2017). Biotoxic impact of heavy metals on growth, oxidative stress and morphological changes in root structure of wheat (*Triticum aestivum* L.) and stress alleviation by *Pseudomonas aeruginosa* strain CPSB1. *Chemosphere, 185*, 942–952. https://doi.org/10.1016/j.chemosphere.2017.07.088.

Rouhani, M., Samih, M. A., & Kalantri, S. (2012). Insecticidal effect of silica and silver nanoparticles on the cowpea seed beetle, *Callosobruchus maculatus* F. (Col.: Bruchidae). *J. Entomol. Res., 4*, 297–305.

Sabir, S., Arshad, M., & Chaudhari, S. K., (2014). Zinc oxide nanoparticles for revolutionizing agriculture, synthesis and applications. *Sci. World J.*, 1–8.

Santoyo, G., Moreno-Hagelsieb, G., Orozco-Mosqueda, M. C., & Glick, B. R., (2016). Plant growth-promoting bacterial endophytes. *Microbiol. Res., 183*, 92–99. http://dx.doi.org/10.1016/j.micres.2015.11.008.

Schloss, P. D., & Handelsman, J., (2003). Biotechnological prospects from metagenomics. *Curr. Opin. Biotechnol., 14*(3), 303–310.

Singh, M., Pant, G., Hossain, K., & Bhatia, A. K., (2016). Green remediation. Tool for safe and sustainable environment: A review. *Appl. Water Sci.* http://dx.doi.org/10.1007/s13201-016-0461-9.

Sinha, B., (2012). Global biopesticide research trends: A bibliometric assessment. *Indian J. Agricul. Sci., 82*(2), 95–101.

Souza, A. Q. L., Astolfi, F. S., Belem, P. M. L., Sarquis, M. I. M., & Pereira, J. O., (2004). Antimicrobial activity of endophytic fungi isolated from toxic plants in the Amazon: *Palicourea longiflora* (Aubl.) Rich and *Strychnos cogens* Bentham. *Acta Amaz., 34*(2), 185–195.

Steffens, C., Azambuja, A. O., Pinto, L. M. N., Oliveira, J. V., Menezes, V. G., & Fiuza, L. M., (2001). Pathogenicity of *Bacillus thuringiensis* in larvae of *Oryzophagus oryzae* (Coleoptera: Curculionidae) In: *II Brazilian Congress of Irrigated Rice and XXIV Meeting of the Irrigated Rice Culture*. Porto Alegre, Annals. Porto Alegre.

Teodoro, S., Micaela, B., & David, K. W., (2010). Novel use of nano-structured alumina as an insecticide. *Pest Manag. Sci., 66*, 577–579.

Thakore, Y., (2006). The biopesticide market for global agricultural use. *Industrial Biotechnology, 2*(3), 194–208.

Thuler, R. B., Barros, R., Mariano, R. L. R., & Vendramim, J. D., (2006). Effect of plant growth-promoting bacteria (BPCP) in developing *Plutella xylostella* (L.) (Lepidoptera: Plutellidae) in cabbage. *Sci., 34*(2), 217–222.

Umar, S., Kaushik, N., Edrada-Ebel, R., Ebel, R., & Proksch, P., (2008). Endophytic fungi for pest and disease management. In: Ciancio, A., & Mukerji, K., (eds.), *Integrated Management of Diseases Caused by Fungi, Phytoplasma and Bacteria; Integrated Management of Plant Pests and Diseases* (p. 3). Springer, Dordrecht.

Viana, C. L. T. P., Bortolo, A. S., Thuler, R. T., Goulart, R. M., Lemos, M. V. F., & Ferraudo, A. S., (2009). Effect of new isolates of *Bacillus thuringiensis* Berliner on *Plutella xylostella* (Linnaeus, 1758) (Lepidoptera: Plutellidae). *Sci., 37*(1), 22–31.

Wani, A. H., & Shah, M. A., (2012). A unique and profound effect of MgO and ZnO nanoparticles on some plant pathogenic fungi. *J. Appl. Pharm. Sci., 2*, 40–44.

Waqas, M., Khan, A. L., Hamayun, M., Shahzad, R., Kang, S. M., Kim, J. G., & Lee, I. J., (2015). Endophytic fungi promote plant growth and mitigate the adverse effects of stem rot: An example of *Penicillium citrinum* and *Aspergillus terreus*. *J. Plant Interactions., 510*, 1.

White, J. F., Kingsley, K. L., Verma, S. K., & Kowalski, K. P., (2018). Rhizophagy cycle: An oxidative process in plants for nutrient extraction from symbiotic microbes. *Microorganisms, 6*, 1–20. http://dx.doi.org/10.3390/microorganisms6030095.

CHAPTER 6

SECONDARY METABOLITES AND THEIR BIOLOGICAL ACTIVITIES FROM *CHAETOMIUM*

KASEM SOYTONG[1] and SOMDEJ KANOKMEDHAKUL[2]

[1]*Department of Plant Production Technology, Faculty of Agricultural Technology, King Mongkut's Institute of Technology Ladkrabang, Bangkok, Thailand*

[2]*Department of Organic Chemistry, Faculty of Science, Khon Khan University, Khon Khan, Thailand*

ABSTRACT

Chaetomium spp. has been searched by many researchers for years and has discovered the secondary metabolites against human, animal, and plant pathogens. Our research findings on *Chaetomium* spp. have been conducted since 1986. There are many species found to produce active metabolites against plant and human pathogens viz., *Ch. brasiliense*, *Chaetomium cochliodes*, *Chaetomium cupreum*, *Chaetomium elatum*, *Chaetomium globosum*, *Chaetomium lucknowense*, *Chaetomium longirostre,* and *Chaetomium siamense* which most of them have been developed to be biofungicide for disease control as agricultural input for organic agricultural production. Therefore, *Chaetomium* spp. viz. *Ch. amygdalisporum*, *Ch. brasiliense*, *Ch. coarctatum*, *Ch. cochliodes*, *Ch. Cupreum*, *Ch. elatum*, *Ch. funicola*, *Ch. globosum*, *Ch. gracile*, *Ch. mollicellum*, *Ch. murorum*, *Ch. olivaceum*, *Ch. quadrangulatum*, *Ch. retardatum*, *Ch. seminudum*, *Ch. siamense*, and *Ch. trilaterale* were reported to produce natural active metabolites against *Plasmodium falciparum*, *Mycobacterium tuberculosis*, antibacterial, antifungal, cytotoxicity against cancer cells, anti-Alzheimer, and anti-inflammatory, human tumor cell lines, human breast cancer

(Bre04), human neuroma (N04) cell lines and human lung (Lu04), cytotoxicity against cholangiocarcinoma cell lines at IC_{50} 3.41–86.95 μM, the human HL-60 leukemia and murine P388 leukemia cell lines. Those metabolites found from several species of *Chaetomium* are reported for antimicrobial activity against anaerobic bacteria, especially *Bacteroides fragillis*, *Propionibacterium acnes*, *Escherichia coli* W3110, *Staphylococcus aureus* 209P, *Cladosporium resinae*, *Bacillus subtilis*, *Trichophyton mentagrophytes*, *Candida albicans*, *Salmonella typhimurium*, *Streptococcus pyogenes*, *Escherichia coli*, *Salmonella choler-aesuis*, *Corynebacterium diphtheriae*, *Streptococcus aureus*, and *Aspergillus fumigatus*. *Chaetomium* species are being discovered as the potential antagonistic fungi against phytopathogens, viz. *Drechslera oryzae* (leaf blight of maize), *Pyricularia oryzae* (rice blast disease), *Cochliobolus lunatus* (perfect stage) which the imperfect stage is known as *Curvularia lunata* (leaf spot disease), *Pythium ultimum* (damping-off disease of sugarbeet), *Botrytis cinerea* (BC) (gray mold disease of grape), *Rhizopus stolonifer* (postharvest disease) and *Coniella diplodiella* (grape white rot disease), *Phytophthora infestans* (late blight of potato), *Phytophthora* spp. and *Pythium* spp. (root rot of plants). Interestingly some *Chaetomium* spp. reported to release the metabolite that showed excellent insecticidal activity against *Plutella xylostella*. Biofungicides for plant disease control have developed *Chaetomium* species and their active metabolites have also investigated to be natural products of nanoparticles used for plant immunity. Those bioproducts are used to promote the non-agrochemical production (NAP) and organic agriculture (OA), which has been contributed for farmers use in many countries, e.g., Thailand, Myanmar, Vietnam, Laos, Cambodia, and China.

6.1 INTRODUCTION

Chaetomium species are one of the richest sources of biologically active compounds. Previous investigation of the authors and other researchers on secondary metabolites of *Chaetomium* spp. recorded to isolate various types of active compounds, e.g., benzoquinone derivatives, tetra-S-methyl derivatives, azaphilones, chaetochalasins, indol-3-yl-[13]cytochalasans, depsidones, anthraquinone-chromanone, globosumones, and orsellinic acid. Some of those secondary metabolites showed their biological activity against human diseases such as *Plasmodium falciparum*, *Mycobacterium tuberculosis*, antibacterial, antifungal, cytotoxicity against cancer cells, anti-Alzheimer, and anti-inflammatory. In addition, some of those compounds were reported as fungal mycotoxins. The studied species, their structures and biological activities are discussed in this chapter.

6.2 BIOACTIVE SECONDARY METABOLITES FROM *CHAETOMIUM*

Our review literature found that *Ch. amygdalisporum* extract from the culture on rice medium of *Ch. amygdalisporum* strain NHL2874 encountered *bis*-(3-indolyl)-dihydroxybenzoquinone, neocochlio-dinol (**1**) and mollicellin G (**2**) (Sekita, 1983). *Chaetomium atrobrunneum* is found by Okeke et al. (1993) who reported a mycotoxin, patulin (**3**) that inhibited the rice disease; *Drechslera oryzae*, *Pyricularia oryzae*, and *Gerlachia oryzae*. Hwang et al. (2000) found a novel metabolite, chaetoatrosin A (**4**) from broth culture of *Ch. atrobrunneum* F449. Compound **4** inhibited chitin synthase and antifungal activities against BC, *Cryptococcus neoformans* and *Trichophyton mentagrophytes*. *Chaetomium aureum* is recorded by Li et al. (2010) who found the compounds named chaetoaurin (**5**), eugenitol (**6**), eugenetin (**7**), chaetoquadrin A (**8**), chaetoquadrin B (**9**), chaetoquadrin G (**10**), and chaetoquadrin H (**11**).

The investigation on the ethyl acetate extract of *Ch. brasiliense* (NRRL 22999) found the new compound, chaetochalasin A (**12**), along with four known compounds: chaetoglobosin D (**13**), chaetoglobosin F_{ex} (**14**), 19-*O*-acetyl-chaetoglobosins A (**15**) and 19-*O*-acetylchaetoglobosins D (**16**) (Oh et al., 1998). Chaetochalasin A (**12**) expressed cytotoxicity against the NCI's panel of 60 human tumor cell lines and antibacterial activity against *Bacillus subtilis* (ATCC 6051) and *Staphylococcus aureus* (ATCC 25923). Later, Li et al. (2008) reported chromone, 2-(hydroxymethyl)-6-methylmethyleugenin (**17**), eugenetin (**18**), *O*-methyl-sterigmatocystin (**19**), sterigmatocystin (**20**), chaetocin (**21**), and depsidones, mollicellin D (**22**) and mollicellins H-J (**23–25**). Mollicellins H and I (**23** and **24**) significantly inhibited the growth of human breast cancer (Bre04), human neuroma (N04) cell lines and human lung (Lu04).

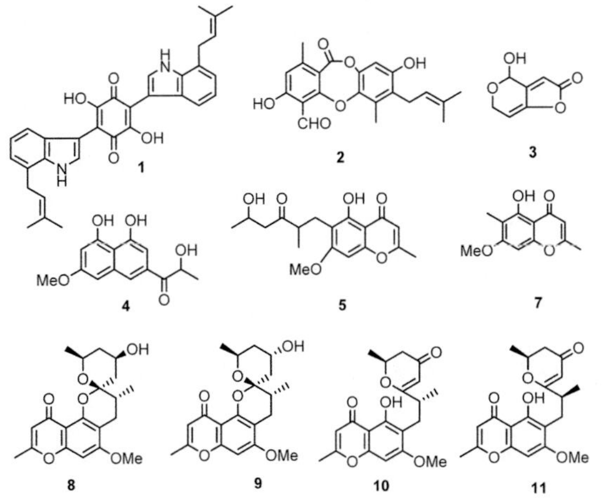

Our research finding revealed that 10 depsidones, five known mollicellins B (**26**), C (**27**), E (**28**), F (**29**), H (**23**), and J (**25**), and four new, mollicellins K-N (**30–33**) from fungus *Ch. brasiliense*. It showed that compounds **25–28** and **30–32** expressed antimalarial activity against *Plasmodium falciparum*. Only compound **30** resulted in antimicrobial activities against *Candida albicans* and *Mycobacterium tuberculosis*. Moreover, all compounds (**23, 25–33)** expressed cytotoxicity to KB, BC1, NCI-H187 and five cholangiocarcinoma cell lines (Khumkomkhet et al., 2009).

Burrows et al. (1975) studied on *Ch. coarctatum* and found two metabolites 2-(buta-1,3-dienyl)-3-hydroxy-4-(penta-1,3-dienyl)tetrahydro-furan, aureonitol (**39**) and 8-ethylidene-7,8-dihydro-4-methoxy-pyrano[4,3-*b*] pyran-2,5-dione, coarctatin (**40**). *Chaetomium cochliodes* is reported by Brewer et al. (1968) who isolated cochliodinol (**41**) from three strains of *Ch. cochliodes* (HLX 374, HLX 577 and HLX 366).

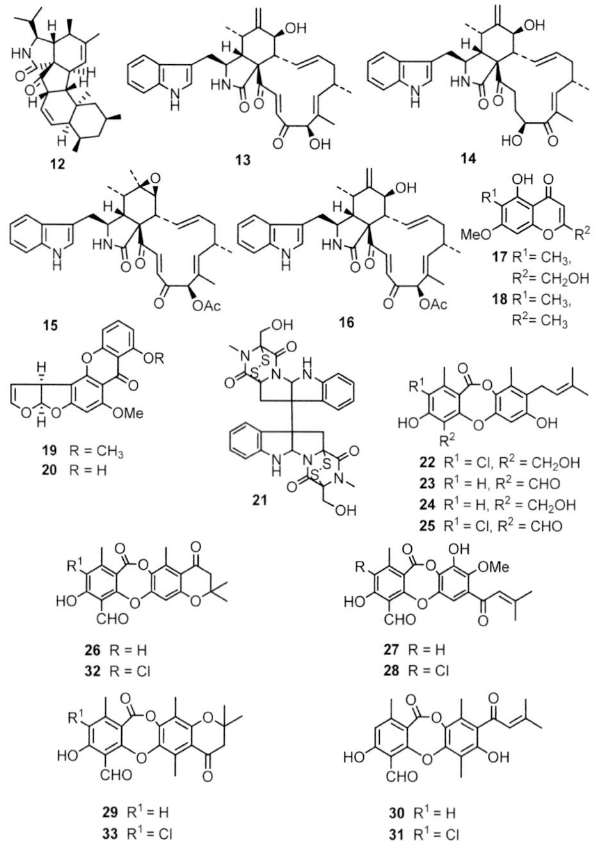

Thereafter, they further recorded that chetomin (**42**) from *Ch. cochliodes* (HLX 440) actively against several gram-positive bacteria and inhibited the mycelial growth of some fungi and protein synthesis in culture of HeLa cells (Brewer et al., 1972). Abraham et al. (1992) investigated the ethyl acetate extract of *Ch. cochliodes* DSM 63353 resulting two tetrahydrofurans, 2-(buta-1,3-dienyl)-3-hydroxy-4-(Penta-1,3-dienyl)tetrahydrofuran, aureonitol (**39**) and a new spiroketal (1*RS*,9*RS*)-3-hydroxymethyl-8*Z*-(2′*E*-pentenylidene)-2,6-dioxa-spiro[4,4]nonanol-9 (**43**).

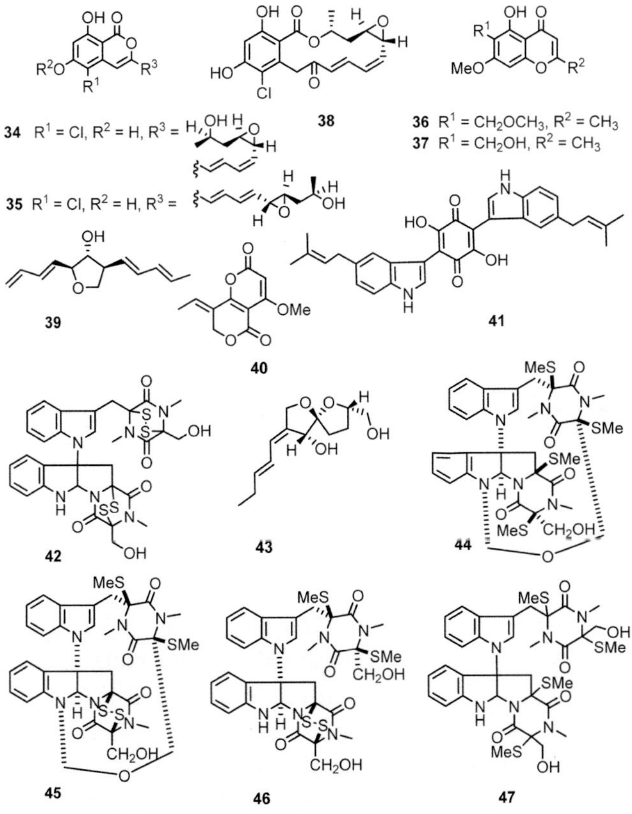

Li et al. (2006a) recorded that the ethyl acetate extract of *Ch. cochliodes* cultured in solid-state fermented rice medium found three new epipolythio-dioxopiperazines, chaetocochins A-C (**44–46**), and dethio-tetra (methylthio) chetomin (**47**), and chetomin (**42**). It expressed those compounds **42**, **44**, and **46** showed significantly cytotoxicity against cancer cell lines, Bre-04, Lu-04, and N-04.

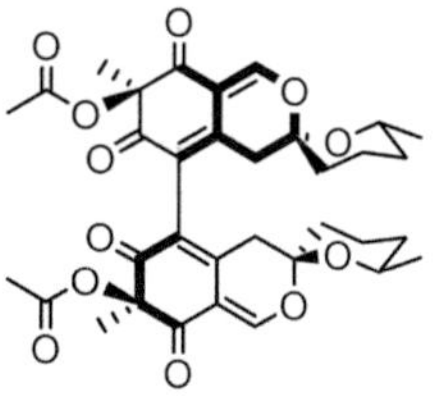

48

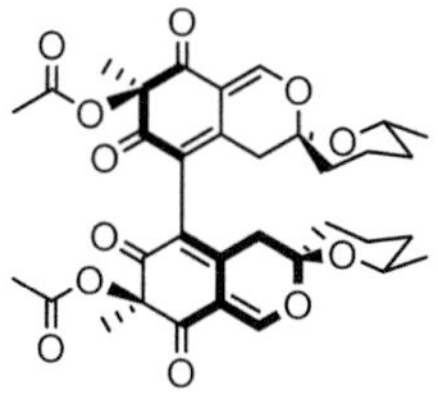

49

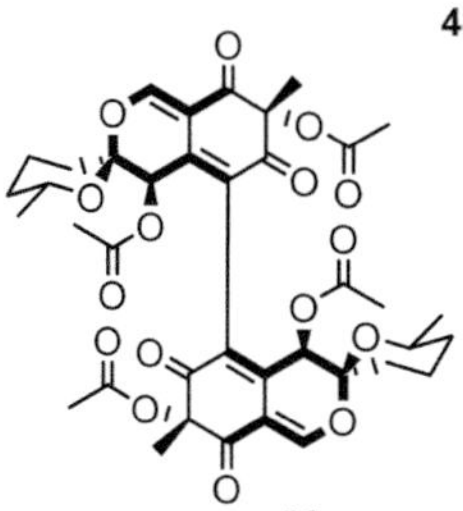

50

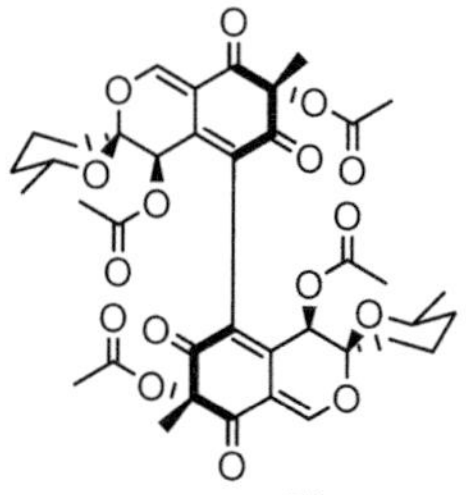

51

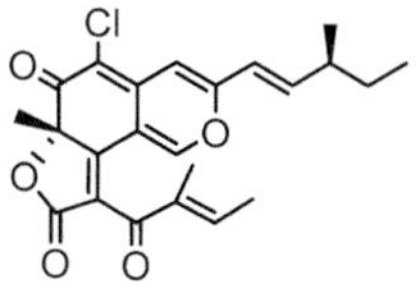

52

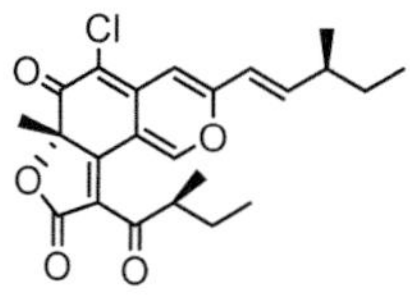

53

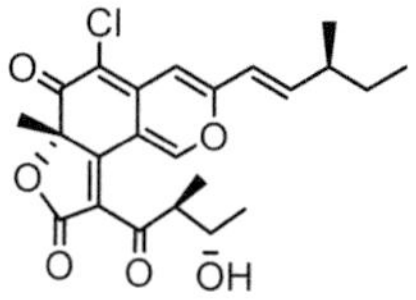

54

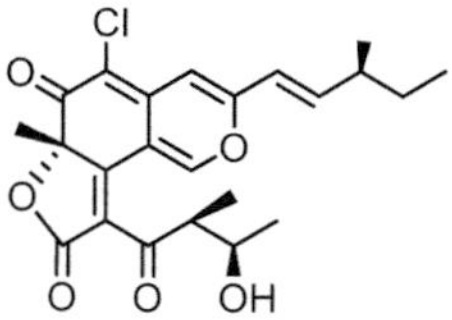

55

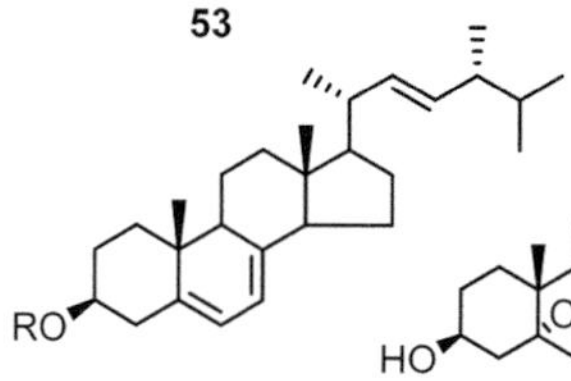

56

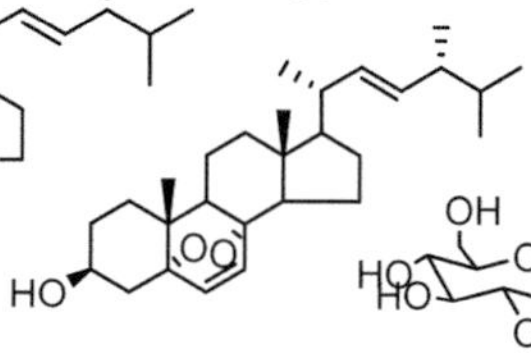

57

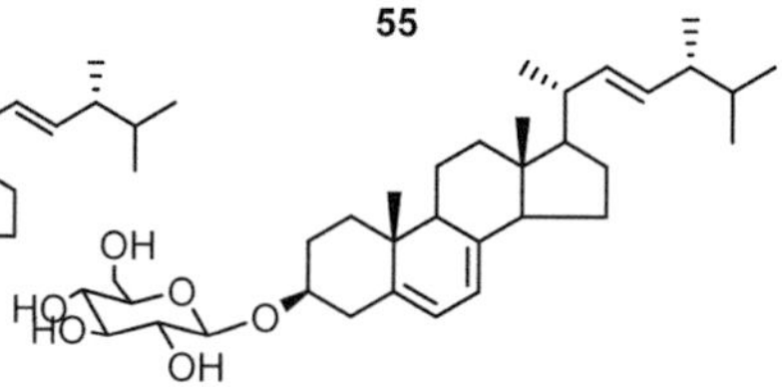

58

Our research investigation found that the extracts of fungal biomass of *Ch. cochliodes* VTh01 and *Ch. cochliodes* CTh05 yielded chaetochalasin A (**12**), two new azaphliones, chaetoviridines E and F (**52** and **53**), four new dimeric spiro-azaplilones named cochliodones A–D (**48–51**), a new *epi*-chaetoviridin A (**54**), chaetoviridin A (**55**), and five known showed that compounds **12**, **53,** and **57** expressed antimalarial activity against *Plasmodium falciparum*. Compounds **12**, **50**, **52**, **53,** and **57** resulted to antimicrobial activity against *Mycobacterium tuberculosis*. Moreover, **52** and **53** recorded to be cytotoxic against the KB, BC1, and NCI-H187 cell lines (Pholkerd et al., 2008).

Chaetomium cupreum CC3003 is first recorded by our investigation that discovered three new azaphilones named rotiorinols A-C (**59–61**), two new stereoisomers, (–)-rotiorin (**62**) and *epi*-isochromophilone II (**63**), and a known compound, rubrorotiorin (**64**). It resulted those compounds **59**, **61**, **62**, and **64** exhibited antifungal activity against *Candida albicans* (Kanokmedhakul et al., 2006).

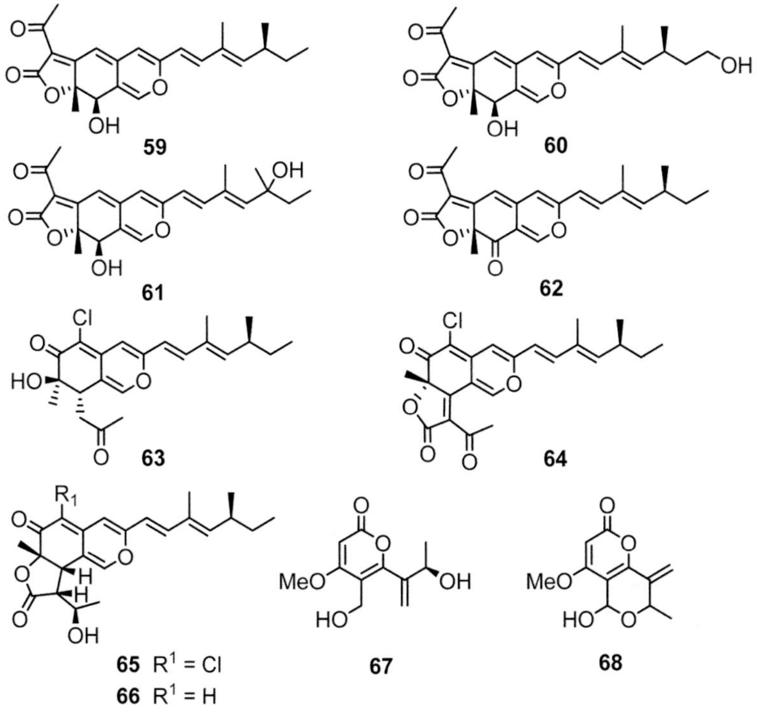

Moreover, we also recorded that *Ch. cupreum* RY202 released two new angular types of azaphilones, sochromophilonol (**65**), ochrephilonol (**66**), and known compounds clearanols A and B (**67** and **68**), rubrorotiorin (**64**), isochromophilone II (**63**), (-)-rotiorin (**62**), rotiorinols A–C (**59–61**). As a result, compounds **65**, **66,** and **68** expressed moderately cytotoxicity toward the KB and NCI-H187 cell lines of IC_{50} 9.63–32.42 µg/mL (Panthama et al., 2015).

Chaetomium elatum ChE01 is further found by our research group as the first recorded to discover the cytochalasans type, chaetoglobosin V (**69**), prochaetoglobosin III (**70**), prochaetoglobosin III_{ed} (**71**), chaetoglobosins B (**72**), C (**73**), D (**13**), F (**74**), G (**75**), and isochaetoglobosin D (**76**). These compounds expressed cytotoxicity against cholangiocarcinoma cell lines at IC_{50} 3.41–86.95 µM and human breast cancer at IC_{50} 2.54–21.29 µM (Thohinung et al., 2010). Our further investigation also found the crude metabolites and nano-particles constructed from *Ch. elatum* gave the effective disease control of rice blast pathogen (*Magnaporthe oryzae*) (Song et al., 2020b).

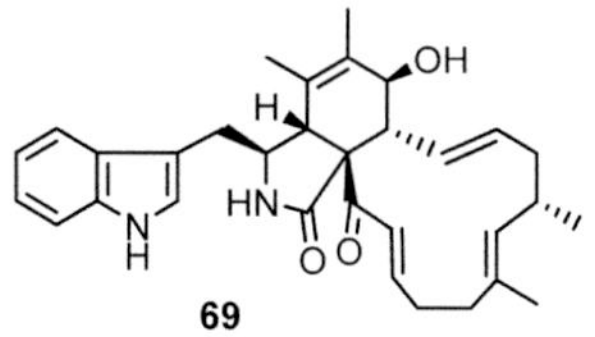

69

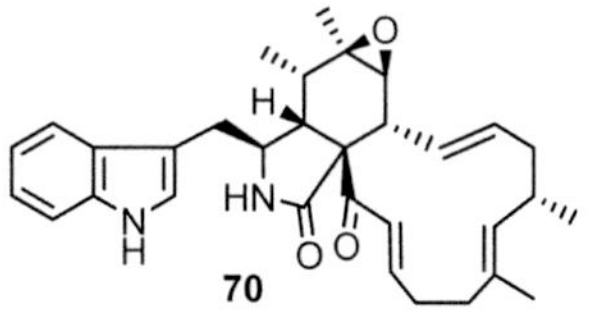

70

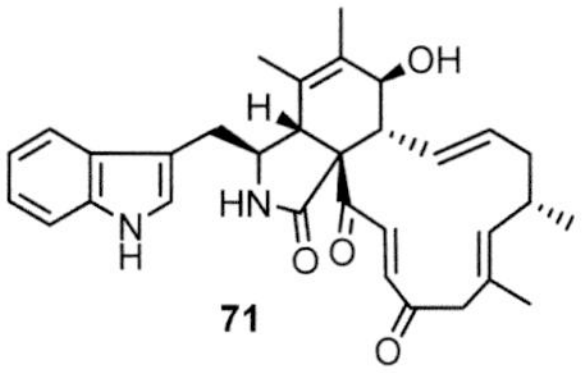

71

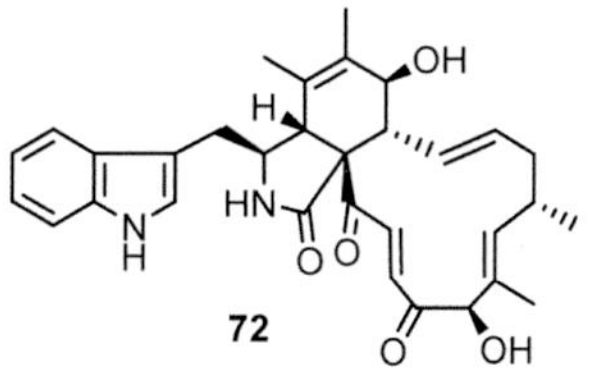

72

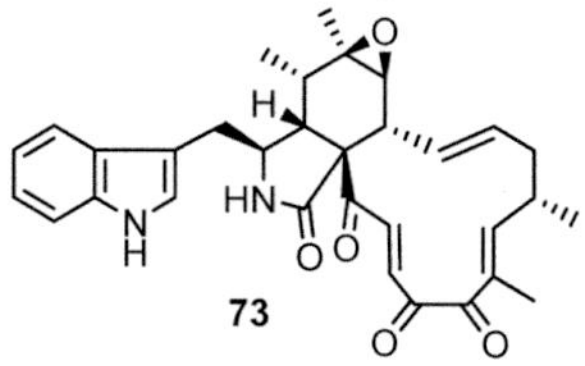

73

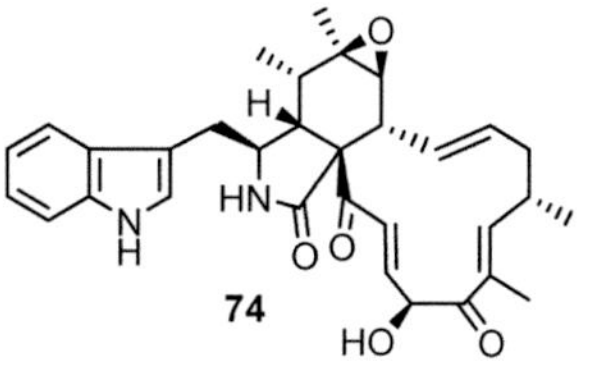

74

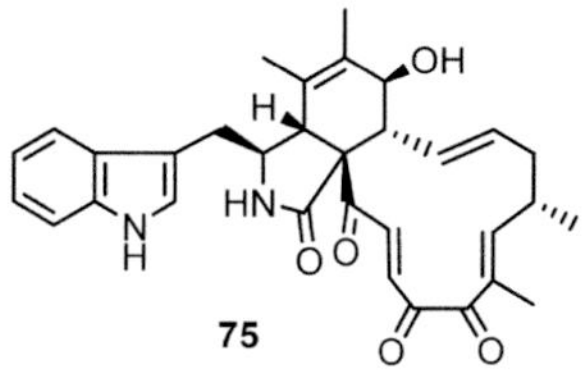

75

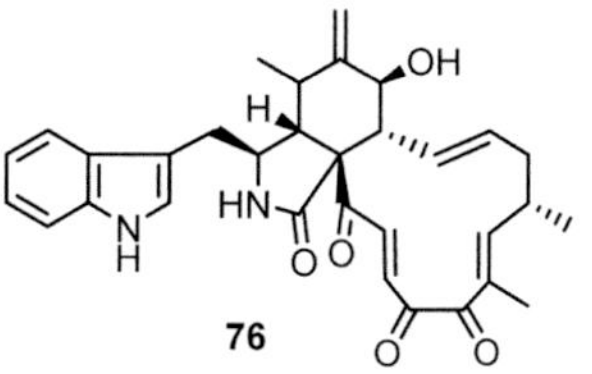

76

Payne et al. (2002) found a series of tricyclic natural products, SB236049 (**77**), SB236050 (**78**), and SB238569 (**79**) from *Ch. funicola*. These compounds exhibited activity against broad-spectrum of metallo-β-lactamases. *Chaetomium globosum* has been investigated by many researchers. Brewer et al. (1968) found that two strains of *Ch. globosum* (HLX 707 and HLX 819) produced a purple pigment, named cochliodinol (**41**). They also further recorded to discover chetomin (**42**) from two strains of *Ch. globosum* (HLX 581 and HLX 707) (Brewer et al., 1972). In 1973, Sekita et al. recorded three cytotoxic indol-3-yl-[13]cytochalasans, chaetoglobosins A (**80**), B (**72**), and C (**73**) from *Ch. globosum* who defined the structures of chaetoglobosins (**73**), D (**13**), E (**81**), and F (**82**) G (**83**) and J (**84**) (Sekita et al., 1976, 1977). Itoh et al. (1980) displayed a new sesquiterpene antibiotic, heptelidic acid (**85**) from *Ch. globosum* (SANK 13379). This compound showed antimicrobial activity against anaerobic bacteria, especially *Bacteroides fragillis* and *Propionibacterium acnes*. In 1981, Probst, and Tamm investigated the dichloromethane extract of *Ch. globosum* (Lederle H-124) and found five cytochalasans, chaetoglobosins A (**80**) and C (**73**), and 19-*O*-acetylchaetoglobosins A (**15**), D (**16**), and B (**86**).

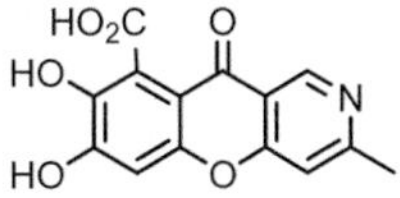

77

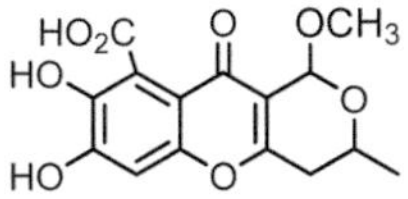

78

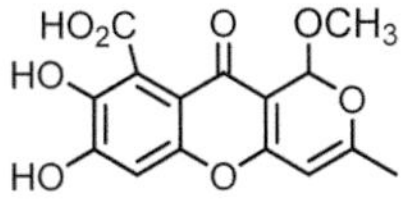

79

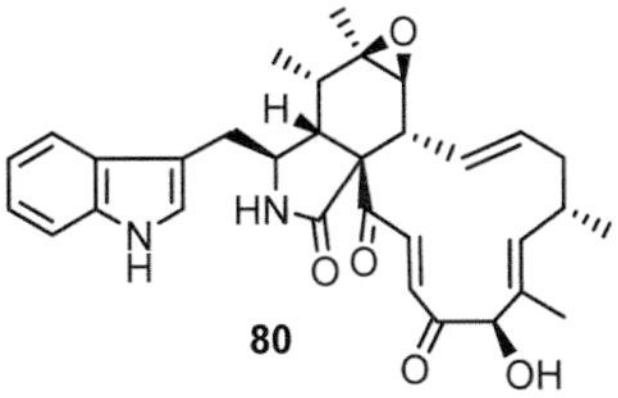

80

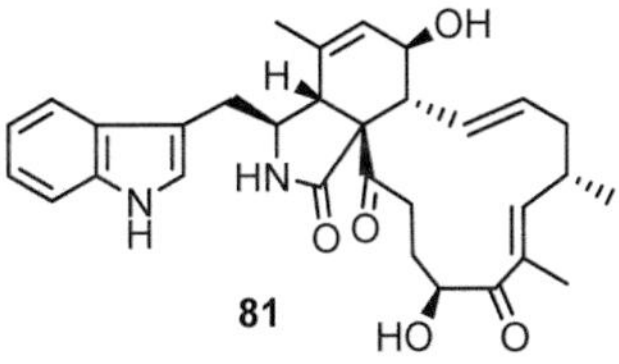

81

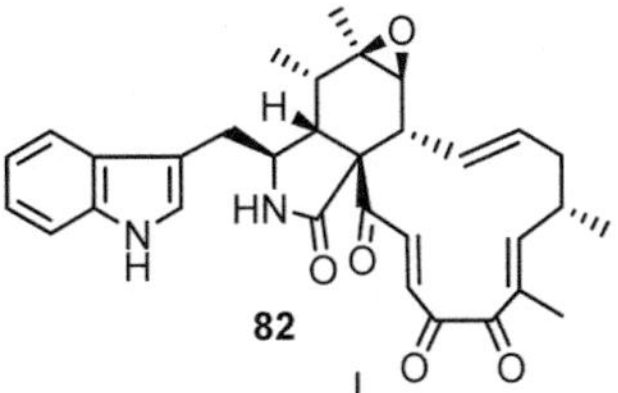

82

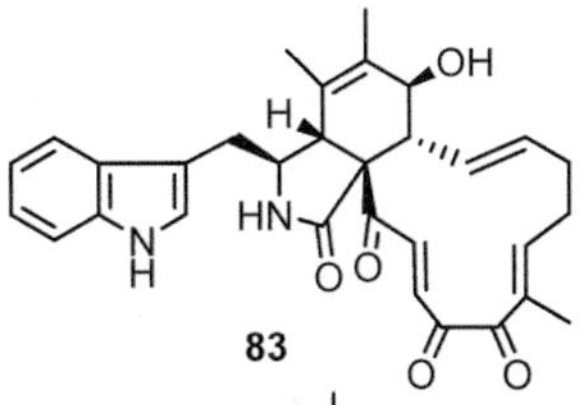

83

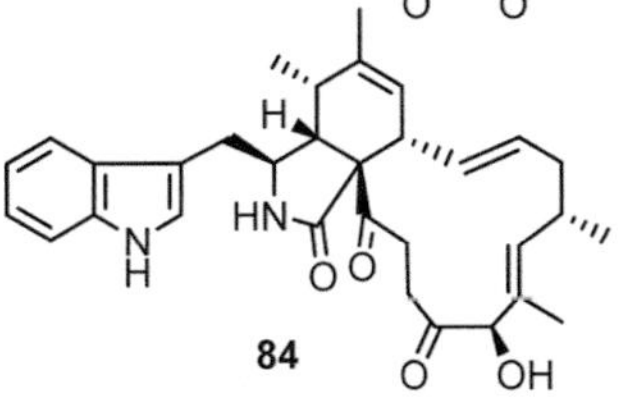

84

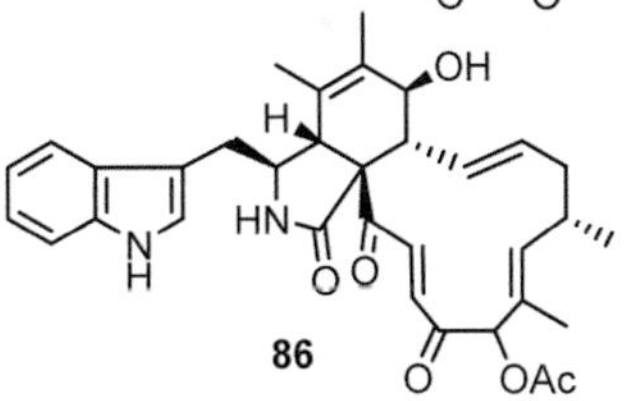

86

Kikuchi et al. (1981) also isolated and structural elucidated a new metabolite, dethio-tetra (methylthio) chetomin (**47**) and a known chetomin (**42**) from the ethyl acetate extract of *Ch. globosum*. These metabolites expressed antimicrobial activity toward *Escherichia coli* W3110 and *Staphylococcus aureus* 209P. Moreover, Sekita et al. (1981) recorded that chaetoglobosin A (**80**) affected to the structure and functions of mammalian cells causing the inhibition of cellular movements and cell division motility, secretion, and phagocytosis, and the cell shape changes. Thereafter, *Ch. globosum* var. *flavo-viridae* (TRTC 66.631a) was reported by Takahashi et al. (1990) found four new azaphilones of angular type, named chaetoviridin A (**55**) as the major compound and chaetoviridins B-D (**87–89**) as the minor congers. It resulted that Chaetoviridin A (**55**) expressed a weak inhibitory activity on monoamine oxidase (MAO), and chlamydospore-like cells of *Cochliobolus lunatus* (perfect stage) which is a plant pathogen causing humans and animals disease but the imperfect stage is known as *Curvularia lunata* (leaf spot disease of plants) and inhibited the growth of *Pyricularia oryzae* (rice blast disease).

Di Pietro et al. (1992) also recorded that *Ch. globosum* produced 2-(buta-1,3-dienyl)-3-hydroxy-4-(Penta-1,3-dienyl)-tetrahydrofuran (**39**) and chetomin (**42**) which expressed activity against *Pythium ultimum* (damping-off disease of sugarbeet). Then, Tanida et al. (1992) reported TAN-1142 (**90**) from EtOAc extract of *Ch. globosum* I-319 (IFO 32395, FERM BP-3429) inhibited the growth of murine tumor cells. Yasukawa et al. (1994) penciled that chaetoviridin A (**55**) showed the inflammatory activity of 12-*O*-tetradecanoylphorbol-13-acetate in mice, a tumor-promoting agent to the mouse ear resulted to induce inflammation.

In 1996, Breinholt et al. found a novel compound of prenisatin (**91**), 5-(3-methyl-2-butenyl)-indole-2,3-dione (5-prenylisatin) which derived from *Ch. globosum*. This compound exhibited antifungal activity *in vitro* against BC (gray mold disease of grape). Further investigation on *Ch. globosum* PF1138 was done by Tabata et al. (2000) found the novel *trans*-epoxysuccinyl peptides named PF1138 A (**92**) and B (**93**) demonstrated inhibitory effect on cysteine proteases, e.g., cathepsin B, papain, and ficin, but no effect on serine protease, e.g., trypsin.

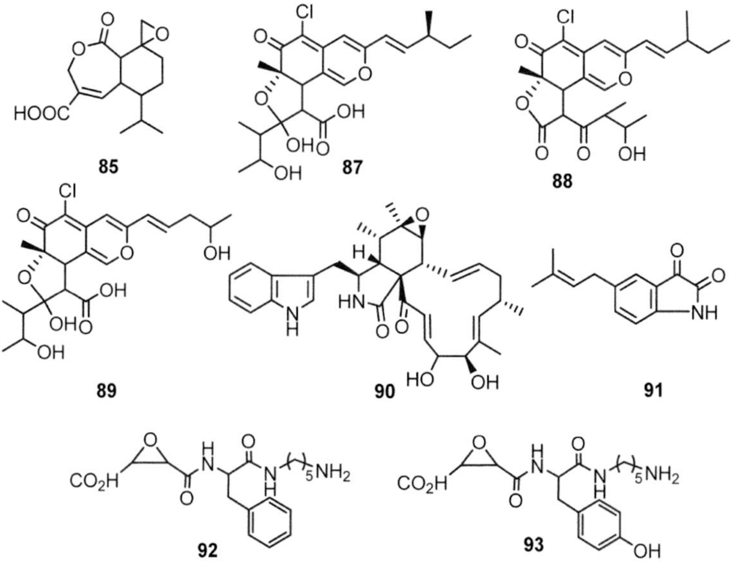

In 2002, our research finding on *Ch. globosum* strain KMITL-N0802 resulted to record a novel anthraquinone-chromanone compound named chaetomanone (**94**) and seven known compounds, ergosterol (**56**), ergosterol palmitate (**95**), chrysophanol (**96**), chaetoglobosin C (**73**), alternariol monoethyl ether (**97**), echinulin (**98**), and isochaetoglobosin D (**76**). Chaetomanone (**94**) and echinulin (**98**) expressed antibacterial against *Mycobacterium tuberculosis* (Kanokmedhakul et al., 2002).

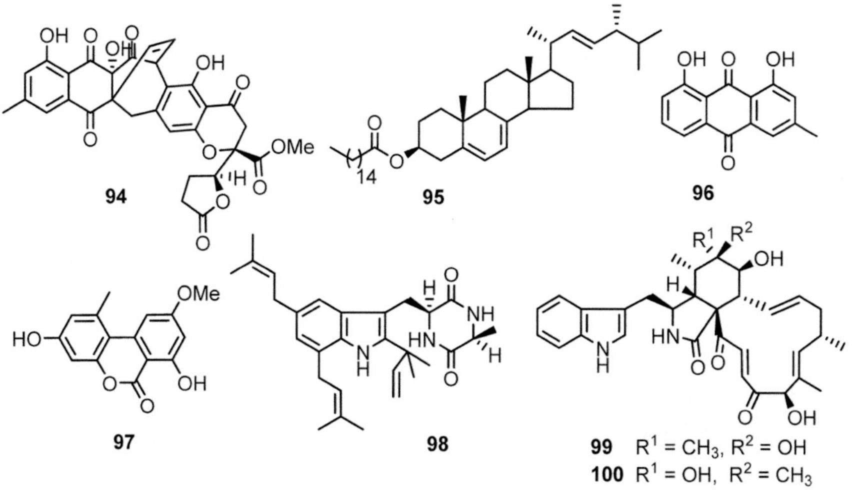

Moreover, Jiao et al. (2004) stated that the metabolites of *Ch. globosum* strain CANU N60 can be found three novel compounds, chaetoglobosins Q (**99**), R (**100**), and T (**101**), six known compounds, chaetoglobosins A (**80**), B (**72**), D (**13**), and J (**84**) and prochaetoglobosins I (**102**) and II (**103**). With this, chaetoglobosins A (**80**), B (**72**), D (**13**), J (**84**), Q (**99**), and T (**101**) and prochaetoglobosins I (**102**) and II (**103**) expressed significantly cytotoxicity toward the P388 murine leukemia cell lines. While, chaetoglobosins A (**80**), B (**72**), D (**13**), and J (**84**) and prochaetoglobosins I (**102**) and II (**103**) showed antimicrobial activities toward *Cladosporium resinae*, *Bacillus subtilis*, and *Trichophyton mentagrophytes*. Bashyal et al. (2005) recorded that *Ch. globosum* released three new esters of orsellinic acid, globosumones A-C (**104–106**) and three known compounds, orsellinic acid (**107**), orcinol (**108**), and trichodion (**109**). They stated that globosumones A and B (**104** and **105**) were moderately inhibited cell proliferation of four cancers cell lines; MCF-7, NCI-H460, SF-268 and MIA Pa Ca-2 (pancreatic carcinoma).

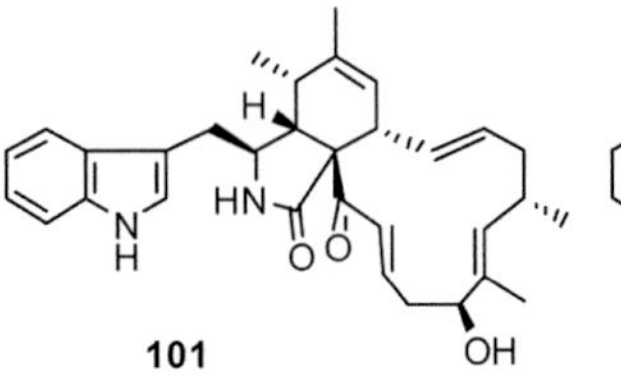

101

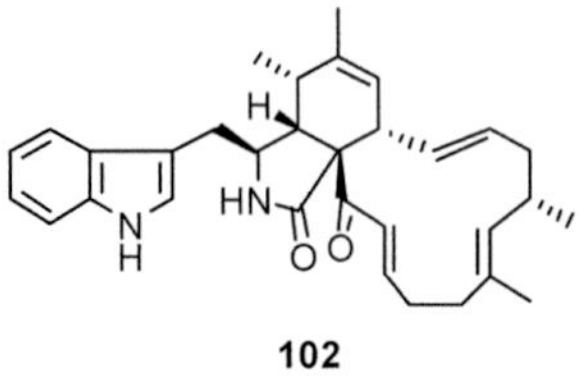

102

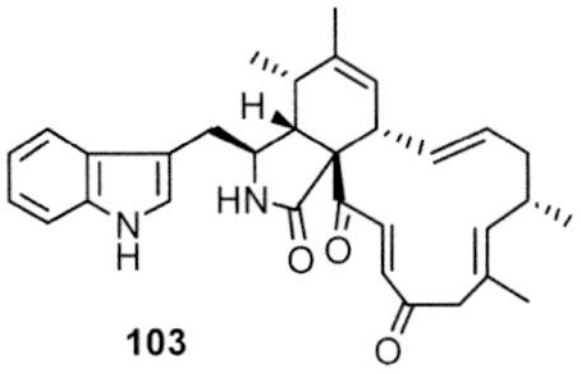

103

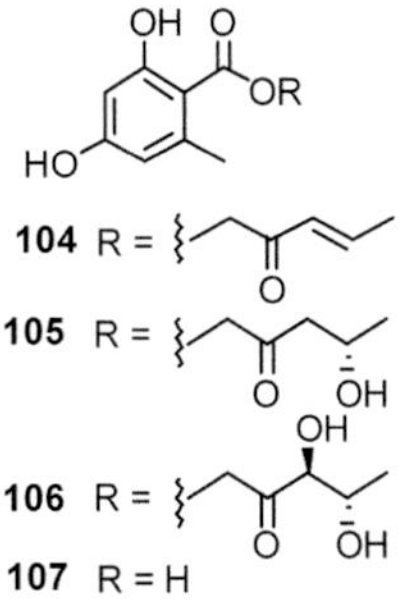

104 R =
105 R =
106 R =
107 R = H

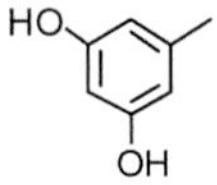

108

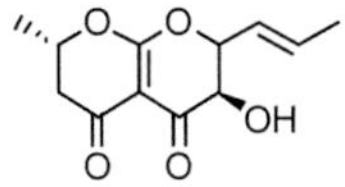

109

In 2006, Ding et al. recorded to isolate the metabolites from *Ch. globosum* IFB-E019 and discovered a new cytotoxic cytochalasin-based alkaloid named chaetoglobosin U (**110**), and four known analogs, chaetoglobosins C (**73**), E (**81**), and F (**82**) and penochalasin A (**111**). With this, Chaetochalasin U (**110**) showed cytotoxic activity against the human nasopharyngeal epidermoid tumor KB cell lines, while **73**, **81**, **82**, and **111** showed moderately against the cell lines. Wijeratne et al. (2006) found dihydroxyxanthenone, globosuxanthone A (**112**), tetrahydroxanthenone, globosuxanthone B (**113**), two xanthones, globosuxanthones C (**114**) and D (**115**), 2-hydroxyvertixanthone (**116**), and two anthraquinones, chrysazin (**117**) and 1,3,6,8-tetrahydoxyanthraquinone (**118**) from *Ch. globosum*. It showed that compound **112** expressed strongly cytotoxicity toward the panel of seven human solid tumor cell lines, induced classic signs of apoptosis and disrupted the cell cycle resulting to accumulate cells in either G_2/M or S phase.

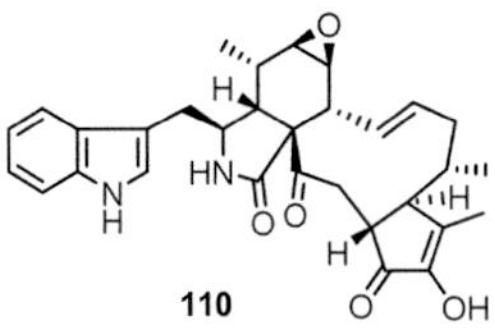

110

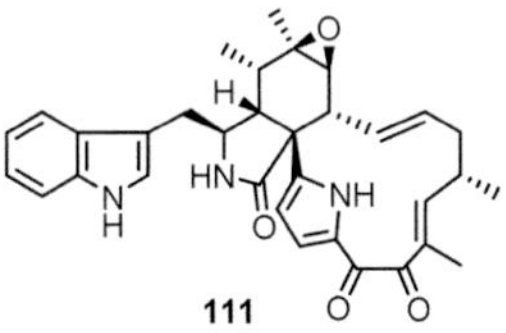

111

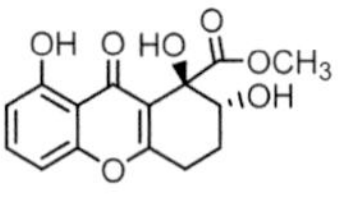

112

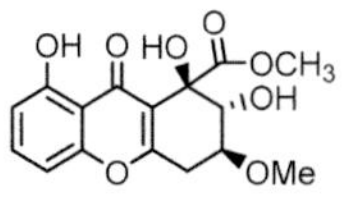

113

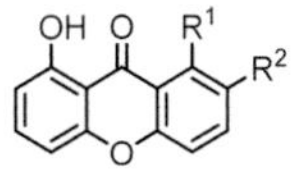

114 R^1 = OH, R^2 = OCH_3
115 R^1 = COOH, R^2 = H
116 R^1 = $COOCH_3$, R^2 = OH

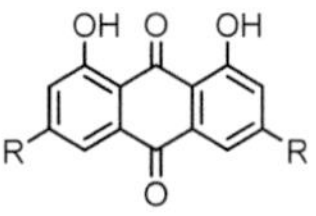

117 R = H
118 R = OH

Lately, Yang et al. (2006) recorded that *Ch. globosum* which isolated from fermentation broth yielded two novel chemokine receptor CCR-5 inhibitors, Sch 210971 (**119**) Sch 210972 (**120**) and Sch 213766 (**121**) and encountered a major component **120** that showed a potent inhibition of the CCR-5 receptor binding. However, Wang et al. (2006) stated that the extracts of endophytic *Ch. globosum* which isolated from the inner tissue of the marine red alga *Polysiphonia urceolata* yielded a new benzaldehyde, chaetopyranin (**122**), 10 known compounds including 2-(2,'3-epoxy-1,'3'-heptadienyl)-6-hydroxyl-5-(3-methyl-2-butenyl) benzaldehyde (**123**) and isotetrahydroauroglaucin (**124**), two anthraquinone derivatives, erythroglaucin (**125**) and parietin (**126**), five asperentin derivatives including asperetin (**127**) which known as cladosporin), 5'-hydroxy-asperentin-8-methylether (**128**), asperentin-8-methylether (**129**), 4'-hydroxyasperentin (**130**), and 5'-hydroxyasperentin (**131**), and the prenylated diketopiperazine congener neoechinulin (**132**). It showed that compound **122** expressed a moderately cytotoxicity towards three tumor cell lines, including human microvascular endothelial cells (HMEC), hepatocellular carcinoma cells (SMMC-7721), and human lung epithelial cells (A549).

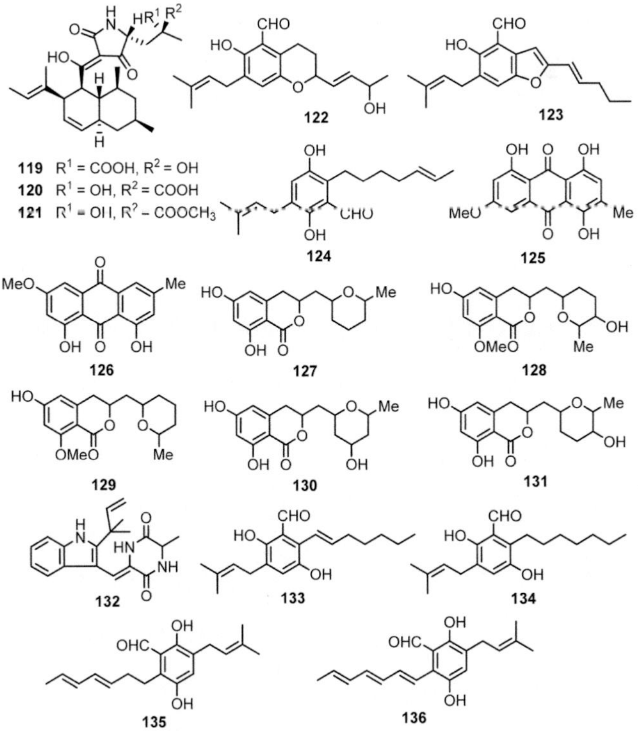

In 2007, Wang et al. recorded four benzaldehyde derivatives from *Ch. globosum* which isolated from a marine-alga and defined as 2-(1-heptenyl)-3,6-dihydroxy-5-(3-methyl-2-butenyl) benzaldehyde (**133**), 2-heptyl-3,6-dihydroxy-5-(3-methyl-2-butenyl)benzaldehyde(**134**),2-(3,5-heptadienyl)-3,6-dihydroxy-5-(3-methyl-2-butenyl) benzaldehyde (**135**), and 2-(1,3,5-heptatrienyl)-3,6-dihydroxy-5-(3-methyl-2-butenyl) benzaldehyde (**136**). Moreover, Yang et al. (2007) discovered that *Ch. globosum* from fermentation broth can be isolated a novel secondary metabolite, Sch 213766 (**121**). It showed activity in the CCR-5 receptor.

In 2008, Yamada et al. isolated three new azaphilones, chaetomugilins A-C (**137–139**), chaetoglobosins A (**80**), and C (**73**) from *Ch. globosum* OUPS-T106B-6 originated from the marine fish *Mugil cephalus*. These metabolites showed significant cytotoxicity against and the human HL-60 leukemia cell line and murine P388 leukemia cell line. In 2009, Qin et al. recorded the endophytic *Ch. globosum* ZY-22 from *Ginkgo biloba* to produce chaetomugilin A (**137**), chaetomugilin D (**140**) globosterol (**139**), chaetoglobosin A (**80**) and C **73**). Further isolation of this fungus yielded globosterol (**141**), tetrahydroxylated ergosterol, (22*E*,24*R*)-ergosta-7,22-diene-3β,5α,6β,9α-tetraol (**142**). In addition, they found more compounds of cerebroside B (**143**), cerebroside C (**144**), and ergosta-4,6,8,22-tetraen-3-one (**145**). Moreover, Yamada et al. (2009) indicated that new azaphilones, seco-chaetomugilins A and D were produced by a marine-fish-derived *Ch. globosum* and found Seco-chaetomugilins A (**146**) and D (**147**).

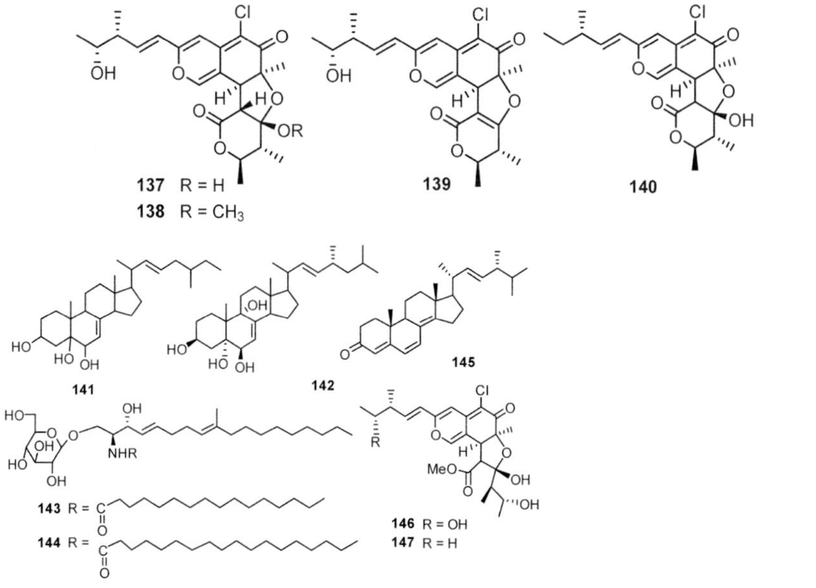

In 2010, a marine-derived endophytic fungus, *Ch. globosum* QEN-14 reported to produce cytoglobosins A-G (**148–154**), isochaetoglobosin D (**76**), chaetoglobosins Fex (**14**), U (**155**), and 20-dihydrochaetoglobosin A (**156**), chaetomugilins A (**137**), 11-*epi*-chaetomugilin A (**157**), and 4′-*epi*-chae-tomugilin A (**158**) (Cui et al., 2010). In the same year, Zhang et al. (2010) further recorded that cytotoxic chaetoglobosins V (**159**), W (**160**) and congeners (**161–166**) from the endophyte *Ch. globosum*.

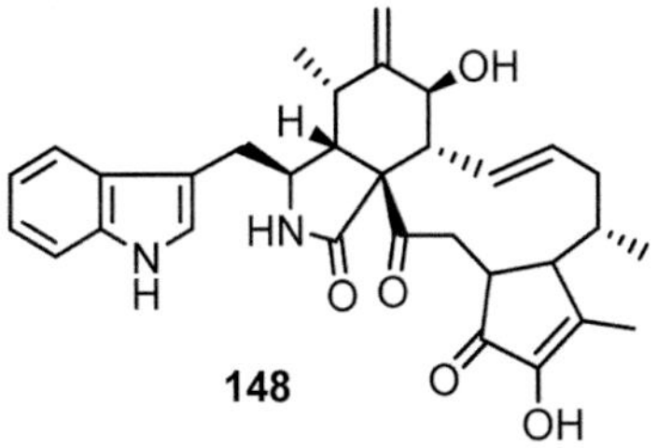

148

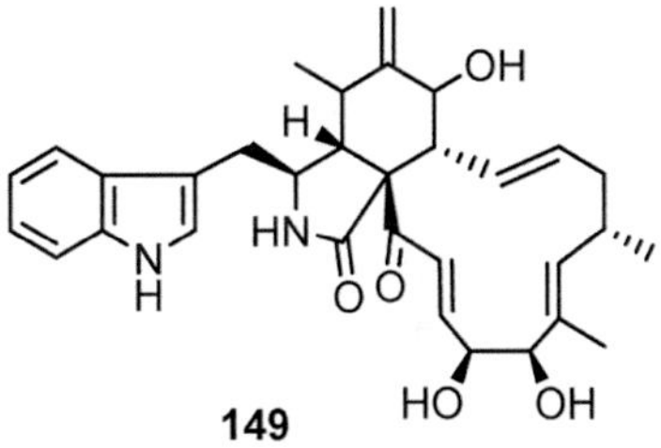

149

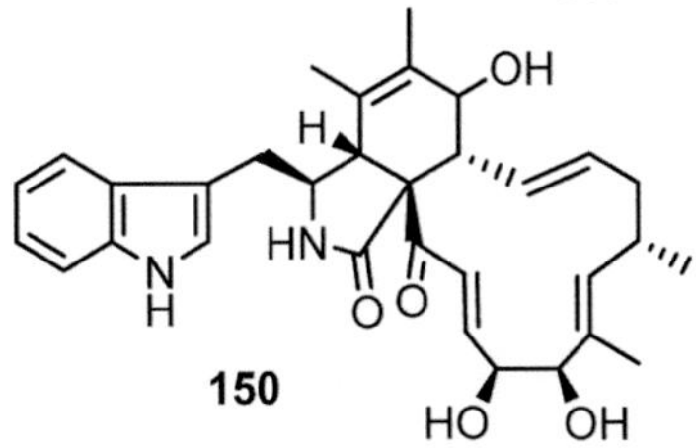

150

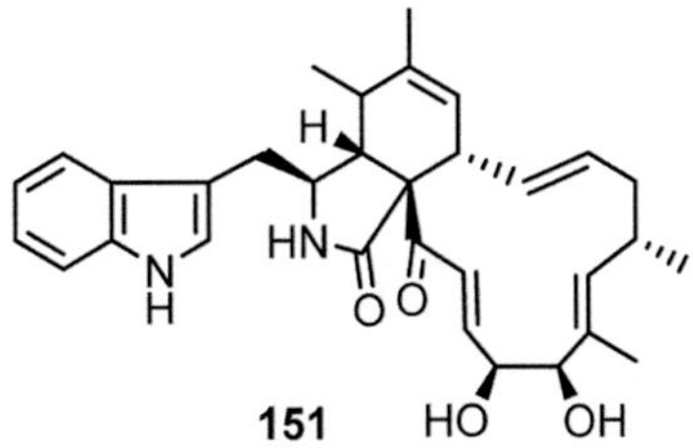

151

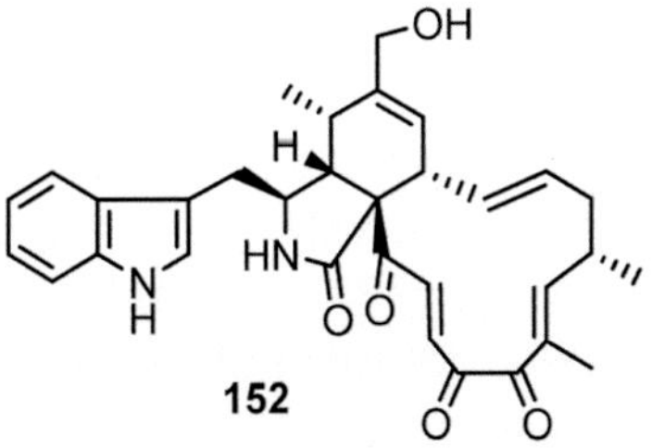

152

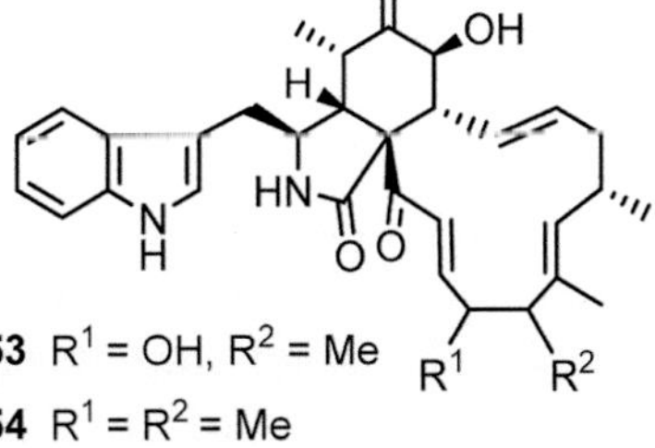

153 R^1 = OH, R^2 = Me

154 $R^1 = R^2$ = Me

In 2011, Yamada et al. discovered chaetomugilins P-R (**167–169**), and 11-*epi*-chaetomugilin I (**170**). While Borges et al. (2011) recorded the compounds chaetoviridins A-F, 5’-*epi*-chaetoviridin A (**54**), 4’-*epi*-chaetoviridin A (**171**), 4’-*epi*-chaetoviridin F (**172**), 12β-hydroxy-chaetoviridin C (**173**), and chaetoviridins G-I (**174–176**) that isolated from the *Ch. globosum*. However, Ge et al. (2011) reported that four new metabolites chaetoglocins A-D (**177–180**) from the endophytic fungus *Ch. globosum* were found.

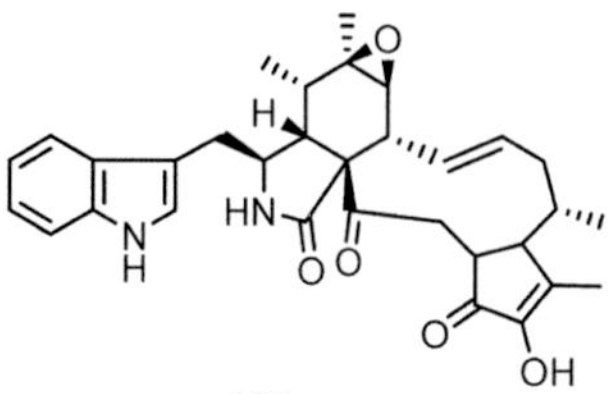

155

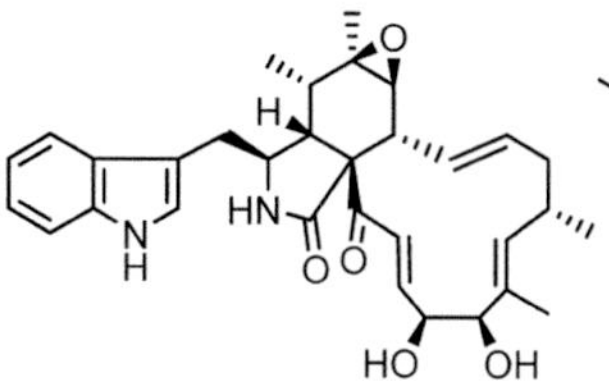

156

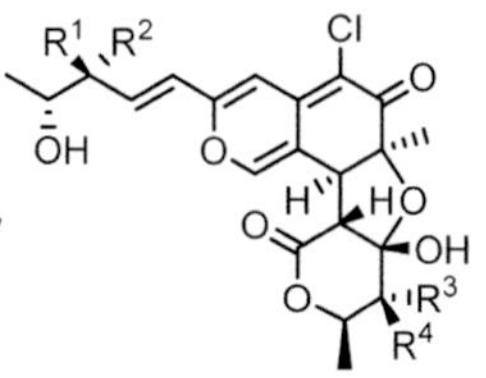

	R^1	R^2	R^3	R^4
157	Me	H	Me	H
158	H	Me	H	Me

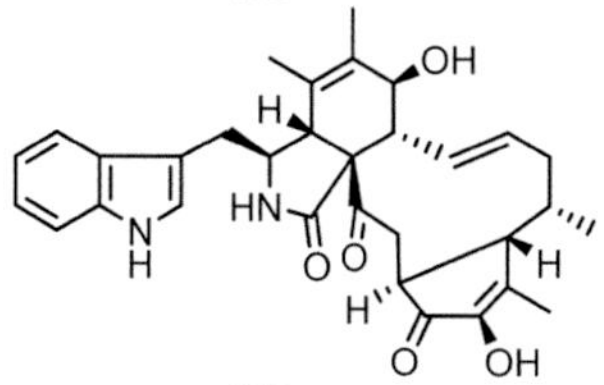

159

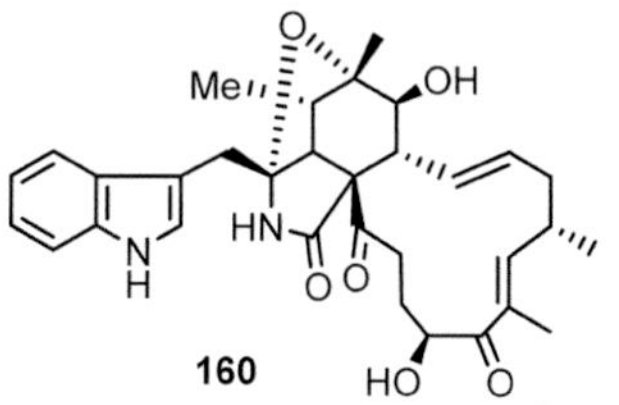

160

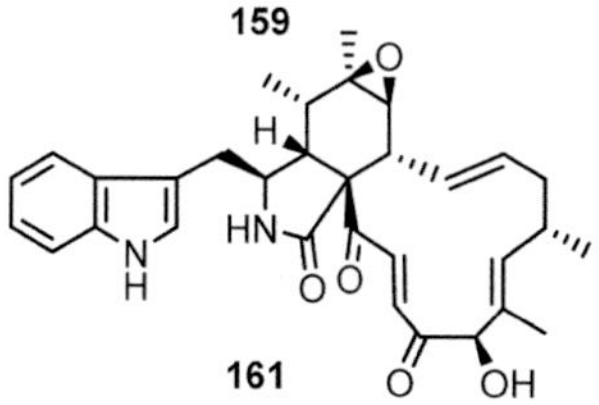

161

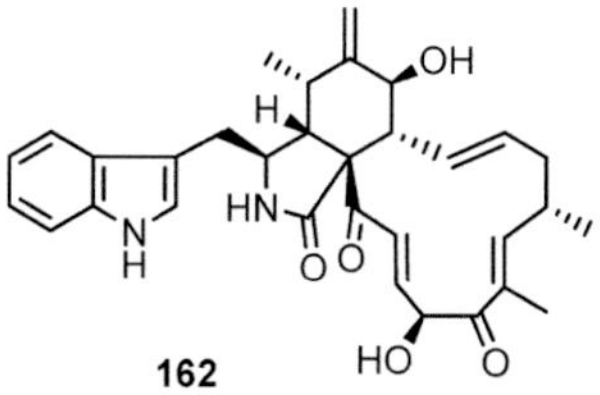

162

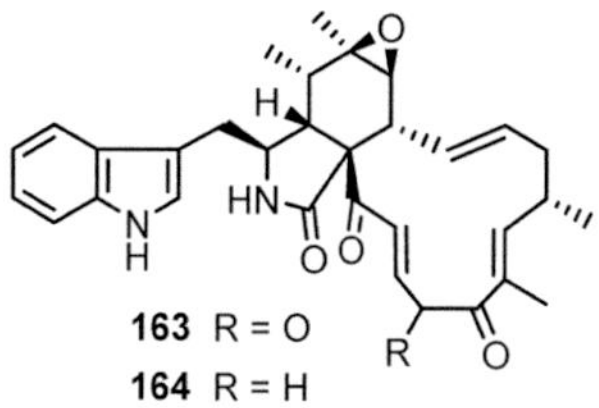

163 R = O
164 R = H

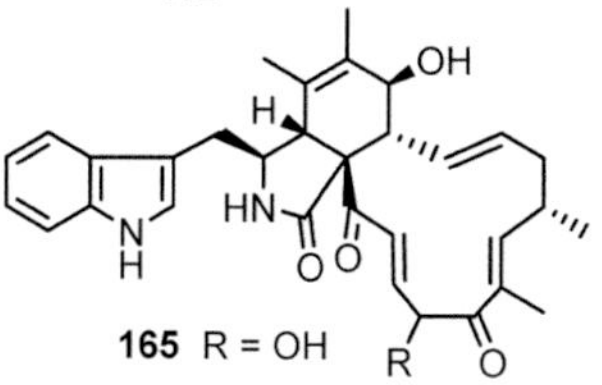

165 R = OH
166 R = O

In 2013, McMullin et al. recorded the new azaphilones nitrogenous, azaphilones; 4′-*epi*-*N*-2-hydroxyethyl-azachaetoviridin A (**181**), *N*-2-butyric-azochaetoviridin E (**182**), and isochromophilone XIII (**183**), chaetoglobosins A (80), C (**73**), and F (**82**), chaetomugilin D (**140**), chaetoviridin A (**54**) from *Ch. globosum* (DAOM 240359). It showed that all compounds were antimicrobial activity by using quantitative growth inhibition assays.

In 2016, Chen et al. found two new azaphilone derivatives, chaephilones A (**198**) and B (**199) and** four structurally related analogs chaetomugilin Q (**168**), chaetomugilin D (**140**), 11-*epi*-chaetomugilin A (**157**), and chaetomugilin S (**200**) which isolated from *Ch. globosum*. It showed that compounds **198** and **199** expressed the cytotoxic activities against five human cancer cell lines (HL-60, SMMC-7721, A-549, MCF-7, and SW480) by the MTS method.

Li et al. (2016) also recorded antifungal metabolites from endophytic *Ch. globosum* which isolated in *Gingko biloba*. Their found structures were elucidated as chaetoglobosin A (**80**), C (**73**), D (**74**), E (**81**), G (**75**), R (**100**) and significantly inhibited the growth of *Rhizopus stolonifer* (postharvest disease) and *Coniella diplodiella* (grape white rot disease) at a concentration of 20 μg/disc.

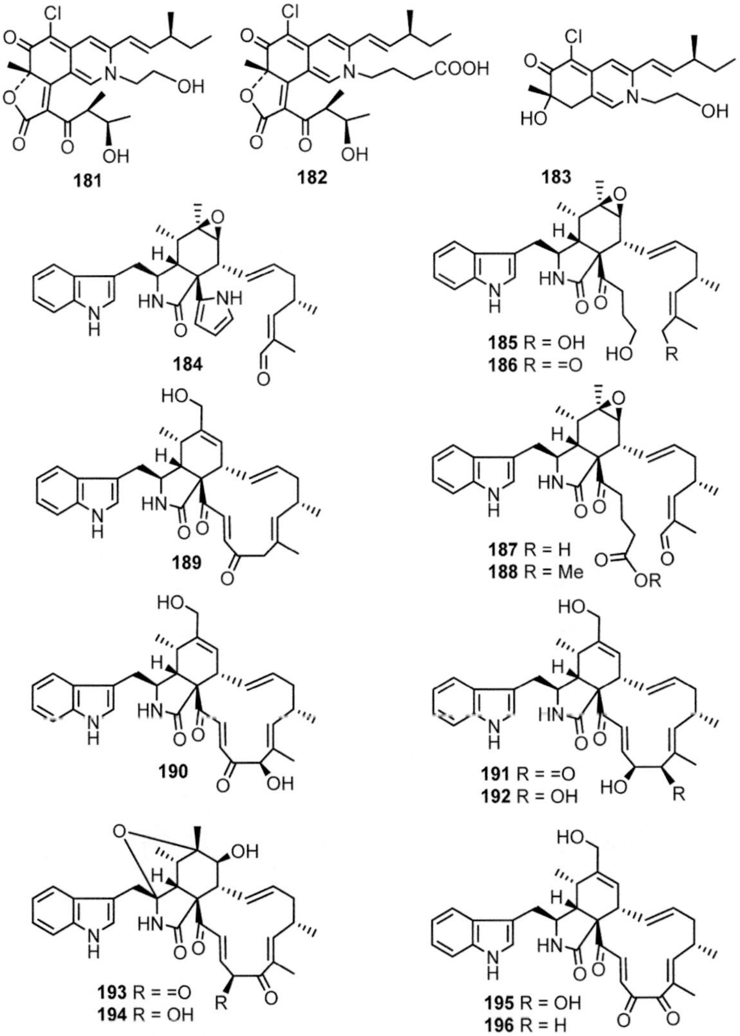

In 2017, Wang et al. reported to isolate two new cytochalasin derivatives, isochaetoglobosin D_b (**201**) and cytoglobosin A_b (**202**) from *Ch. globosum* SNSHI-5 which has taken from extreme environment. They found isochaetoglobosin D_b (**201**), which expressed a potent cytotoxicity with IC_{50} value of 3.5 μM, while cytoglobosin A_b was inactive (IC_{50} > 10 μM).

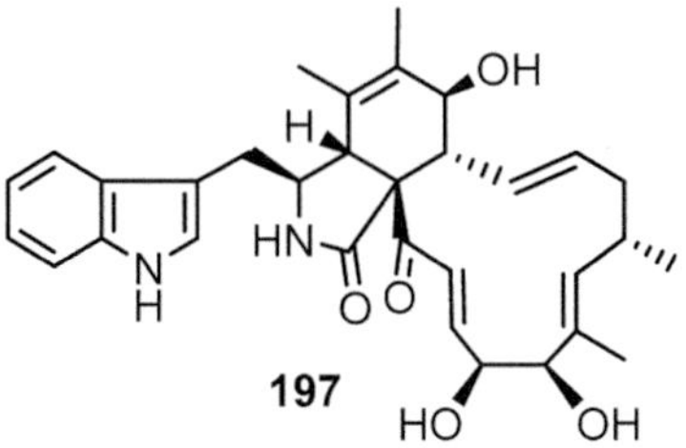

197

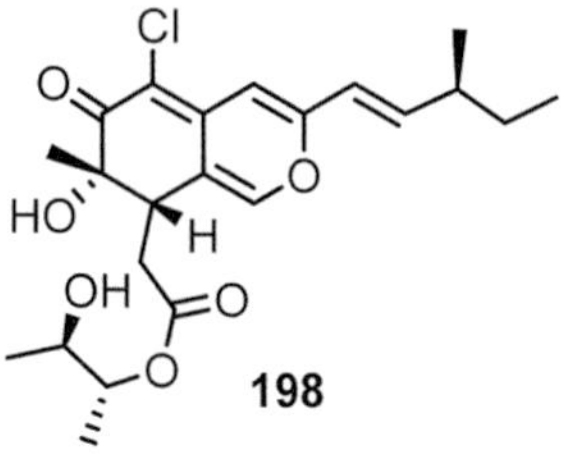

198

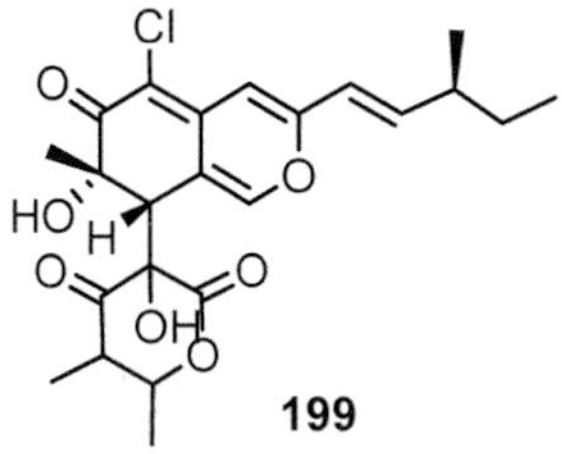

199

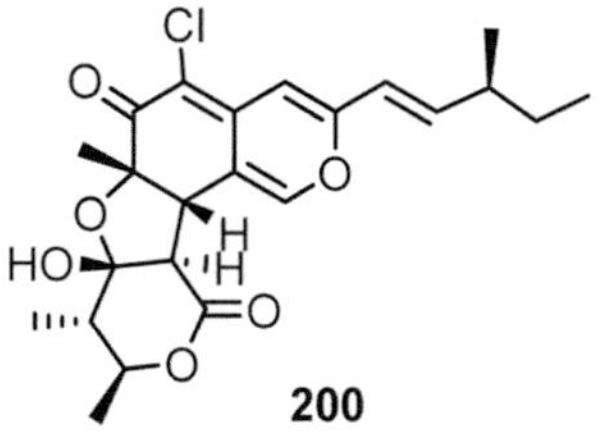

200

In 2018, *Ch. globosum* isolated from the fermented Chinese yam (*Dioscorea opposita*) found two new oxidation products-related aureonitol and cytochalasin and defined as 10,11-dihydroxyl-aureonitol (**203**) and yamchaetoglobosin A (**204**). It showed that compound **204** significant inhibited the nitric oxide production in LPS-activated macrophages, anti-acetylcholinesterase activity with inhibition ratio 92.5, 38.2% at concentration of 50 μM, and cytotoxicity to HL-60, A-549, SMMC-7721, MCF-7 and SW480 with the inhibition of 51–96% at concentration 40 μM (Ruan et al., 2018).

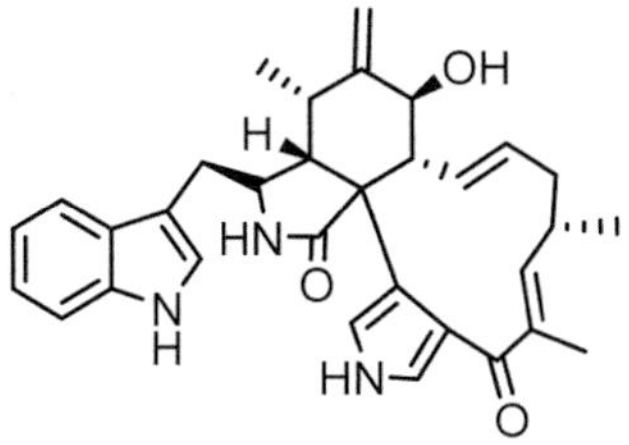

201

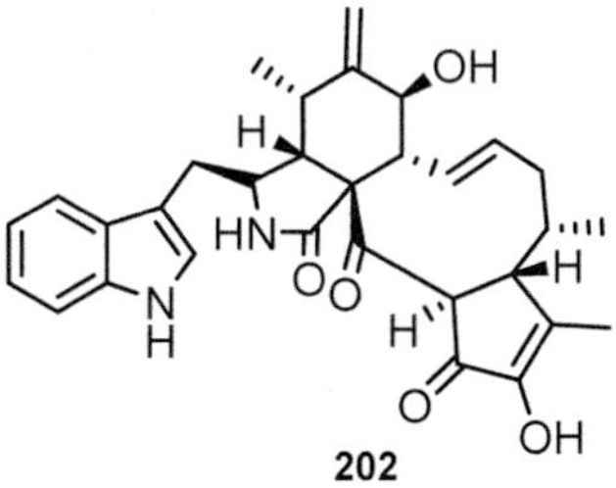

202

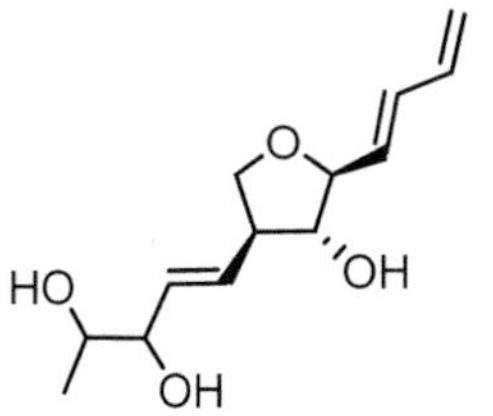

203

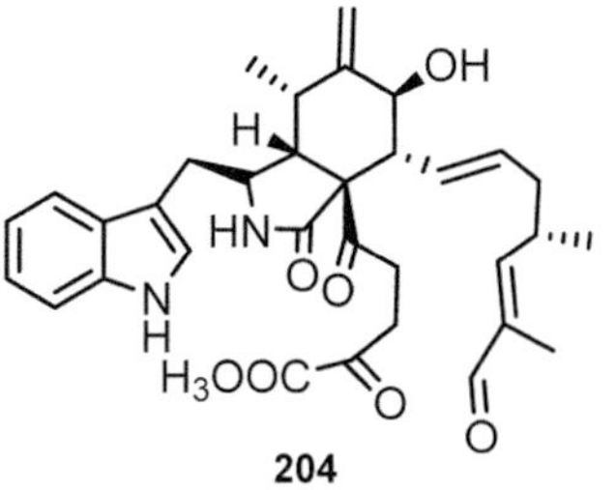

204

Recently of our research investigation on *Chaetomium* species, the endophytic *Ch. globosum* 7s-1 isolated from a plant species, *Rhapis cochinchinensis* yielded a new xanthoquinodin B9 (**205**), three epipolythiodioxopiperazines, chetomin (**42**), chaetocochin C (**46**), dethio-tetra(methylthio) chetomin (**47**), four other compounds, chrysophanol (**96**), two known xanthoquinodins, xanthoquinodin A1 (**206**) and xanthoquinodin A3 (**207**), emodin (**208**), alatinone (**209**) and ergosterol. It demonstrated that compounds **205–207**, **42**, **46,** and **47** expressed the antimicrobial activity toward Gram positive bacteria at concentration of 0.02 *p*M to 10.81 μM. Moreover, these metabolites further expressed the cytotoxicity against a normal cell line (*Vero* cell) at IC_{50} values of 0.04−3.86 μM and cytotoxicity toward KB, MCF-7, NCI-H187 cancer cell lines at IC_{50} 0.04−18.40 μM (Tantapakul et al., 2019).

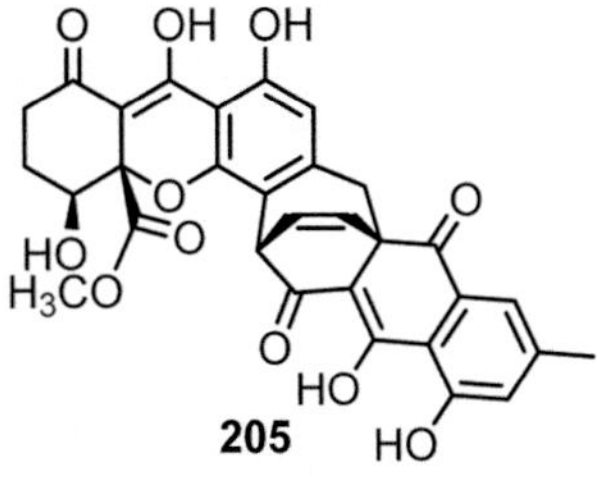

205

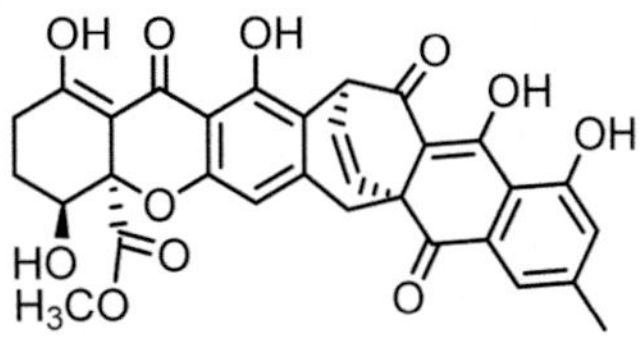

206

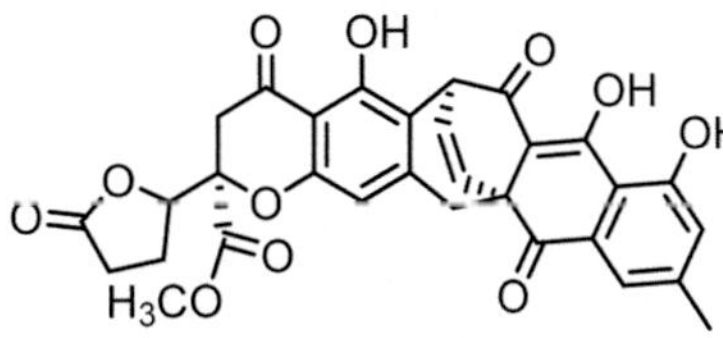

207

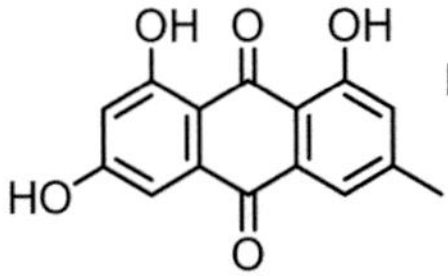

208

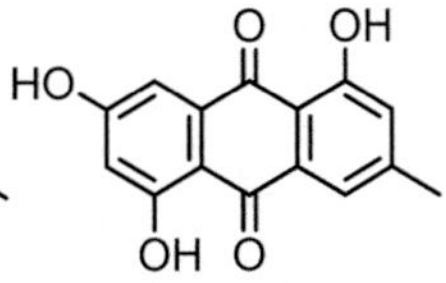

209

Chaetomium gracile is reported by Koyama and Natori (1987) who found the dichloromethane and *bis*(naphtho-γ-pyrone) derivatives named chaetochromins A-D (**210–213**). *Chaetomium indicum* is studied by Li et al. (2006b) who stated that the isoquinolines with novel skeletons named chaetoindicins A-C (**214–216**) was isolated from the solid-state fermented culture.

In 2011, our research group investigation on ethyl acetate extract of *Ch. longirostre* found four new azaphilones, longirostrerones A-D (**217–220)** and three known sterols. Those compounds of **217–220** expressed strongly cytotoxicity toward KB cancer cell lines at IC_{50} 0.23–6.38 μM. Only compound

217 resulted the effective cytotoxicity against MCF7 and NCI-H187 cell lines at IC_{50} 0.24 and 3.08 μM, respectively. Moreover, compounds **217–219** expressed the antimalarial activity against *Plasmodium falciparum* at IC_{50} 0.62–3.73 μM (Panthama et al., 2011).

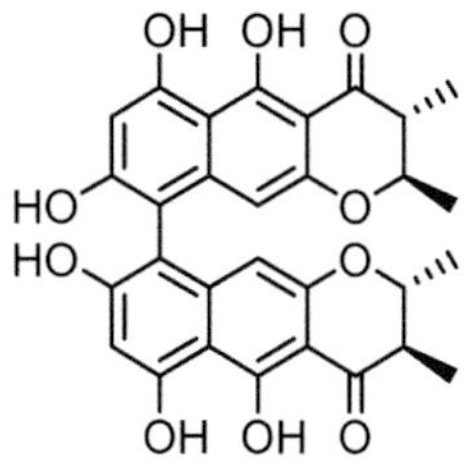

210

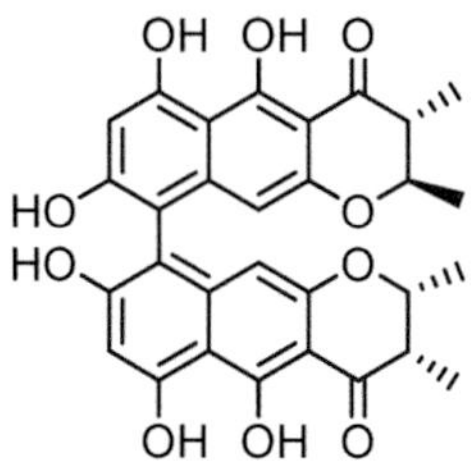

211

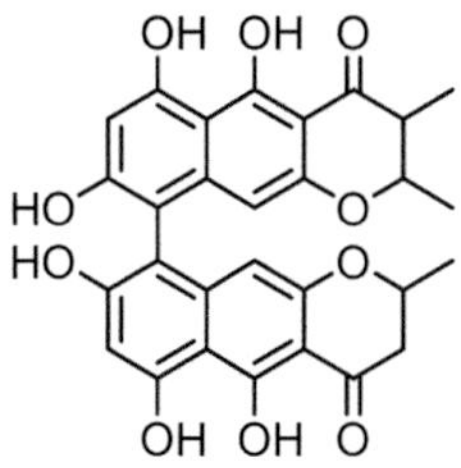

212

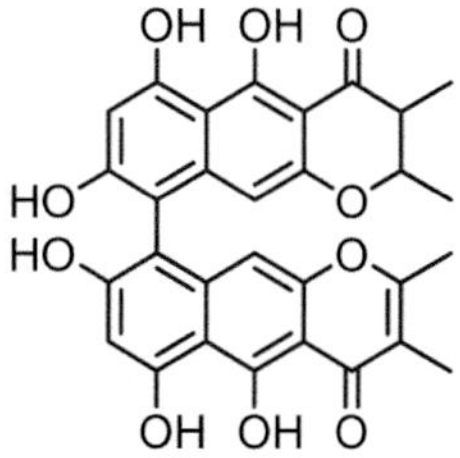

213

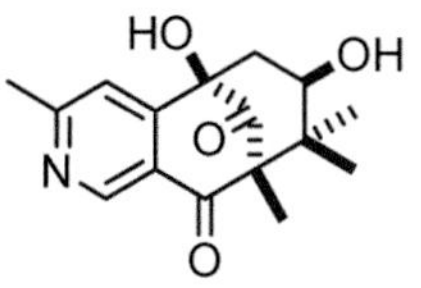

214

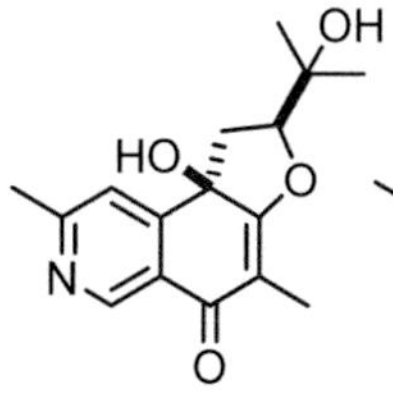

215

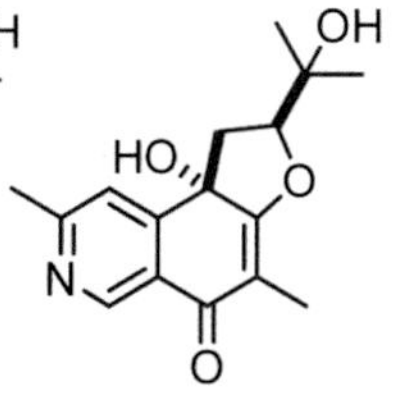

216

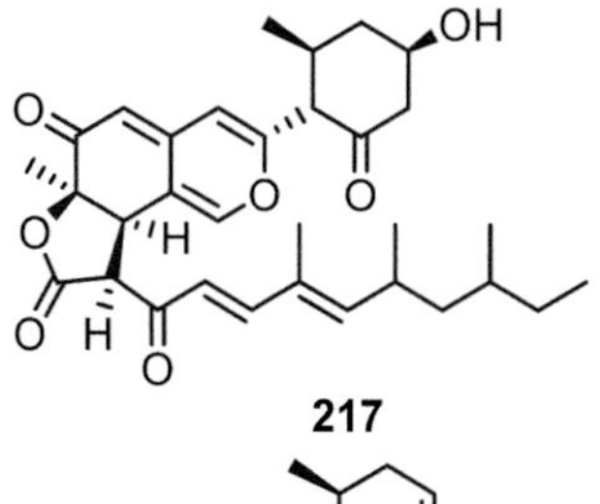

217

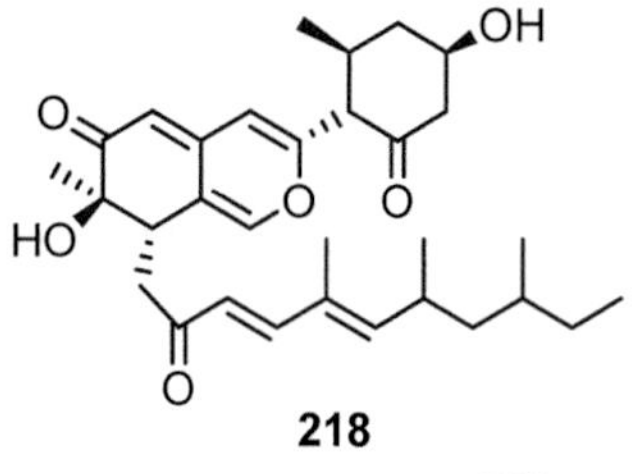

218

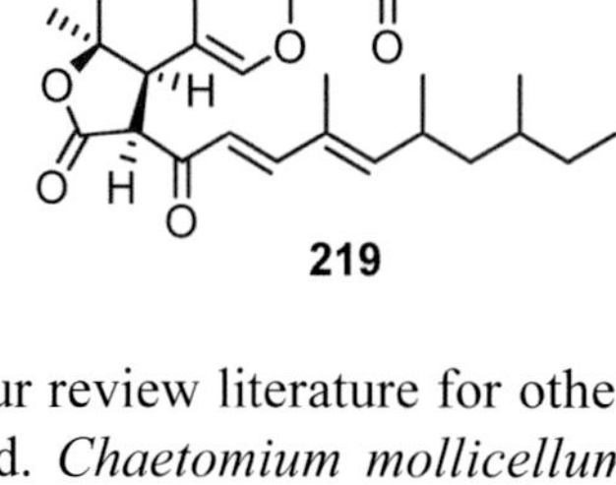

219

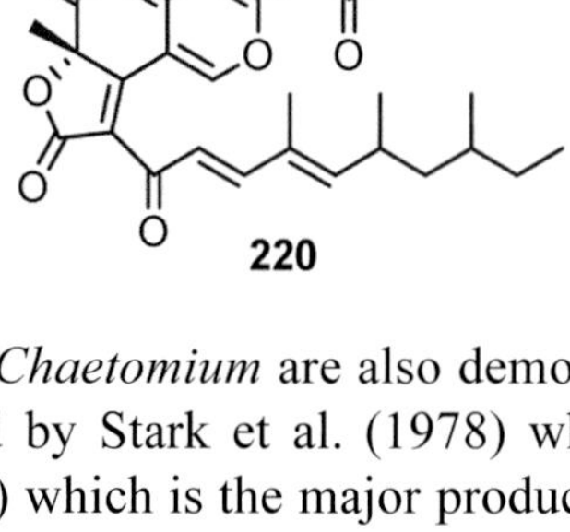

220

Our review literature for other species of *Chaetomium* are also demonstrated. *Chaetomium mollicellum* is reported by Stark et al. (1978) who discovered the eight mollicellins (depsidones) which is the major products of *Ch. mollicellum* MIT M-37, mollicellins A (**221**), B (**26**), C (**27**), D (**22**),

E (**28**), F (**29**), G (**2**), and H (**23**). It showed that Mollicellins C (**27**) and E (**28**) were mutagenic and bactericidal to *Salmonella typhimurium* in the absence of microsomes, while mollicellins D (**22**) and F (**29**) which each contained a chlorine atom, were bactericidal effect but not mutagenic. Sekita et al. (1983) reported the structural elucidation of isocochliodinol (**222**), a metabolite of *Ch. murorum* NHL (78-SH-271-4) and NHL 2240. Saito et al. (1988) reported that ethyl acetate extract from *Ch. nigricolor* which cultured in rice medium found a dimeric epipolythiodioxopiperazine named chetracin A (**223**) with two tetrasulfide bridges and a known compound cochliodinol (**41**).

Smetanina et al. (2001) recorded a pentacyclic triterpenoid, 3-*β*-methoxyolean-18-ene (miliacin) (**224**) for the first time from the marine *Ch. olivaceum*. Moreover, Fujimoto et al. (2002) found that ethyl acetate extract of *Ch. quadrangulatum* strain 71-Ng-22 yielded the five novel chromones (1,4-benzopyran-4-ones), three were tetracyclic and one contained a sulfonyl group, named chaetoquadrins A (**8**), B (**9**), C (**225**), D (**226**) and E (**227**) chaetoquadrins F (**228**), G (**10**), H (**11**), I (**229**), J (**230**) and K (**231**). These metabolites actively inhibited mouse liver MAO. One year later, they recorded six new constituents: chaetoquadrins F-K. It showed that chaetoquadrins G (**10**) and H (**11**) displayed appreciable MAO inhibitory activity (Fujimoto et al., 2003). *Chaetomium retardatum* is recorded in 1988, Saito's group that the isolation of chetracin A (**223**) and 11α,11′α-dihydroxychaetocin (**232**) from *Ch. retardatum* TRTC 66.1778b. *Chaetomium seminudum* is further investigated by Fujimoto et al. (2004) who stated that the ethyl acetate extract of *Ch. seminudum* found a known epipolythiodioxopiperazine, chetomin (**42**), three new chetomin-related metabolites named chetoseminudins A-C (**233–235**). These four natural metabolites, **232** and **233**, inhibited the blastogenesis of mouse splenic lymphocytes which stimulated by mitogens, concanavalin A (Con A) and lipopolysaccharide (LPS).

The further of our research investigation is discovered a new species of *Chaetomium* in Thailand. In 2011, the soil planted to pineapple was isolated by baiting technique and identified as a new species, *Ch. siamense* sp. nov. We found the secondary metabolites of a new chaetoviridin G (**236**) and seven known compounds, ergosterol (**56**), 24(*R*)-5α,8α-epidioxyergosta-6-22-diene-3β-ol, ergosterylplamitate (**95**), cochliodone D (**51**), chaetoviridin A (**55**), chaetoviridin F (**53**), chrysophanol (**96**) (Pornsuriya et al., 2011). This species is reported to be actively against some phytopathogens, e.g., *Phytophthora* spp. and *Pythium* spp.

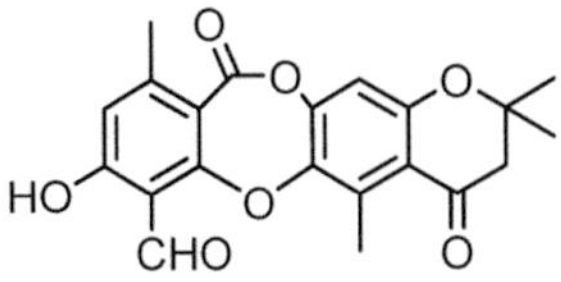

221

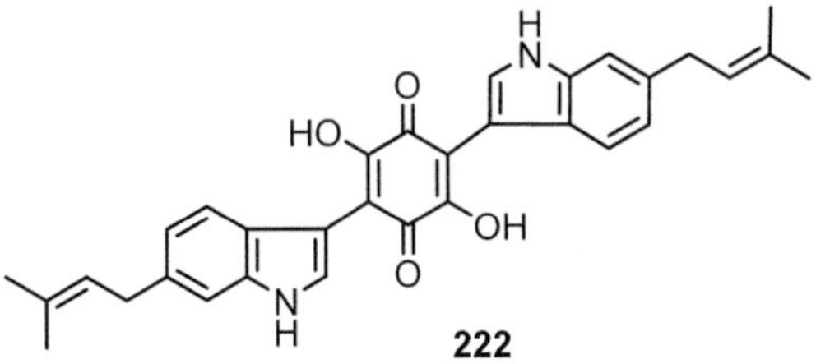

222

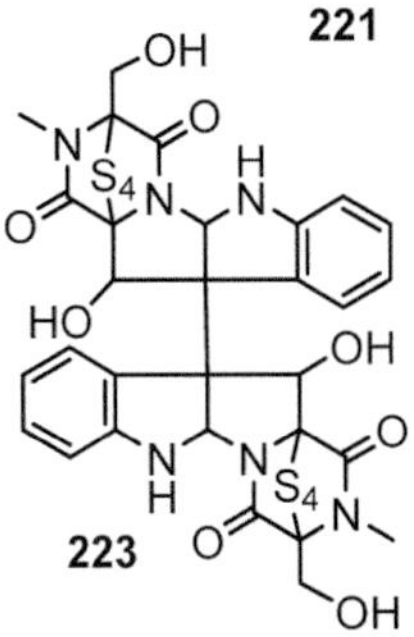

223

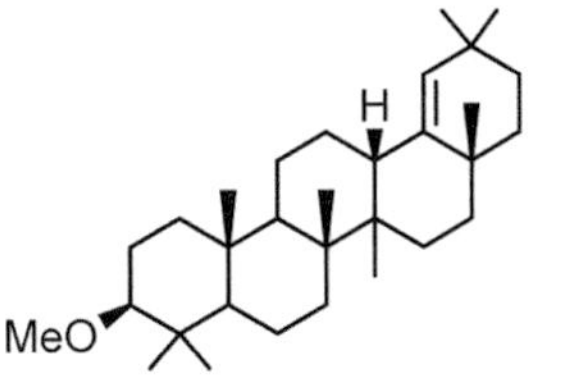

224

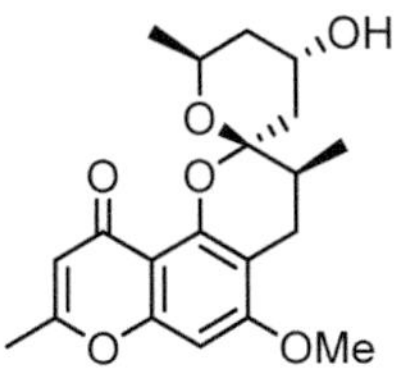

225

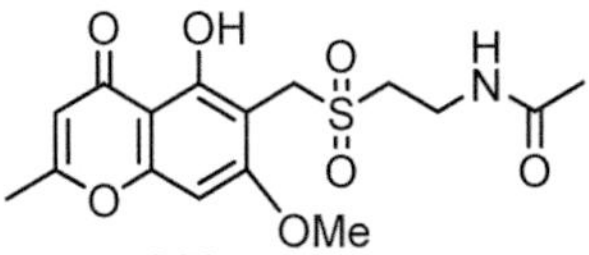

226

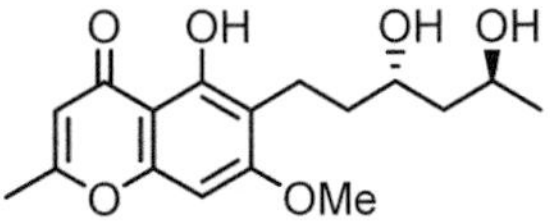

227

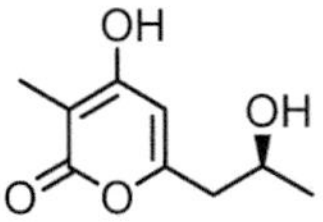

228

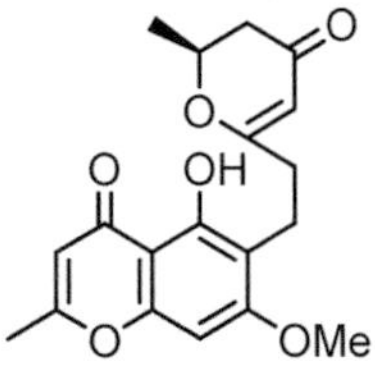

229

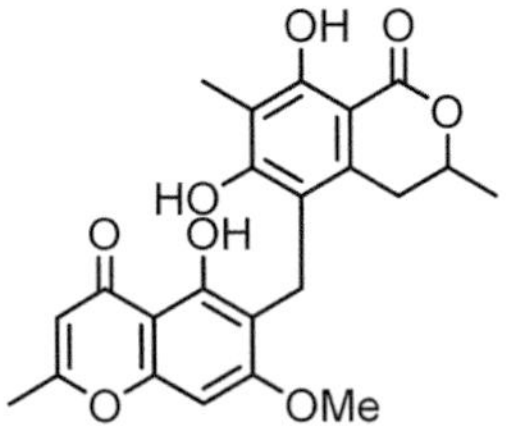

230

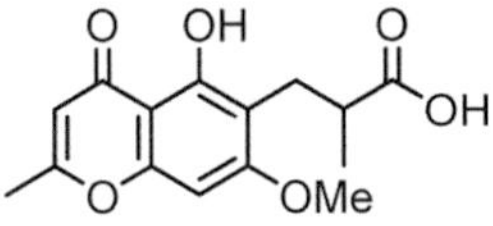

231

Chaetomiumn subaffine is recorded by Oikawa et al. (1933) who found two new metabolites of chaetoglobosins, named chaetoglobosin F_{ex} (**14**) and 20-dihydrochaetoglobosin A (**237**). *Chaetomium subspirale* is reported by Rether et al. (2004) isolated oxaspirodion (**238**) which was a new inhibitor of inducible TNF-α expression. *Chaetomium thielavioideum* is investigated by Sekita et al. (1980) who found the metabolites of *Ch. thielavioideum* NHL 2829 and isolated a new phenolic chaetochromin A (**210**), known compounds eugenitin (**17**), *O*-methyl-sterigmatocystin (**19**), sterigmatocystin (**20**), chaetocin (**21**), and ergosterol (**56**). In 1988, Saito's group also isolated chaetocins B (**239**) and C (**240**) which were strongly inhibited *S. aureus* FDA 209P from *Ch. thielavioidium* NHL 2827. *Chaetomium trilaterale* is recorded by Cole et al.

(1974) who isolated a dibenzoquinone, oosporein (3,3′,6,6′-tetrahydroxy-5,5′-dimethyl-2,2′-bi-benzoquinone) (**241**). *Chaetomium trilaterale* (Chivers, 1915) ATCC 24912. The compound was a moderately oral lethal dose in day-old cockerels and inhibited the plant growth and phytotoxic effects in some plant species.

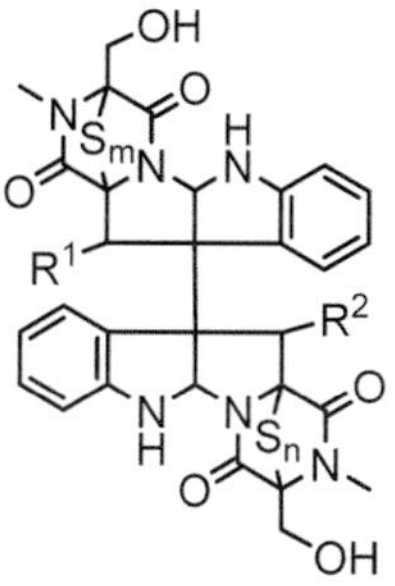

232 $R^1 = R^2 = OH$, m = n = 2
233 $R^1 = R^2 = H$, m = n = 2
239 $R^1 = R^2 = H$, m = 3, n = 2
240 $R^1 = R^2 = H$, m = n = 3

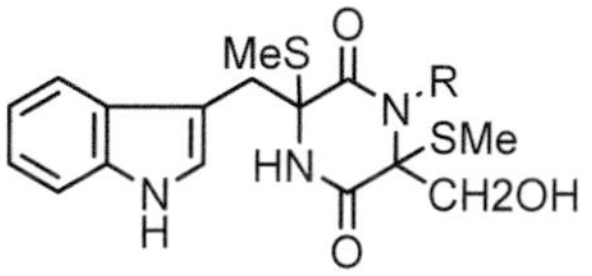

234 R = CH_3
235 R = H

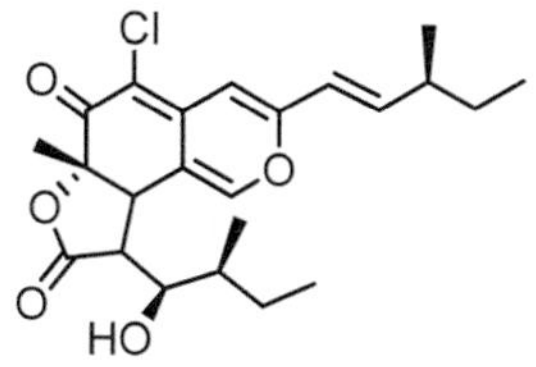

236

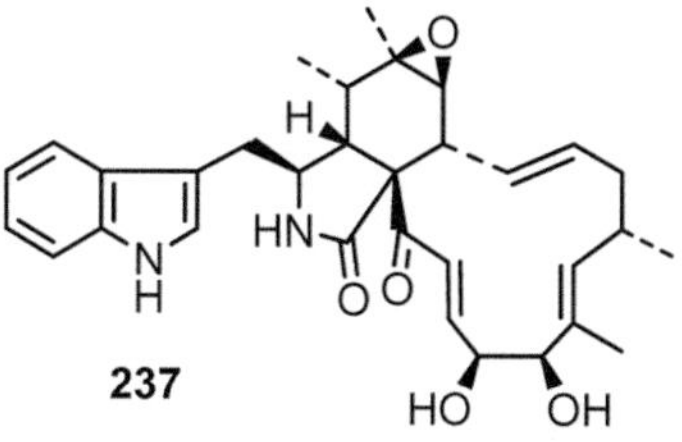

237

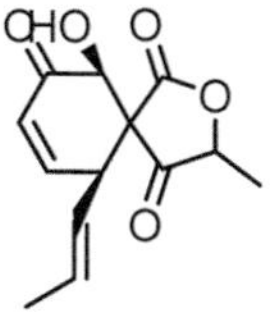

238

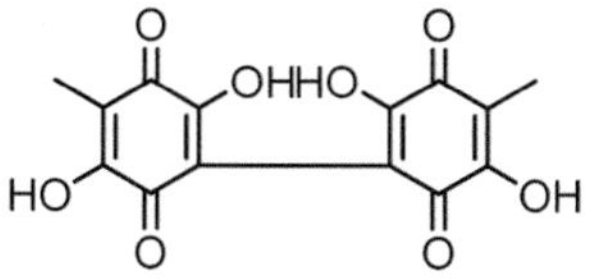

241

The studies on secondary metabolites of *Chaetomium* species are continuously conducted by several researchers. Some literature reviews have not been stated the specific epithet (species) which recorded only genus *Chaetomium*. There were many reports on unidentified *Chaetomium* species. Oka et al. (1985) found differanisole A (**242**) as a new defense inducing substance against leukemia cells from EtOAc extract from *Chaetomium* sp. RB-00. Imamura et al. (1993) reported a novel insecticidal, PF1093 (**243**), from *Chaetomium* sp. PF1093 (FERM P-12541) and the metabolite showed excellent insecticidal activity against *Plutella xylostella*. Kobayashi et al. (2005) found that *Chaetomium* sp. No. 217 produced a triterpene glucoside, FR207944 (**244**) which exhibited activities against *Aspergillus fumigatus* and *Candida albicans*. Moreover, Schlörke and Zeeck (2006) reported that the antibacterial orsellides A-E (**245–249**), novel esters consisting of orsellinic acid (**107**), and a 6-deoxyhexose (**251**) from *Chaetomium* sp. (strain Gö

100/9) together with the known metabolites globosumones A (**104**) and B (**105**). However, Jiao et al. (2006) recorded chaetominine (**251**), an alkaloidal metabolite from *Chaetomium* sp. IFB-E015, an endophytic fungus on the healthy leaves of *Adenophora axilliflora*. It showed that Chaetominine was more cytotoxic than 5-fluorouracil against the human leukemia K562 and colon cancer SW1116 cell lines.

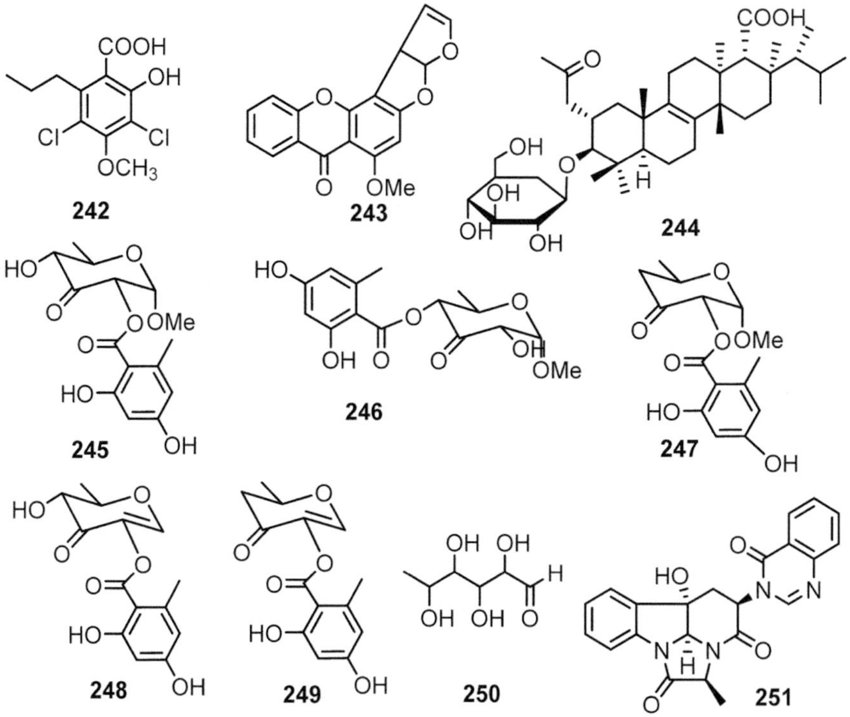

Lösgen et al. (2007) discovered three new fungal polyketide metabolites, chaetocyclinones A-C (**252–254**), two known compounds, SB238569 (**79**) and anhydrofulvic acid from *Chaetomium* sp. (strain Gö 100/2) which was isolated from a marine algae. It showed that chaetocyclinone A (**252**) inhibited the growth of selected phytopathogenic fungus, *Phytophthora infestans*, causing late blight of potato. Marwah et al. (2007) recorded a new furano-polyene, (–)-musanahol (**255**), a known furano-polyene, 3-*epi*-aureonitol (**256**), and a fatty acid, linoleic acid (**257**) from *Chaetomium* sp. isolated from tomato fruits and grown on YMG medium (yeast extract, glucose, malt extract and water). The (–)-musanahol (**255**) and

3-*epi*-aureonitol (**257**) were presented in the culture filtrate. The 3-*epi*-aureonitol (**256**) completely inhibited the growth of *Streptococcus pyogenes*, *Escherichia coli*, *Staphylococcus aureus*, *Salmonella choleraesuis*, and *Corynebacterium diphtheriae*, whereas (–)-musanahol (**255**) was no effect on the antimicrobial activity of compound **256** even if its similarity in their structures. Moreover, linoleic acid (**257**) resulted in inhibiting the growth of *S. aureus* and *Bacillus subtilis*.

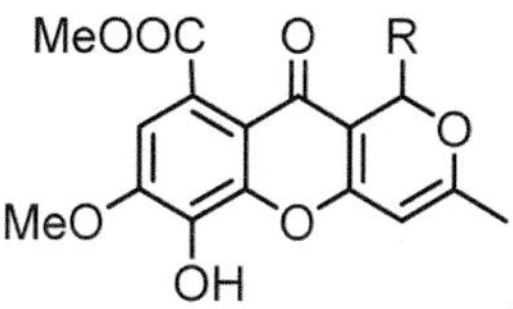

252 R = CH_3
253 R = H

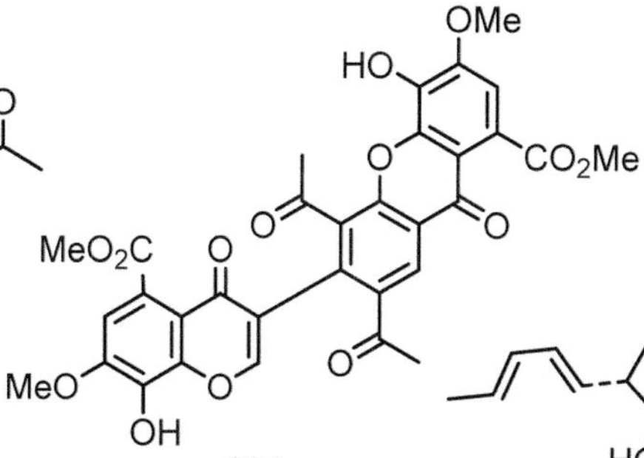

254

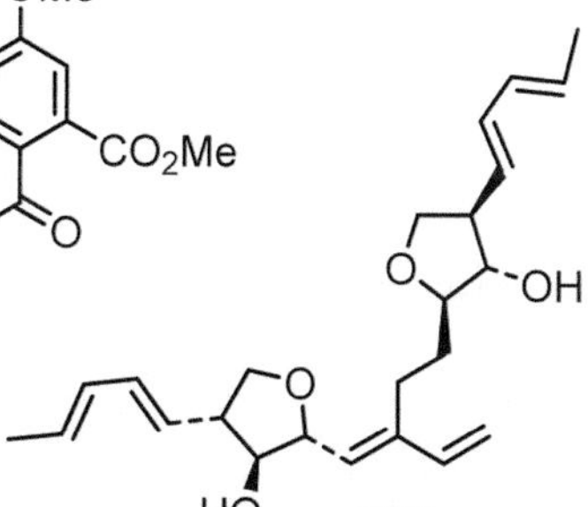

255

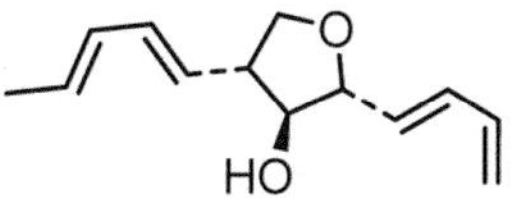

256

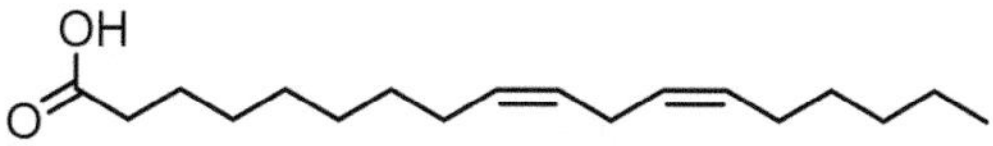

257

6.3 CONCLUSION AND FUTURE TRENDS

It is concluded that *Chaetomium* spp. has been investigated by many researchers from years to find out the secondary metabolites against human, animal, and plant diseases. Our research investigation on *Chaetomium* spp. has been started from 1986 to find out the antagonistic effect against phytopathogens as well as human pathogens. It is confirmed that further ongoing research has been developed on the selected active metabolites to formulate biofungicide for plant disease control to promote non-agrochemicals-based agriculture and organic agriculture and has been contributed for farmers use in many countries, e.g., Thailand, China, Myanmar, Vietnam, and Laos. The ongoing research is developed on the active metabolites from *Chaetomium* spp. which considered as natural products and nanoparticles thereof for plant immunity induction.

KEYWORDS

- **biofungicide**
- ***Chaetomium***
- **lipopolysaccharide**
- **monoamine oxidase**
- **organic agriculture**
- **plant disease control**
- ***Pyricularia oryzae***

REFERENCES

Abraham, W. R., & Arfmann, H. A., (1992). Rearranged tetrahydrofurans from *Chaetomium cochlioides*. *Phytochemistry, 31*, 2405–2408.

Bashyal, B. P., Wijeratne, E. M. K., Faeth, S. H., & Gunatilaka, A. A. L., (2005). Globosumones A-C, cytotoxic orsellinic acid esters from the Sonoran Desert endophytic fungus *Chaetomium globosum*. *Journal of Natural Products, 68*, 724–728.

Borges, W. S., Mancilla, G., Guimaraes, D. O., Duran-Patron, R., Collado, I. G., & Pupo, M. T., (2011). Azaphilones from the *Chaetomium globosum*. *Journal of Natural Products, 74*, 1182–1187.

Breinholt, J., Demuth, H., Heide, M., Jensen, G., Moller, I., Nielsen, R., et al., (1996). Prenisatin (5-(3-methyl-2-butenyl)indole-2,3-dione) an antifungal isatin derivative from *Chaetomium globosum*. *Acta Chemica Scandinavica, 50*, 443–445.

Brewer, D., Duncan, J. M., Jerram, W. A., Leach, C. K., Safe, S., Taylor, A., & Christensen, C. M., (1972). Ovine ill-thrift in Nova Scotia. 5. The production and toxicology of chetomin, a metabolite of *Chaetomium* spp. *Canadian Journal of Microbiology, 18*, 1129–1137.

Brewer, D., Jerram, W. A., & Taylor, A., (1968). The production of cochliodinol and a related metabolite by *Chaetomium* species. *Canadian Journal of Microbiology, 14*, 861–866.

Burrows, B. F., Turner, W. B., & Walker, E. R. H., (1975). 8-ethylidene-7,8-dihydro-4-methoxypyrano[4,3-*b*]pyran-2,5-dione (Coarctatin), a metabolite of *Chaetomium coarctatum*. *Journal of the Chemical Society, Perkin Transactions, 1*, 999–1000.

Chen, C., Jing, W., Hucheng, Z., Jianping, W., Yongbo, X., Guangzheng, W., et al., (2016). Chaephilones A and B, two new azaphilone derivatives isolated from *Chaetomium globosum*. *Chem Biodivers., 13*(4), 422–426.

Chivers, A. H., (1915). *A Monograph of the Genera Chaetomium and Ascotricha* (Vol. 14, pp. 155–240). Memoirs of the Torrey Botanical Club.

Cole, R. J., Kirksey, J. W., Gutter, H. G., & Davis, E. E., (1974). Toxic effects of oosporein from *Chaetomium trilaterale*. *Journal of Agricultural and Food Chemistry, 22*, 517–520.

Cui, C. M., Li, X. M., Li, C. S., Proksch, P., & Wang, B. G., (2010). Cytoglobosins A-G, cytochalasans from a marine-derived endophytic fungus, *Chaetomium globosum* QEN-14. *Journal of Natural Products, 73*, 729–733.

Di Pietro, A., Gut-Rella, M., Pachlatko, J. P., & Schwinn, F. J., (1992). Role of antibiotics produced by *Chaetomium globosum* in biocontrol of *Pythium ultimum*, a causal agent of damping-off. *Phytopathology, 82*, 131–135.

Ding, G., Song, Y. C., Chen, J. R., Xu, C., Ge, H. M., Wang, X. T., et al., (2006). Chaetoglobosin U, a cytochalasan alkaloid from endophytic *Chaetomium globosum* IFB-E019. *Journal of Natural Products, 69*, 302–304.

Fujimoto, H., Nozawa, M., Okuyama, E., & Ishibashi, M., (2002). Five new chromones possessing monoamine oxidase inhibitory activity from an ascomycete, *Chaetomium quadrangulatum*. *Chemical and Pharmaceutical Bulletin, 50*, 330–336.

Fujimoto, H., Nozawa, M., Okuyama, E., & Ishibashi, M., (2003). Six new constituents from an ascomycete, *Chaetomium quadrangulatum*, found in a screening study focused on monoamine oxidase inhibitory activity. *Chemical and Pharmaceutical Bulletin, 51*, 247–251.

Fujimoto, H., Sumino, M., Okuyama, E., & Ishibashi, M., (2004). Immunomodulatory constituents from an ascomycete, *Chaetomium seminudum*. *Journal of Natural Products, 67*, 98–102.

Ge, H. M., Zhang, Q., Xu, S. H., Guo, Z. K., Song, Y. C., Huang, W. Y., et al., (2011). Chaetoglocins A-D. Four new metabolites from the endophytic fungus *Chaetomium globosum*. *Planta Medica, 77*, 277–280.

Hwang, E. I., Yun, B. S., Kim, Y. K., Kwon, B. M., Kim, H. G., Lee, H. B., et al., (2000). Chaetoatrosin A, a novel chitin synthase II inhibitor produced by *Chaetomium atrobrunneum* F449. *The Journal of Antibiotics, 53*, 248–255.

Imamura, K., Gomi, S., Yaguchi, T., Moryama, C., & Iwata, M., (1993). Novel insecticidal PF1093 and its manufacture with *Chaetomium*. *Jpn. Kokai Tokkyo Koho,* 1–8.

Itoh, Y., Takahashi, S., Haneishi, T., & Arai, M., (1980). Structure of heptelidic acid, a new sesquiterpene antibiotic from fungi. *The Journal of Antibiotics, 33*, 525–526.

Jiao, R. H., Xu, S., Liu, J. Y., Ge, H. M., Ding, H., Xu, C., et al., (2006). Chaetominine, a cytotoxic alkaloid produced by endophytic *Chaetomium* sp. IFB-E015. *Organic Letters, 8*, 5709–5712.

Jiao, W., Feng, Y., Blunt, J. W., Cole, A. L. J., & Munro, M. H. G., (2004). Chaetoglobosins Q, R, and T, three further new metabolites from *Chaetomium globosum*. *Journal of Natural Products, 67*, 1722–1725.

Kanokmedhakul, S. K., Nasomjai, P., Loungsysouphanh, S., Soytong, K., Isobe, M., Kongsaeree, K., Prabpai, S., & Suksamran, A., (2006). Antifungal Azaphilones from the fungus, *Chaetomium cupreum* CC3003. *Journal of Natural Products, 69*, 891–895.

Kanokmedhakul, S., Kanokmedhakul, K., Phonkerd, N., Soytong, K., Kongsaeree, P., & Suksamrarn, A., (2002). Antimycobacterial anthraquinone-chromanone compound and diketopiperazine alkaloid from the fungus *Chaetomium globosum* KMITL-N0802. *Planta Medica, 68*, 834–836.

Khumkomkhet, P., Kanokmedhakul, S., Kanokmedhakul, K., Hahnvajanawong, C., & Soytong, K., (2009). Antimalarial and cytotoxic depsidones from the fungus *Chaetomium brasiliense*. *Journal of Natural Products, 72*, 1487–1491.

Kikuchi, T., Kodata, S., Nakamura, K., Nishi, A., Taga, T., & Kaji, T., (1981). Dethio-tetra (methylthio) chetomin, a new antimicrobial metabolite of *Chaetomium globosum* KINZE ex FR: Structure and partial synthesis from chetomin. *Chemical and Pharmaceutical Bulletin, 30*, 3846–3848.

Kobayashi, M., Kanasaki, R., Sato, I., Abe, F., Nitta, K., Ezaki, M., et al., (2005). FR207944, an antifungal antibiotic from *Chaetomium* sp. No. 217 I. Taxonomy, fermentation, and biological properties. *Bioscience, Biotechnology, and Biochemistry, 69*, 515–521.

Koyama, K., & Natori, S., (1987). Chaetochromins B, C, and D; *bis*(naphtho-γ-pyrone) derivatives from *Chaetomium gracile*. *Chemical and Pharmaceutical Bulletin, 35*, 578–584.

Li, G. Y., Li, B. G., Yang, T., Liu, G. Y., & Zhang, G. L., (2006b). Chaetoindicins A-C, three isoquinoline alkaloids from the fungus *Chaetomium indicum*. *Organic Letters, 8*, 3613–3615.

Li, G. Y., Li, B. G., Yang, T., Liu, G. Y., & Zhang, G. L., (2008). Secondary metabolites from the fungus *Chaetomium brasiliense*. *Helvetica Chimica Acta, 91*, 124–129.

Li, G. Y., Li, B. G., Yang, T., Yan, J. F., Liu, G. Y., & Zhang, G. L., (2006a). Chaetocochins A-C, epipolythiodioxopiperazines from *Chaetomium cochliodes*. *Journal of Natural Products, 69*, 1374–1376.

Li, L. M., Zou, Q., & Li, G. Y., (2010). Chromones from an ascomycete, *Chaetomium aureus*. *Chinese Chemical Letters, 21*, 1203–1205.

Li, W., Yang, X., Yang, Y., Duang, R., Chen, G., Li, X., Li, Q., et al., (2016). Anti-phytopathogen, multi-target acetylcholinesterase inhibitory and antioxidant activities of metabolites from endophytic *Chaetomium globosum*. *Natural Product Research, 30*, 2616–2619.

Lösgen, S., Schlörke, O., Meindl, K., Herbst-Irmer, R., & Zeeck, A., (2007). Structure and biosynthesis of chaetocyclinones, new polyketides produced by an endosymbiotic fungus. *European Journal of Organic Chemistry, 2191*–2196.

Marwah, R. G., Fatope, M. O., Deadman, M. L., Al-Maqbali, Y. M., & Husband, J., (2007). Musanahol: A new aureonitol-related metabolite from a *Chaetomium* sp. *Tetrahedron, 63*, 8174–8180.

McMullin, D. R., Sumarah, M. W., Blackwell, B. A., & Miller, J. D., (2013). New azaphilones from *Chaetomium globosum* isolated from the built environment. *Tetrahedron Letters, 54*, 568–572.

Oh, H., Swenson, D. C., Gloer, J. B., Wicklow, D. T., & Dowd, P. F., (1998). Chaetochalasin A: A new bioactive metabolite from *Chaetomium brasiliense*. *Tetrahedron Letters, 39*, 7633–7636.

Oikawa, H., Murakami, Y., & Ichihara, A., (1933). 20-Ketoreductase activity of chaetoglobosin A and prochaetoglobosins in a cell-free system of *Chaetomium subaffine* and the isolation of new chaetoglobosins. *Bioscience, Biotechnology, and Biochemistry, 57*, 628–631.

Oka, H., Asahi, K., Morishima, H., Sanada, M., Shiratori, K., Iimura, Y., et al., (1985). Differanisole A, a new differentiation inducing substance. *The Journal of Antibiotics, 38*, 1100–1102.

Okeke, B., Seiglemurandi, F., Steiman, R., Benoitguyod, J., & Kaouaji, M., (1993). Identification of mycotoxin-producing fungal strains: A step in the isolation of compounds active against rice fungal disease. *Journal of Agricultural and Food Chemistry, 41*, 1731–1735.

Panthama, N., Kanokmedhakul, S., Kanokmedhakul, K., & Soytong, K., (2011). Cytotoxic and antimalarial azaphilones from *Chaetomium longirostre*. *J. Nat. Prod., 74*(11), 2395–2399.

Payne, D. J., Hueso-Rodríguez, J. A., Boyd, H., Concha, N. O., Janson, C. A., Gilpin, M., Bateson, J. H., et al., (2002). Identification of a series of tricyclic natural products as potent broad-spectrum inhibitors of metallo-beta-lactamases. *Antimicrob Agents Chemother., 46*, 1880–1886.

Pholkerd, N., Kanokmedhakul, S., Kanokmedhakul, K., Soytong, K., Prabpai, S., & Kongsearee, P., (2008). Bis-spiro-azaphilones and azaphilones from the fungi *Chaetomium cochliodes* VTh01 and *C. cochliodes* CTh05. *Tetrahedron, 64*, 9636–9645.

Pornsuriya, C., Soytong, K., Poeaim, S., Kanokmedhakul, S., Khumkomkhet, P., Lin, F. C., Wang, H. K., & Hyde, K. D., (2011). *Chaetomium siamense* sp. nov., a soil isolate from Thailand, produces a new chaetoviridin, G. *Mycotaxon, 115*, 19–27.

Probst, A., & Tamm, C., (1981). 19-*O*-acetylchaetoglobosin B and 19-*O*-acetylchaetoglobosin D, two new metabolites of *Chaetomium globosum*. *Helvetica Chimica Acta, 64*, 2056–2064.

Qin, J. C., Gao, J. M., Zhang, Y. M., Yang, S. X., Bai, M. S., Ma, Y. T., et al., (2009). Polyhydroxylated steroids from an endophytic fungus, *Chaetomium globosum* ZY-22 isolated from *Ginkgo biloba*. *Steroids, 74*, 786–790.

Rether, J., Erkel, G., Anke, T., & Sterner, O., (2004). Oxaspirodion, a new inhibitor of inducible TNF-α expression from the Ascomycete *Chaetomium subspirale*. Production, isolation and structure elucidation. *The Journal of Antibiotics, 57*, 493–495.

Ruan, B. H., Yu, Z. F., Yang, X. Q., Yang, Y. B., Hu, M., Zhang, Z. X., Zhou, Q. Y., et al., (2018). New bioactive compounds from aquatic endophyte *Chaetomium globosum*. *Natural Products Research, 32*, 1050–1055.

Saito, T., Suzuki, Y., Koyama, K., Natori, S., Iitaka, Y., & Kinoshita, T., (1988). Chetracin A and chaetocins B and C, three new epipolythiodioxopiperazines from *Chaetomium* spp. *Chemical and Pharmaceutical Bulletin, 36*, 1942–1956.

Schlörke, O., & Zeeck, A., (2006). Orsellides A-E: An example for 6-deoxyhexose derivatives produced by fungi. *European Journal of Organic Chemistry*, 1043–1049.

Sekita, S., (1983). Isocochliodinol and neocochliodinol, bis(3-indolyl)-benzoquinones from *Chaetomium* spp. *Chemical and Pharmaceutical Bulletin, 31*, 2998–3001.

Sekita, S., Yoshihira, K., & Natori, S., (1973). Structures of chaetoglobosin A and B, cytotoxic metabolites of *Chaetomium globosum*. *Tetrahedron Letters, 23*, 2109–2112.

Sekita, S., Yoshihira, K., & Natori, S., (1976). Structures of chaetoglobosin C, D, E and F, cytotoxic indol-3-yl-[13]cytochalasans from *Chaetomium globosum*. *Tetrahedron Letters, 17*, 1351–1354.

Sekita, S., Yoshihira, K., & Natori, S., (1980). Chaetochromin, a bis(naphthodihydropyran-4-one) mycotoxin from *Chaetomium thielavioideum*: Application of ^{13}C-^{1}H long-range coupling to the structure elucidation. *Chemical and Pharmaceutical Bulletin, 28*, 2428–2435.

Sekita, S., Yoshihira, K., & Natori, S., (1983). Chaetoglobosins, cytotoxic 10 (indol-3-yl)-[13] cytochalasans from *Chaetomium* spp. IV. ^{13}C-nuclear magnetic resonance spectra and their application to a biosynthetic study. *Chemical and Pharmaceutical Bulletin, 31*, 490–498.

Sekita, S., Yoshihira, K., Natori, S., & Chaetoglobosins, G. J., (1977). Cytotoxic indol-3-yl[13]-cytochalasans from *Chaetomium globosum*. *Tetrahedron Letters, 32*, 2771–2774.

Sekita, S., Yoshihira, K., Natori, S., Udagawa, S. I., Sakabe, F., Kurata, H., et al., (1981). Chaetoglobosins, cytotoxic 10-(indol-3-yl)-[13]cytochalasans from *Chaetomium* spp. I. production, isolation and some cytological effects of chaetoglobosins A-J. *Chemical and Pharmaceutical Bulletin, 30*, 1609–1617.

Smetanina, O. F., Denisenko, V. A., Pivkin, M. V., Khudyakova, Y. V., Gerasimenko, A. V., Popov, D. Y., et al., (2001). 3β-methoxyolean-18-ene (miliacin) from the marine fungus *Chaetomium olivaceum*. *Russian Chemical Bulletin, 50*, 2463–2465.

Song, J. J., Soytong, K., Kanokmedhakul, S., Kanokmedhakul, K., & Poeaim, S., (2020b). Antifungal activity of microbial nanoparticles derived from *Chaetomium* spp. against *Magnaporthe oryzae* causing rice blast. *Plant Protection Science, 56*, 180–190.

Stark, A. A., Kobbe, B., Matsuo, K., Buchi, G., Wogan, G. N., & Demain, A. L., (1978). Mollicellins: Mutagenic and antibacterial mycotoxins. *Applied and Environmental Microbiology, 36*, 412–420.

Tabata, Y., Miiko, N., Yaguchi, T., Hatsu, M., Ishii, S., & Imai, S., (2000). PF1138A and B, novel trans-epoxysuccinate-type cysteine protease inhibitors produced by *Chaetomium globosum*. *Meiji Seika Kenkyu Nenpo, 39*, 55–64.
Takahashi, M., Koyama, K., & Natori, S., (1990). Four new azaphilones from *Chaetomium globosum* var. *flavo-viridae*. *Chemical and Pharmaceutical Bulletin, 38*, 625–628.
Tantapakul, C., Promgool, T., Kanokmedhakul, K., Soytong, K., Song, J., Hadsadee, S., Jungsuttiwon, S., & Kanokmedhakul, S., (2019). Bioactive xanthoquinodins and epipolythiodioxopiperazines from *Chaetomium globosum* 7s-1, an endophytic fungus isolated from *Rhapis cochinchinensis* (Lour.) Mart. *Natural Product Research, 34*, 494–502.
Thohinung, S., Kanokmedhakul, S., Kanokmedhakul, K., Kukongviriyapan, V., Tusskorn, O., & Soytong, K., (2010). Cytotoxic 10-(indol-3-yl)-[13] cytochalasans from the fungus *Chaetomium elatum* ChE01. *Arch Pharm. Res., 33*, 1135–1141.
Wang, S., Li, X. M., Teuscher, F., Li, D. L., Diesel, A., Ebel, R., et al., (2006). Chaetopyranin, a benzaldehyde derivative, and other related metabolites from *Chaetomium globosum*, an endophytic fungus derived from the marine red alga *Polysiphonia urceolata*. *Journal of Natural Products, 69*, 1622–1625.
Wang, S., Zhang, Y., Li, X. M., & Wang, B. G., (2007). Benzaldehydes from endophytic fungus *Chaetomium globosum* separated from marine red alga *Polysiphonia urceolata*. *Haiyang Yu Huzhao, 38*, 131–135.
Wang, X. Y., Yan, X., Fang, M. J., Wu, Z., Wang, D., & Qiu, Y. K., (2017). Two new cytochalasan derivatives from *Chaetomium globosum* SNSHI-5, a fungus derived from extreme environment. *Natural Product Research, 31*, 1669–1675.
Wijeratne, E. M. K., Turbyville, T. J., Fritz, A., Whitesell, L., & Gunatilaka, A. A. L., (2006). A new dihydroxanthenone from a plant-associated strain of the fungus *Chaetomium globosum* demonstrates anticancer activity. *Bioorganic and Medicinal Chemistry, 14*, 7917–7923.
Yamada, T., Doi, M., Shigeta, H., Muroga, Y., Hosoe, S., Numata, A., et al., (2008). Absolute stereostructures of cytotoxic metabolites, chaetomugilins A-C, produced by *Chaetomium* species separated from a marine fish. *Tetrahedron Letters, 49*, 4192–4195.
Yamada, T., Muroga, Y., & Tanaka, R., (2009). New azaphilones, seco-chaetomugilins A and D, produced by a marine-fish-derived *Chaetomium globosum*. *Marine Drugs, 7*, 249–257.
Yamada, T., Muroga, Y., Jinno, M., Kajimoto, T., Usami, Y., Numata, A., et al., (2011). New class azaphilone produced by a marine fish-derived *Chaetomium globosum*. The stereochemistry and biological activities. *Bioorganic and Medicinal Chemistry, 19*, 4106–4113.
Yang, S. W., Mierzwa, R., Terracciano, J., Patel, M., Gullo, V., Wagner, N., et al., (2007). Sch 213766, a novel chemokine receptor CCR-5 inhibitor from *Chaetomium globosum*. *The Journal of Antibiotics, 60*, 524–528.
Yang, S. W., Mierzwa, R., Terrcciano, J., Patel, M., Gullo, V., Wagner, N., et al., (2006). Chemokine receptor CCR-5 inhibitors produced by *Chaetomium globosum*. *Journal of Natural Products, 69*, 1025–1028.
Yasukawa, K., Takahashi, M., Natori, S., Kawai, K., Yamazaki, M., Takeushi, M., et al., (1994). Azaphilones inhibit tumor promotion by 12-O-tetradecanoylphorbol-13-acetate in 2-stage carcinogenesis in mice. *Oncology, 51*, 108–112.
Zhang, J., Ge, H. M., Jiao, R. H., Li, J., Peng, H., Wang, Y. R., et al., (2010). Cytotoxic chaetoglobosins from the endophyte *Chaetomium globosum*. *Planta Medica, 76*, 1910–1914.

PART II

Organic Amendments and Sustainable Practices for Plant and Soil Management

CHAPTER 7

GARLIC PRODUCTS FOR SUSTAINABLE ORGANIC CROP PROTECTION

ANJORIN TOBA SAMUEL[1] and ADENIRAN LATEEF ARIYO[2]

[1]Department of Crop Protection, Faculty of Agriculture, University of Abuja, PMB 117, Abuja, Nigeria

[2]Department of Veterinary Physiology and Biochemistry, Faculty of Veterinary Medicine, University of Abuja, Nigeria

ABSTRACT

The crop protection potential and pesticidal efficacy of garlic (*Allium sativum* L.) has been attributed to its production of several biologically active defense compounds. The plant is rich in organo-sulfur content, which has great potential to prevent and treat many diseases of plants and animals. Garlic phytochemical contains alliinase which is released when it is chopped. Allinase is involved in catalyzing the formation of S-allyl cysteine sulfoxide (allin) which is the main active component of garlic. It is possible that constituents from garlic could be used to develop alternatives to conventional pesticides for the management of crop pests and disease-causing pathogens. This chapter describes the biochemical compounds found in garlic bulb, preparatory procedures of garlic pesticidal products from fresh and aged bulbs, and the discovery process of bioactive phytochemical from garlic bulbs. It further explains the insecticidal, fungicidal, bactericidal, and nematicidal usage of garlic products and discusses issues on natural pesticides formulation from garlic. Usage of garlic herbal preparations by growers can reduce non-target exposure to hazardous pesticides and curb resistance development in pests, thus enabling sustainable organic crop protection.

7.1 INTRODUCTION

Concerns regarding the potential health and environmental impacts of synthetic chemical pesticides have led to increased interest in the development and use of safer pest control alternatives such as plant-based pesticides. Botanical pesticides are generally developed from naturally and locally available plant renewable resources, instead of purchased chemical inputs, and hence reduce the need for synthetic pesticides (IFOAM, 2019). This enables food production with minimal or no harm to ecosystems, animals, or humans (Brooklyn Botanic Garden, 2000; Agama, 2015). Also, the involvement of plant resources in crop protection can increase incomes and generation of employment for the populace (Auerbach et al., 2013). *Allium sativum* L.-family Alliaceae, has been used worldwide for medicinal and culinary purposes and currently been exploited as an organic pesticide (Petropoulos et al., 2018). Garlic bulb is made up of several concentric bulblets characterized by acid taste, pungent-smelling malicious odor.

Garlic is native to central Asia and was first encountered by the man about some 7,000 years ago (Ellis and Bradley, 1992). Farmers have cultivated garlic in farms with other crops as companion crops to prevent insect and pest attacks like aphids, caterpillars on their farm crops (Lalla et al., 2013). Garlic products have been effectively used in the treatment of several plant pathogens and insects at different stages in their life cycle. The products have successfully been used against downy mildew, fruit rots and blight (Davidson, 1997; Sallam et al., 2012). The pesticidal efficacy of garlic has been attributed to its production of several biologically active defense compounds.

Silva et al. (2001) attributed the biological activity of garlic to the presence of organosulfur compounds in the plant. The organosulfur compounds elucidated in garlic include S-2-carboxypropylglutathione, S-allyl-L-cysteine-sulfoxide, S-(trans-1-propenyl)-L-cysteine (Lalla et al., 2013; Lanzotti et al., 2013; Beni et al., 2018). Synthesis of allicin may be categorized as the first line of defense against attack by pests and other pathogens (Kodera et al., 2002). Other minerals and microelements found in garlic like iodine, zinc, copper, selenium, tocopherols, and metabolites like ascorbic acid, protein content and polyphenol are the components responsible for its medicinal properties (Borlinghaus et al., 2014).

Release of allinase occurs when garlic clove is crushed. This enzyme quickly changes alliin to allicin. Separation of the allinase and alliin occur in garlic bulb membrane *in-situ* (Miron et al., 2000). Allicin is an unstable organosulfur compound which can be stabilized by adding oil-soluble polysulfide's like dially (tetrasulfide, trisulfide, disulfide, and sulfide) (Block et al., 2017).

In light of the significant pesticidal potential of garlic and with the advent of modern formulation techniques, there is a need for an extensive understanding of the plant especially on how best to extract its biologically active defense compounds. In this chapter, the biochemical components of the garlic bulb, the procedures for preparing the extracts for pesticidal purposes by the farmers, and the scientific procedures for isolating and formulating garlic-based pesticides are described. Update on critical issues on the current usage of garlic products for pest control are also provided.

7.2 GARLIC PREPARATIONS AND THEIR PESTICIDAL PRODUCTS

There are many pesticidal garlic products that are sold in the market, usually as raw garlic powder, oil, and homogenate which are the major garlic formulation. The homogenate is made up of S-allylmercaptocysteine (SMAC) and S-allycysteine (SAC). Alliin is found in heat-treated and powdered garlic (Figure 7.1). The highly odoriferous oil and powder from garlic is used as pesticide while the supplement could be made from the odorless aged garlic product (Plata-Rueda et al., 2017).

7.2.1 GARLIC HOMOGENATE FROM FRESH BULB

Garlic homogenate can be achieved by pulverizing fresh garlic. One gram of garlic (peeled) is blended by adding 0.01 L of water, and the homogenate produced is left for between 2 and 5 min at 25°C, then a 0.01 L of methanol is added to the filtered homogenate to cause precipitation of the carbohydrate and protein present in the extraction. Allicin is a stable compound at 4°C for up to 36 h (Bayan et al., 2014).

7.2.2 AGED GARLIC EXTRACT (AGE)

Aged garlic extract (AGE) is usually prominent out of many other garlic extracts. Safe and stable sulfur compounds are usually extracted using 15–20% ethanol at 25°C. This stable sulfur compound confers garlic the characteristics flavor and smell. S-allyl-mercaptocysteine and all other stable water-soluble, little oil-soluble allyl sulfides and minimal allicin are found in AGE (Ryu et al., 2017). All these compounds have appreciable antioxidant activities (Corzo-Martínez et al., 2007; Pérez-Torres et al., 2016).

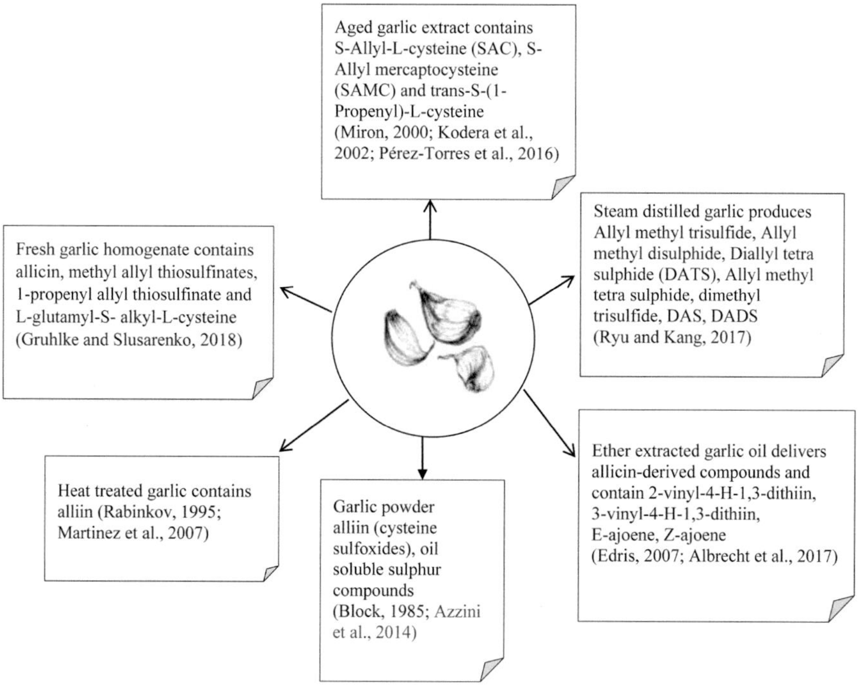

FIGURE 7.1 Garlic products and their major pesticidal organo-sulfur compounds.

7.2.3 BLACK GARLIC

Black garlic is obtained from fermented garlic. Fermentation of garlic at high humidity and temperature or (80–90% and 60–90°C) respectively reduces allicin and thus the offensive odor found in fresh garlic. Black garlic activity is as a result of the change in the physiological properties and this enhances its biological activities (Tan et al., 2019).

7.2.4 GARLIC POWDER

Garlic powder is made up of alliin and little oil-soluble sulfides. It is made from garlic cloves that were dehydrated and grounded. Allinase enzyme will be deactivated through the processing heat dehydration of the garlic cloves. In order to prepare garlic powder, the garlic cloves are first separated, and then the skin is peeled away with fingers or cracked with a sterile knife. Each clove is thinly sliced and placed on a sheet lined with parchment paper. For it to dry, the sheets of garlic are placed into the oven and cooked for about 1 h 30 mins at 65°C. The sheet of garlic is removed from the oven and allowed to cool before been ground into powder (Rahman et al., 2007).

If a dehydrator is to be used to dry out garlic, set the temperature to about 51°C and allowed to dehydrate for about 8–12 h, when the garlic bits should be brittle indicating it has dried well (Iciek et al., 2009). In order to grind the dried garlic, a grinder, blender, food processor, spice mill or mortar and pestle could be used. The garlic powder should be stored in an airtight container somewhere cool, dry, and out of direct sunlight. Mason jars are a good storage container for locally-made garlic powder. Freezing your garlic powder is also an option.

7.2.5 GARLIC OIL

Garlic oil is mostly obtained by extracting it with ether followed by steam distillation, with a yield around 2.5–3.0 g/kg fresh garlic. In garlic oil, diallyl sulfide (DAS), diallyl disulfide (DADS), and dimethyl trisulfide (DATS), differing in their number of sulfur atoms, and allylmethyl sulfide are the four most abundant volatile allyl sulfides (Block et al., 2017). However, the composition of these ingredients may vary slightly based on the extraction method and garlic cultivar (Pérez-Torres et al., 2016). The types of commercially available garlic products and their major pesticidal organo-sulfur compounds are shown in Figure 7.1.

7.3 PREPARATION OF PESTICIDE FROM CRUDE GARLIC EXTRACT

For production of garlic extract, healthy cloves of garlic should be selected. The garlic bulbs are then stored in a cool, dry shady place. Garlic extract are obtained by first blending the cloves, followed by filtration of the mixture. Most of the times, garlic extract is prepared by thoroughly mixing garlic cloves with water in a blender in a ratio 1:2 w/w. The mixture is then filtered with the filtrate collected as garlic extract before being mixed with oil or organic solvents (Kaluwa and Kruger, 2012). Clean utensils should be used and washed immediately after use, in order to avoid cross-contamination. In the preparation and application process, it should be ensured that there is no direct contact with the crude extract. Care must be taken not to spray the plant with garlic extract close to harvest time because sprayed garlic remains long on the plant. A potash-based soap like laundry soap can be added to the extract as an emulsifier. Appropriate clothing should always be worn while applying the extract and the hands should always be washed after handling the plant extract (Tafadzwa et al., 2016). Before going for large scale spraying of crops, it is advisable to first try the extract formulated on few plants infested with the disease. Two out of various traditional preparation protocols are as follows:

1. About 80 grams of garlic bulbs are chopped into small pieces by a clean and sterile sharp knife and ground with a grinder. About 600 ml water and 3–4 drops of liquid soap are added to the mixture, stirred, and allowed to stand for 24 h and filtered through a fine cloth. As garlic components are volatile, the extract is to be stored in a tightly sealed container or vessel like bottle before use (Mikaili, 2013). In order to use it, 1 part of the emulsion is diluted with 9 parts of water and shaken well before spraying. Spraying on the infested plants should be thorough, preferably applied in early in the morning.
2. About 85 grams of chopped or crushed garlic are added to 50 ml of mineral oil (vegetable oil) and allowed to stand for 24 h. The ratio of garlic extract to essential oil used could range from 0–70% garlic extract to 90–30% oil. 950 ml of water is then added and stirred in 10 ml of liquid dish soap and filtered through a fine cloth (Tafadzwa et al., 2016). Such a mixture needed to be stored in a non-transparent container such as a bottle in a cool place before use in order to check thermo- and photo-degradation. For its use, 1 part of the emulsion is diluted with 19 parts of H_2O and shaken well prior to spraying. Spraying early in the morning of the infested plant thoroughly is advocated (Vijayalakshmi et al., 1999). A synergistic effect is often observed when garlic extract is combined with essential or mineral oils. These often result in improved natural insecticidal, bactericidal, and fungicidal qualities than when garlic extract only is used (Curtis et al., 2004). The essential volatile oils could be obtained from plants or seeds of cottonseed oil, soybean oil, castor oil, lemongrass oil, sesame oil, ginger oil (Edris, 2007).

7.4 DISCOVERY PROCESS OF PESTICIDAL BIOACTIVE PHYTOCHEMICAL FROM GARLIC BULBS

7.4.1 EXTRACTION AND PURIFICATION OF MOLECULE FROM GARLIC

The current use of crude extracts of garlic for pesticide formulation by the farmers is usually substandard. A standard and scientific stepwise procedure for formulating pesticides from garlic bulb via bioassay evaluation is as shown in Figure 7.2. Garlic efficacy optimization depends on the extraction techniques and conditions. The crude extract can be reconstituted with all the organic solvent while the organic layer can be separated by drying over sodium sulfate and evaporated at 40°C on a Rota evaporator. The crude extract obtained from this step could then be utilized for various biological activities (Atanasov et al., 2011). The contents could be subjected to a defatting process by pouring them into a solution of 10 ml brine, 10 ml methanol and 20 ml hexane. This solution is then stirred for 15 min following which methanol/water layer is extracted with ethyl acetate. Extracted content is then subjected to flash chromatography and each fraction checked for activity (Kamleh et al., 2008).

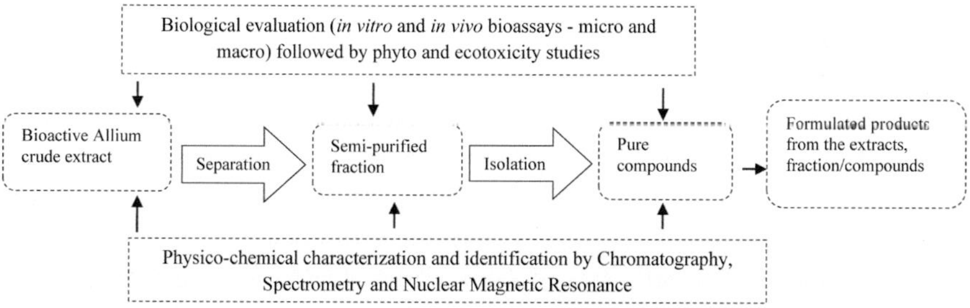

FIGURE 7.2 Bio-guided discovery process of pesticidal bioactive phytochemical from garlic bulbs.
Source: Adapted from: Pino et al. (2013); http://scielo.sld.cu/pdf/rpv/v28n2/rpv01213.pdf.

7.4.2 ISOLATION AND PURIFICATION OF GARLIC BIOACTIVE COMPOUNDS

The purification and subsequent isolation of biologically active compound can be achieved through the use of column chromatography from garlic before obtaining a standardized formulation. Highly valued separation can be accomplished through increase polarity of the mobile phases. However, a rapid analysis of the different compound present can be achieved through the use of thin-layer chromatography (TLC). Accelerating this process can occur through the use of high-pressure liquid chromatography (HPLC) accelerate the process of extracting the target structurally-related compound or purification of the bioactive molecule. Each compound exhibits a peculiar peak when subjected to a definite chromatographic condition: the pure compound obtained from the garlic extract can then be used for structural elucidation and pesticidal activities (Pino et al., 2013).

Spectroscopic data obtained from one- and two-dimensional nuclear magnetic resonance (NMR) mass spectrometry (MS), ultraviolet (UV) visible light, and infrared spectroscopy can be deployed in identifying isolated compound. Compounds that have powerful chromophore can be carried out using UV-visible spectroscopy (Ziegler et al., 2013).

Immunoassays, monoclonal antibodies (Mabs) are part of non-chromatographic technique required to identify pure isolate of garlic extract. Fourier transform infrared spectroscopy (FTIR) offers a rapid and non-destructive investigation to fingerprint herbal extracts or powders (Zhang et al., 2018).

The purity and characterization of active fractions that are obtained from the flash chromatography can be determined by HPLC using C-18 column. The preparative HPLC is carried out using a C-18 column, elution is performed using acetonitrile/water (90/10) at a flow rate of 3 ml/min and detection when performed at λ254. These are performed using the procedure of Adetunji et al. (2019). HPLC system is made up of the solvent reservoir, or multiple reservoirs, a high-pressure Agilent 1200 series preparative pump, a column, injector system and the UV diode array detector. In vitro and in vivo evaluation of the activities of the crude extract is needed before isolation and elucidation is carried out.

The discovery and development of new compounds from garlic is extremely intensive but necessary. It is imperative to search for new and potent compound from garlic. This is an important strategy in order to overcome the challenge of widespread synthetic pesticide resistance. That is why additional phytochemical or essential oils are often added in the formulation of garlic-based products. Also, it is increasingly desirable to substitute some existing pesticides that are highly toxic with safe compounds that are only toxic to the pest and are environmentally safe (Clough and Godfrey, 1998).

7.5 PESTICIDAL USAGE OF GARLIC PRODUCTS

The increasing level of appreciation for the value of garlic as pesticides has been recorded, especially among organic farmers. The potency of garlic products has been reported in the management of several crop insect pests (Badoo, 2016).

7.5.1 GARLIC USAGE AGAINST INSECT PESTS

Garlic has been severally reported to have properties for controlling insect pests. These include African bollworm, African armyworm, onion thrips, whiteflies, termites, through their antifeedant properties and repellent properties (Prowse et al., 2006; Denloye, 2010; Kiani et al., 2012). The level of insecticidal activity of garlic components depends on the extracting solvent, concentration, and species susceptibility of target insect pest. Concentration dependent mortalities on pulse beetle has been recorded for the use of Sativum oil extract (Chaubey, 2017). The garlic extracted using methanol as a solvent has shown 76% mortality of maize weevil mortality at higher dosage treatment and 35.9% mortality at lower concentration (Ibrahim and Garba, 2011; Lalla et al., 2013). Other studies carried out by Mobki (2014) documented high fumigant toxicities of garlic extracts against red flour beetle-*T. casteneum* and *S. zeamais*. Use of 7.6 µl/L of garlic extract causes 30% mortalities of Japanese termite (Park and Shin, 2005). *D. domestica* and *D. radium* larva mortality occurs upon exposure to garlic juice (Gareth et al., 2006). The larvicidal effects of garlic juice were comparable to insecticidal activities of organophosphate pesticides. Dailly (disulfide and trisulfide) are the phyto-constituent with insecticidal properties which affect a large range of insect pest (Ouko et al., 2017). These organosulfur compound causes pathology in the respiratory organ of the insect (Yang, 2012). Additionally, products of ethanol extract identified include vinydithiins, allicin, ajoene, and diallyl methyl trisulfide which may be responsible for the insecticidal activities of the extract (Nwachuku and Asawalam, 2014).

Low toxicity resulting in 24% mortality of *S. zeamais* has been associated with hexane extract of garlic bulb. The low polarity of hexane has been described to be the reason for the poor activity of the hexane extract since most of the compound with insecticidal activities has higher polarity. Ouko et al. (2017) recorded the use of higher polarity solvent for the extraction of *Allium sativum* since most of the Alky substance found in the plant is soluble in highly polar solvents. This could be attributed to the presence of potent compounds such as phenolic, flavonoids, and saponins in low concentration.

The use of methanol/hexane in extracting garlic improves its activities again maize weevils than when the individual solvent is used. The extract showed 96% mortality while 88% mortality occurs within 24 h. This activity is comparable to actellic synthetic insecticide used in the treatment of insect pest. The activities displayed by methanol and hexane extract may be attributed to increase solubility of more alkyl compound and allicin in the garlic extract (Hamed et al., 2012).

There were reports of high efficacy of garlic essential oils on rice and bean bruchid weevils (Lalla et al., 2013). Organosulfur compound from Alliaceae has an effect on several insect species. These compounds affect the respiratory cells of the insects (Adedire and Ajayi, 1996). The compounds induced inflammation through the activation of transient receptor potential Ankyrin-1 ion channel in the insect neural tube; this may be responsible for the mortality of insect observed (Nwachukwu and Asawalam, 2014).

The presence of allicin and diallyl sulfide which are majorly organosulfur compounds responsible for the pungent smell of garlic extract. The odor repels feeding insects (Karavina et al., 2014). The presence of diallyl trisulfide also deters insect oviposition on stored grains (Nwachukwu and Asawalam, 2014). The larvicidal and ovicidal activities of these biologically active constituent of garlic extract on larva and egg of beetles have been demonstrated (Ali et al., 2014).

Effect of garlic extract on leaf miner disease which infest tomato showed that spraying the garlic extract thrice at 2-week interval starting from day 40 significantly reduced the population of the leaf miner. Also, it increased tomato yield (Hussein et al., 2014). Also, the extract from garlic has activities against blights, downy mildew, rust, and fruit rots (Sallam et al., 2012).

7.5.2 GARLIC USAGE AGAINST PATHOGENIC FUNGI

Chips, extracts, and oil of garlic have been reported to inhibit the growth of fungi such as *Phytophthora* spp., *Penicillium* spp., *Saccharomyces* spp., *Aspergillus fumigatus* and *Aspergillus parasiticus* (Singh et al., 2001; Barile et al., 2007; Bayili et al., 2011; Perello et al., 2013). The antifungal activity of garlic was tested for controlling tomato early blight caused by *Alternaria solani, in vitro* and *in vivo* (Tagoe et al., 2011). In greenhouse experiments, the highest reduction of disease severity was achieved by the extracts of garlic at 5% concentration at this level, the fruit yield was increased by 76.2%

and 66.7% compared to the infected control. *In vitro* and *in vivo* experiments have shown that garlic extract inhibited cyst growth, germination of sporangia and growth of germ tube of *Phytophthora* infestation of leaf. Also, allicin at concentration ranging from 50 μg to 1 g/mL causes appreciable reduction (50–100%) in cucumber downy mildew, a disease caused by *Pseudoperonospora cubensis* (Davidson, 1997).

Garlic extract containing allicin has shown great activity against *Drechslera tritici-repentis* and *Bipolaris sorokiniana,* which are fungal contaminant of wheat seeds. Result obtained from the use of garlic extract to correct wheat seed poor germination is comparable to result from the use of synthetic pesticide. Garlic extract induced structural modification of the conidia and hyphae of *Bipolaris sorokiniana* and *Drechslera tritici-repentis,* thus restricting these fungal colonies radial growth for up to 10 days. Significant antifungal activities have been associated with the essential oil extracted from garlic (Onyeagba et al., 2004).

The essential oils extracted from garlic have been shown to possess significant antifungal properties. In separate studies by Salim (2011); and Tagoe (2011) documented broad-spectrum antifungal activities of garlic. The extract has shown between 60% and 82% growth inhibition of seed borne fungi of *Penicillium* and *Aspergillus* genera. The wide range antifungal effect may be due to the presence of allicin which can be converted to diallyl (trisulfide, disulfide, sulfide, and allymethyltrisulfide) ajoene and dithiins (Naganawa et al., 1996; Cowan, 1999).

Garlic extract was found to completely inhibit the mycelial growth of *Aspergillus ochraceus* and OTA biosynthesis by *A. ochraceus* (Reddy et al., 2010). Inhibition of *Fusarium* species found in cucumber by garlic tablet has been documented (Amin et al., 2009). Also, the allicin from garlic extract potently inhibited the growth of *Botrytis* and *Alternaria* (Perello et al., 2013). Rice seedling treated with garlic extract show resistance to rice blast pathogen (*M. grisea*) attack, and the treated plant showed fever symptoms (Hubert et al., 2015). Use of garlic extract as the curative measure is less effective compared to its use as a prophylactic measure.

Plant extracts including garlic reduced the activity of cucumber soil-borne pathogens. Similarly, spraying of garlic extract has shown appreciable inhibition of downy mildew and cucumber soil borne disease which encourages more cucumber yield. As well as controlling disease the sprays had a beneficial effect on cucumber yields (Hajano et al., 2012). The results indicated that the combination of garlic extracts with either cottonseed oil or cinnamon oil inhibited the growth of powdery mildew disease more effectively than applications not in combination.

Exposure of *Fusarium* species, *Alternaria* species and *Colletotrichum* species to garlic extract causes inhibition of their spore germination and halt their mycelial growth, especially when the extracts were applied after the spore has germinated (Singh et al., 2001). This may be due to the exchange reaction of the thiol-disulfide which elicit the antimicrobial action (Fass and Thorpe, 2018).

7.5.3 GARLIC USAGE AGAINST PATHOGENIC BACTERIA AND NEMATODES

Rahman et al. (2007) determined the anti-microbial activity of different garlic products, which include garlic oil, dried garlic powder, commercially produced garlic products were tested against different selected bacterial pathogen of plant (Rahman et al., 2007; Mikaili, 2013). The result indicated that the highest growth inhibition occurs in the lactic culture of *Thermophilus*, *Streptococcus*, *Lactobacillus delbrueckii* subsp. *bulgaricus*. Moderate growth inhibition was observed in *Staphylococcus aureus*. While poor growth inhibition of garlic extract was seen in *Bacillus cereus* and *Salmonella typhimurium*. It was observed that the antimicrobial activities of the garlic products tested decreases with decreasing concentration of garlic powder. The temperature of the preparation of garlic products affects the antimicrobial properties. Increase temperature of mode of preparation decreases the antimicrobial activities of garlic products.

Garlic extract showed comparable result in control of nematode causing root-knot in groundnut. Garlic extract showed appreciable effect through suppression of nematode, thus stimulating the growth of the plant (Fatema, 2005). A 100% mortality was recorded in *Heterodera cajanis* second stage juvenile which is the cause of pigeon pea root-knot. This result was observed after 14 h exposure to garlic extract. Similarly, after exposure of *Terranchus urticae* to garlic extract, 100% mortality of the helminth was recorded (Singh et al., 2001).

7.6 ISSUES CONCERNING NATURAL PESTICIDES FORMULATION FROM GARLIC BULB

Ordinarily, purified pesticidal natural compounds from garlic can either be used in pesticide formulation, as such compounds constitute the biologically active principle in the formulation mixture, sometime the mixture may include synthetic material. Alternatively, such pure compounds obtained may be explored as a leading drug candidate in the design of novel synthetic drug (Gordon et al., 2013).

The use of plant natural product as pesticide has been filled with a lot of problems. Many natural products are unstable under sunlight and high temperature that may be encouraged in the field; thus, they are not able to exert any toxic effect on plant pathogens. Also, due to lack of selectivity, many of the natural products exert their toxicity on plants and mammals. The suggestion of adding synthetic material to the natural products to overcome this limitation can definitely add to the overall cost of producing the natural products (Clough and Godfrey, 1998; Atanasov, 2015).

In spite of the arguments above expressed and the difficulties to obtain a natural substance which can have pesticidal activity and stability at the same time, the researchers are continually searching new substances naturally occurring in nature with pesticidal properties so that in the future and after optimization could be used as commercial proposes.

The use of garlic extracts to improve plant health and increase disease/pest resistance are currently and largely unregulated. Thus, legislative control by various governments should be applied to ensure regulatory compliance of wherever pesticidal claims are made. When growing garlic for pest control, it has been observed that heavy doses of fertilizer can reduce the concentration of the effective substances in the garlic (Amagase, 1993). Apart from avoiding the use of large amounts of fertilizers, other factors that could influence the efficacy of garlic should be studied.

7.7 CONCLUSION

Research reports support the use of garlic extract as a useful, simple, cost effective and environmental-safe crop pest and diseases management strategy with the purpose to minimize the use of synthetic insecticides and fungicides. The combination of garlic natural ingredients when combined with other plant essential oils, have superior antifungal and anti-bacterial qualities, than if applied separately. However, most of the pesticidal studies of garlic in developing countries mostly involved crude extracts. Further studies on the isolation of pure compounds of *A. sativum* and determine the mode of action to obtain standard formulation is imperative. Also, because garlic has a broad-spectrum effect and might be non-selective, it should be used with caution or used as directed so that non-target beneficial organisms will be safe.

Although new pesticides based on natural plant extracts are continually developing, more research is necessary for optimizing applications and become a safe alternative for eliminating the chemical pesticides from agriculture. Though garlic products are safe pesticide, for the sake of IPM strategy, it should be applied together with low residual synthetic pesticides. Due to the easy preparation of garlic products, systemic action after application, biodegradability products, and possible low-cost alternatives to agrochemicals, it is hoped that garlic-based pesticides would no doubt play a significant role in achieving sustainable crop protection in modern agriculture.

KEYWORDS

- **bactericidal activity**
- **bioactive phytochemicals**
- **formulation procedure**
- **fungicidal activity**
- **garlic products**
- **insecticidal activity**
- **nematicidal activity**
- **organic crop protection**
- **organo-sulfur compounds**

REFERENCES

Adedire, C. O., & Ajayi, T. S., (1996). Assessment of insecticidal properties of some plants as grain protectants against the maize weevil, *Sitophilus zeamais* (Motsch.). *Nig. J. Ent., 13*, 93–101.

Adetunji, C. O., Olaniyi, O. O., & Ogunkunle, A. T. J., (2013). Bacterial activity of crude extracts of *Vernonia amygdalina* on clinical isolates. *Journal of Microbiology and Antimicrobials, 5*(6), 60–64.

Agama, J., (2015). Africa: Latest development in organic agriculture in Africa. In: Willer, H., & Lernoud, J., (eds.), *The World of Organic Agriculture* (p. 15). Statistics and Emerging Trends FiBL-IFOAM Report. FiBL, Frick, and IFOAM-Organics International, Bonn.

Ali, S., Sagheer, M., Hassan, M., Abbas, M., & Hafeez, F., (2014). Insecticidal activity of turmeric (*Curcuma longa*) and garlic (*Allium sativum* L.) extracts against red flour beetle, *Tribolium castaneum*: A safe alternative to insecticides in stored commodities. *J. Entomol. Zool. Stud., 2*, 201–205.

Amagase, H., & Milner, J. A., (1993). Impact of various sources of garlic and their constituents on 7,12-dimethylbenz[α]anthracene binding to mammary cell DNA. *Carcinogenesis, 14*(8), 1627–1631.

Amin, R. A. B. M., Rashid M. M., & Meah, M. B., (2009). Efficacy of garlic to control seed-borne fungal pathogens of cucumber. *J. Agric. Rur. Dev., 7*(1–2), 135–138.

Atanasov, A. G., Waltenberger, B., & Pferschy-Wenzig, E. M., (2011). Discovery and resupply of pharmacologically active plant-derived natural products: A review. *Biotechnology Advances, 33*(8), 1582–1614.

Auerbach, R., Rundgren, G., & Scialabba, N. E., (2013). *Organic Agriculture: African Experiences in Resilience and Sustainability*. (p. 219) FAO Natural Resources Management and Environment Department.

Baidoo, P. K., & Mochiah, M. B., (2016). Comparing the Effectiveness of garlic (*Allium sativum* L.) and hot pepper (*Capsicum frutescens* L.) in the management of the major pests of cabbage (*Brassica oleracea* (L.). *Sustainable Agriculture Research, 5*(2), 83–91.

Barile, E., Bonanomi, G., Antignani, V., Zolfaghari, B., Sajjadi, S. E., Scala, F., & Lanzotti, V., (2007). Saponins from *Allium minutiflorum* with antifungal activity. *Phytochemistry, 68*, 596.

Bayan, L., Koulivand, P. H., & Gorji, A., (2014). Garlic: A review of potential therapeutic effects. *Avicenna J. Phytomed., 4*(1), 1–14.

Bayili, R. G., Abdoul-Latif, F., Kone, O. H., Diao, M., Bassole, I. H. N., & Dicko, M. H., (2011). Phenolic compounds and antioxidant activities in some fruits and vegetables from Burkina Faso. *African Journal of Biotechnology, 10*, 62.

Beni, C., Casorri, L., Masciarelli, E., Ficociello, B., Masetti, O., Rinaldi, S., & Cichelli, A., (2018). Characterization of garlic (*Allium sativum* L.) aqueous extract and its hypothetical role as biostimulant in crop protection. *Journal of Food Agriculture and Environment, 16*(3, 4), 38–44.

Block, E., Bechand, B., Gundala, S., Vattekkatte, A., Wang, K., Mousa, S., Godugu, K., et al., (2017). Fluorinated analogs of organosulfur compounds from garlic (*Allium sativum*): Synthesis, chemistry and anti-angiogenesis and antithrombotic studies. *Molecules, 22*, 2081.

Borlinghaus, J., Albrecht, F., Gruhlke, M. C. H., Nwachukwu, I. D., & Slusarenko, A. J., (2014). Allicin: Chemistry and biological properties. *Molecules, 19*, 12591–12618.

Brooklyn Botanic Garden, (2000). *Natural Disease Control*: *A Commonsense Approach to Plant First Aid*. Handbook No. 164. Brooklyn Botanic Garden, Inc., NY.

Chaubey, M. K., (2017). Biological activities of *Allium sativum* essential oil against pulse beetle, *Callosobruchus chinensis* (Coleoptera: Bruchidae). *Herba Polonica, 60*, 41–55.

Clough, J. M., & Godfrey, C. R. A., (1998). The strobilurin fungicides. In: Hutsun, D., & Miyamoto, J., (eds.), *Fungicidal Activity, chemical and Biological Approaches to Plant Protection* (p. 254). John Wiley and Sons Ltd., West Sussex, England.

Corzo, M., Corzo, N., & Villamiel, M., (2007). Biological properties of onions and garlic. *Trends in Food Science and Technology, 18*(12), 609–625.

Cowan, M. M., (1999). Plant products as antimicrobial agents. *Clinical Microbiology Reviews, 12*(4), 564–582.

Curtis, H., Noll, U., Störmann, J., & Slusarenko, A. J., (2004). Broad-spectrum activity of the volatile phytoanticipin allicin in extracts of garlic (*Allium sativum* L.) against plant pathogenic bacteria, fungi and oomycetes. *Physiological and Molecular Plant Pathology, 65*, 79–89.

Davidson, P. M., (1997). Chemical preservatives and natural antimicrobial compounds. In: Doyle, M. P., Beuchat, L. R., & Montville, T. J., (eds.), *Food Microbiology: Fundamentals and Frontiers* (pp. 520–556). Washington D.C., ASM Press.

Denloye, A. A., (2010). Bioactivity of powder and extracts from garlic, *Allium sativum* L. (Alliaceae) and spring onion, *Allium fistulosum* L. (Alliaceae) against *Callosobruchus maculatus* F. (Coleoptera: Bruchidae) on cowpea, *Vigna unguiculata* (L.) Walp. (Leguminosae) seeds. *Entomology*, 1–5.

Edris, A., (2007). Pharmaceutical and therapeutic potentials of essential oils and their individual volatile constituents: A review. *Phytotherapy Research, 21*(4), 308–323.

Ellis, B. W., Bradley, F. M., & Atthowe, H., (1996). *The Organic Gardener's Handbook of Natural Insect and Disease Control: A Complete Problem-Solving Guide to Keeping Your Garden and Healthy Without Chemicals*. Emmaus: Rodale Press.

Fatema, S., & Ahmad, M. U., (2005). Comparative efficacy of some organic amendments and a nematicide (Furadan-3G) against root-knot on two local varieties of groundnut. *Plant Pathology Journal, 4*, 54–57.

Gareth, M. P., Tamara, S. G., & Andrew, F., (2006). Insecticidal activity of garlic juice in two dipteran pests. *Agricultural and Forest Entomology, 8*(1), 1–6.

Gordon, M. C., & Newman, D. J., (2013). Natural products: A continuing source of novel drug leads. *Biochim. Biophys. Acta, 1830*(6), 3670–3695.

Gruhlke, M. C. H., & Slusarenko, A. J., (2018). The chemistry of alliums. *Molecules, 23*(1),143.

Hajano, J. A., Lodhi, A. M., &. Pathan, M. A., & Shah, G. S., (2012). *In-vitro* evaluation of fungicides, plant extracts and bio-control agents against rice blast pathogen *Magnaporthe oryzae* Couch. *Pakistan Journal of Botany, 44*(5), 1775–1778.

Hamed, R. K. A., Ahmed, S. M. S., Abotaleb, A. O. B., & El-Sawaf, B. M., (2012). Efficacy of certain plant oils as grain protectants against the rice weevil, *Sitophilus oryzae* (Coleoptera: Curculionidae) on wheat. *EAJBS, 5*, 49–53.

Hubert, J., Mabagala, R. B., & Mamiro, D., (2015). Efficacy of selected plant extracts against *Pyricularia grisea*, causal agent of rice blast disease. *American Journal of Plant Sciences, 6*(5), 602–611.

Hussein, N. M., Hussein, M. I., Gadel-Hak, S. H., & Hammad, M. A., (2014). Effect of two plant extracts and four aromatic oils on population and productivity of tomato cultivar gold stone. *Nature and Science, 12*(7).

Ibrahim, N. D., & Garba, S., (2011). Use of garlic powder in the control of maize weevil. *Proceeding of the 45th Annual Conference of Agricultural Society of Nigeria* (pp. 177–181).

Iciek, M., Kwiecien, I., & Wlodek, I., (2009). Biological properties of garlic and garlic-derived organosulfur compounds. *Environmental and Molecular Mutagenesis, 50*(3), 24765.

IFOAM, (2019). *The IFOAM Norms for Organic Production and Processing, Version*. The International Federation of Organic Agriculture Movements.

Kaluwa, K., & Kruger, E., (2012). *Natural Pest and Disease Control Handbook*. KwaZulu-Natal Department of Agriculture and Rural Development, Pietermaritzburg.

Kamleh, A., Barrett, M. P., Wildridge, D., Burchmore, R. J., Scheltema, R. A., & Watson, D. G., (2008). Metabolomic profiling using Orbitrap Fourier transform mass spectrometry with hydrophilic interaction chromatography: A method with wide applicability to analysis of biomolecules. *Rapid Commun. Mass Spectrom., 22*, 1912–1918.

Karavina, C., Mandumbu, R., Zivenge, E., & Munetsi, T., (2014). Use of garlic (*Allium sativum*) as a repellent crop to control diamondback moth (*Plutella xylostella*) in cabbage (*Brassica oleraceae* var. *capitata*). *Agric. Res., 52*, 615–621.

Kiani, L., Yazdanian, M., Tafaghodinia, B., & Sarayloo, M. H., (2012). Control of western flower thrips, *Frankliniella occidentalis* (Pergande) (Thysanoptera: Thripidae), by plant extracts on strawberry in greenhouse conditions. *Mun. Ent. Zool., 7*(2), 12–22.

Kodera, Y., Ayabe, M., Ogasawara, K., Yoshida, S., Hayashi, N., & Ono, K., (2002). Allicin accumulation with long-term storage of garlic. *Chem. Pharm. Bull., 50*, 405–407.

Lalla, F. D., Ahmed, B., Omar, A., & Mohieddine, M., (2013). Chemical composition and biological activity of *Allium sativum* essential oils against *Callosobruchus maculatus. IOSR-JESTFT, 3*, 30–36.

Lanzotti, V., Bonanomi, G., & Scala, F., (2013). What makes *Allium* species effective against pathogenic microbes? *Phytochem. Rev., 12*, 751–772.

Mikaili, P., Maadirad, S., Moloudizargari, M., Aghajanshakeri, S., & Sarahroodi, S., (2013). Therapeutic uses and pharmacological properties of garlic, shallot, and their biologically active compounds. *Iran J. Basic Med. Sci., 16*(10), 1031–1048.

Miron, T., Rabinkov, A., Mirelman, D., Wilchek, M., & Weiner, L., (2000). The mode of action of allicin: Its ready permeability through phospholipid membranes may contribute to its biological activity. *Biochim. Biophys. Acta, 1463*, 20–30.

Mobki, M., Safavi, S. A., Safaralizadeh, M. H., & Panahi, O., (2014). Toxicity and repellency of garlic (*Allium sativum* L.) extract grown in Iran against *Tribolium castaneum* (Herbst) larvae and adults. *Arch. Phytopathol. Plant Protect., 47*, 59–68.

Naganawa, R., Iwata, N., Ishikawa, K., Fukuda, H., Fujino, T., & Suzuki, A., (1996). Inhibition of microbial growth by ajoene, a sulfur-containing compound derived from garlic. *Appl. Environ. Microbiol., 62*, 4238–4242.

Nwachukwu, I. D., & Asawalam, E. F., (2014). Evaluation of freshly prepared juice from garlic (*Allium sativum* L.) as a biopesticide against the maize weevil, *Sitophilus zeamais* (Motsch.) (Coleoptera: Curculionidae). *JPPR., 54*, 132–138.

Onyeagba, R. A., Ugbogu, O. C., Okeke, C. U., & Iroakasi, O., (2004). Studies on the antimicrobial effects of garlic (*Allium sativum* Linn.), ginger (*Zingiber officinale* Roscoe) and lime (*Citrus aurantifolia* Linn.). *African Journal of Biotechnology, 3*, 552–554.

Ouko, R. O., Koech, S. C., Arika, W. M., Osano, K. O., & Ogola, P. E., (2017). Bioefficacy of organic extracts of *A. sativum* against *S. zeamais* (Coleoptera; Dryophthoridae). *Biol Syst., 6*(1), 174.

Park, I. K., & Shin, S. C., (2005). Fumigant activity of plant essential oils and components from garlic (*Allium sativum*) and clove bud (*Eugenia caryophyllata*) oils against the Japanese termite (*Reticulitermes speratus* Kolbe). *J. Agric. Food Chem., 53*, 4388–4392.

Perelló, A., Gruhlke, N., & Slusarenko, A. J., (2013). *In vitro* efficacy of garlic extract to control fungal pathogens of wheat. *Journal of Medicinal Plants Research, 7*(24).

Pérez-Torres, I., Torres-Narváez, J., Pedraza-Chaverri, J., Rubio-Ruiz, M., Díaz-Díaz, E., Valle-Mondragón, L. D., Martínez-Memije, R., et al., (2016). Effect of the aged garlic extract on cardiovascular function in metabolic syndrome rats. *Molecules, 21, 1425.*

Petropoulos, S. A., Fernandes, Â., Ntatsi, G., Petrotos, K., Barros, L., & Ferreira, I. C. F. R., (2018). Nutritional value, chemical characterization and bulb morphology of Greek garlic landraces. *Molecules, 23*(2), 319.

Pino, O., Sánchez, Y., & Rojas, M. M., (2013). Plant secondary metabolites as an alternative in pest management. I: Background, research approaches and trends. *Rev. Protección Veg., 28*(2), 81–94.

Plata-Rueda, A., Martínez, L. C., Dos, S. M. H., Fernandes, F. L., Wilcken, C. F., Soares, M. A., Serrão, J. E., & Zanuncio, J. C., (2017). Insecticidal activity of garlic essential oil and their constituents against the mealworm beetle, *Tenebrio molitor* Linnaeus (Coleoptera: Tenebrionidae). *Sci. Rep., 7*, 46406.

Prowse, G. M., Galloway, T. S., & Foggo, A., (2006). Insecticidal activity of garlic juice in two dipteran pests. *Agricultural and Forest Entomology, 8*, 1–6.

Rahman, M. S., Houd, I. A. S., Al-Riziqi, M. H., Mothershaw, A., Guizani, N., & Bengtsson, G., (2007). Assessment of the anti-microbial activity of dried garlic powders produced by different methods of drying. *International Journal of Food Properties, 9*(1), 503–513.

Reddy, K. R. N., Nurdijati, S. B., & Salleh, B., (2010). An overview of plant-derived products on control of mycotoxigenic fungi and mycotoxins. *Asian J. Plant Sci., 9*(3), 126–133.

Ryu, P., & Kang, D., (2017). Physicochemical properties, biological activity, health benefits, and general limitations of aged black garlic: A review. *Molecules, 22*, 919.

Salim, A. B., (2011). Effect of some plant extracts on fungal and aflatoxin production. *Int. J. Acad. Res. 3*(4), 116–120.

Sallam, M. A., Nashwa, K., & Abo-Elyousr, K. A. M., (2012). Evaluation of various plant extracts against the early blight disease of tomato plants under greenhouse and field conditions. *Plant Protect. Sci., 48*(2), 74–79.

Singh, U. P., Prithiviraj, B., Sarma, B. K., Mandadavi, S., & Ray, A. B., (2001). Role of garlic (*Allium sativum* L.) in human and plant diseases. *Indian Journal of Experimental Biology, 39*, 310–322.

Tafadzwa, M., Svotwa, E., & Katsaruware, R. D., (2016). Evaluating the efficacy of garlic (*Allium sativum*) as a bio-pesticide for controlling cotton aphid (*Aphis gossypii*). *Sci. Agri., 16*(2), 54–60.

Tagoe, D., Baidoo, S., Dadzie, I., Kangah, V., & Nyarko, H. A., (2011). Comparison of the antimicrobial (antifungal) properties of onion (*Allium cepa*), Ginger (*Zingiber officinale*) and garlic (*Allium sativum*) on *Aspergillus flavus*, *Aspergillus niger* and *Cladosporium herbarum* using organic and water based extraction methods. *Res. J. Med. Plant, 5*, 281–287.

Tan, B. G. T., Pham, V., & Nam, T. N., (2019). *Studies in Garlic*. Intech Open, London.

Vijayalakshmi, K., Subhashini, B., & Koul, S., (1999). *Plants in Pest Control: Garlic and Onion* (p. 15). Center for Indian Knowledge Systems. Chennai, India.

Yang, F. L., Zhu, F., & Lei, C. L., (2012). Insecticidal activities of garlic substances against adults of grain moth, *Sitotroga cerealella* (Lepidoptera: Gelechiidae). *Insect Science, 19*, 205–212.

Zhang, L., Dou, X. W., Zhang, C., Logrieco, A. F., & Yang, M. H., (2018). A review of current methods for analysis of mycotoxins in herbal medicines. *Toxins, 10*(65), 1–39.

Ziegler, S., Pries, V., Hedberg, C., & Waldmann, H., (2013). Target identification for small bioactive molecules: Finding the needle in the haystack. *Angew. Chem. Int. Ed. Engl., 52*, 2744–2792.

CHAPTER 8

EFFICACY OF ORGANIC SUBSTRATES FOR MANAGEMENT OF SOIL-BORNE PLANT PATHOGENS

MALAVIKA RAM AMANTHRA KELOTH, MEENAKSHI RANA, and AJAY TOMER

Department of Plant Pathology, School of Agriculture, Lovely Professional University, Phagwara–144411, Punjab, India

ABSTRACT

Organic matter is obtained from living organisms which are returned to the soil, providing a large source of carbon-based compounds. It can improve the microbial diversity of the soil, nutrient availability to plants, and maintain soil texture and structure. Organic matters are delivered to the soil in various forms like cattle manure, swine manure, or poultry manure (PM) in the solid or liquid phase, S-H mixture, seed cakes, slurry, composts, and many others which help in disease suppression and reduction in the phytopathogenic population. They adopt various mechanisms in controlling plant diseases like competition, hyper-parasitism, ineffective pathogen proliferation, induced resistance, and also through physicochemical properties of matter. They enable plant protection through their rich microbial community, regarded as bio-control agents that can ensure the elimination of the pathogen without harming other living organisms and the ecosystem. Amendment of soil using organic matter along with other agricultural practices is also giving good results in plant disease management.

8.1 INTRODUCTION

Since a few years back, the farming community has been found to rely on synthetic pesticides as it proved to give them the most immediate and successful results in controlling agricultural pests. But nobody was cautious regarding the after-effects of the unlimited use of these commercial pesticides that has resulted in polluting soil as well as the natural resources. The practice of using animal manures and crop rotations were forsaken after World War II. In replacement to this, farmers practiced monoculturing, intensive tillage, use of synthetic pesticides and fertilizers, shortened rotations, etc. (Katan, 1996). Before this scenario, our ancestors believed in implementing cultural practices and using organic matter, which might not have given them immediate results and required more labor. Eventually, they manipulated the soil ecology consciously and unconsciously (Bailey and Lazarovits, 2003). The realization of the advantages of organic matter and further studies related to it are advanced in the recent times.

Organic matter is capable of influencing the microbial ecology, nutrient availability, drainage, water holding capacity, soil aeration, and soil structure (Davey, 1996). It directly affects the crop productivity and plant health and provides carbon to the beneficial organisms in the soil as an energy source. It also helps in minimizing the infections caused by soilborne pathogens and improves root health. The use of organic matter maintains resident microbial populations and makes the soil immune to various plant diseases (Cook, 1990). It can also replace crop rotation in some instances, sanitize the soil and nourish the farming system by influencing the biological, physical, and chemical processes (Mathre et al., 1999). According to a recent report United States generates about 1 billion inorganic and organic agricultural by-products like crop residue (400 million), cattle manure (50 million), municipal solid wastes (150 million), pulp, and paper industry by-products (10 million) and swine and poultry manure (PM) (30 million) every year (Edwards and Someshwar, 2000). Many research works have also represented organic matters to be an important entity competent against a broad category of plant pathogens and pests to reduce its incidence (Cook, 1986; Abawi and Widmer, 2000; Akhtar and Malik, 2000; Conn and Lazarovits, 2000; Gamliel et al., 2000; Lazarovits et al., 2001).

8.2 MECHANISM OF DIFFERENT TYPES OF ORGANIC AMENDMENTS AND THEIR EFFICIENCY

High nitrogen amendments are one such organic matter used which includes soymeal, blood meal, bone meal, fishmeal, meat meal, and PM (Figure 8.1). Blood meal and fishmeal in pot trials are meant to have efficiency in reducing the pathogen causing Verticillium wilt of tomato (Wilhelm, 1951).

PMs, soymeal, bone meal and meat meal in field conditions at 37 tons/ha, are capable of reducing the incidence of pathogens like *Streptomyces scabies* causing common scab of potato, *Verticillium dahliae* or *Verticillium albo-atrum* causing Verticillium wilt and few nematodes parasitizing plants (Conn and Lazarovits, 1999; Lazarovits et al., 1999). It also inhibited the germination of microsclerotia of *Verticillium dahliae* harboring the sandy soil in lab conditions (Hawke and Lazarovits, 1994) and field conditions of potato crop (Conn and Lazarovits, 1999). Few reports indicated that the harm caused to the phytopathogens was due to the accumulation of ammonia after the application of high nitrogen-based amendments which has a low C:N ratio that is less than 10 (Gilpatrick, 1969; Huber and Watson, 1970; Mian and Rodriquez-Kabana, 1982; Stirling, 1991; Shiau et al., 1999). Thus, microsclerotia of *Verticillium dahliae* was killed due to the liberation of ammonia (Tenuta, 2001). The amount of application of organic amendments differs according to the type of soil. The amount of meat and bone meal required for loamy soil is twice the amount required for sandy loam soil to generate equal amounts of ammonia which shows that sandy loam soil can readily accumulate ammonia. It was observed that the various amendments used to mitigate pathogens or pathogen inoculum like microsclerotia in sandy soils, has urea as a major component which reacts with urease enzyme in soil to liberate ammonia (Rodriguez-Kabana et al., 1989; Stirling, 1991; Tenuta and Lazarovits, 2002). The efficacy of these amendments was confirmed by removing urea from them and observing that the latter mixture did not give supporting results (Huang and Huang, 1993). Soymeal and PM along with heated soil ensured an effective control over the plant diseases, even the heat-tolerant pathogens, as increasing temperatures declines the equilibrium point of ammonia or ammonium (Gamliel et al., 2000). Studies reported that nitrification products like nitrate and nitrite can also be the cause for the death of microsclerotia in Verticillium wilt disease. The fact is that the amendments applied at low rates became effective only after 2 weeks, along with the change in the soil pH to acidic (Tenuta and Lazarovits, 2002). At this pH ammonia cannot be present but the formation of nitrous acid (HNO_2) from nitrite can result in the depletion of the pathogen (Tsao and Oster, 1981; Loffler et al., 1986; Michel and Mew, 1998). Mineral soils comprising of less than 0.5% bone and meat meal, rapid nitrification rate and acidic soil tends to be the ideal condition for the presence of nitrous acid (Bailey and Lazarovits, 2003).

Organic amendments like crab shell, soybean stalk, PM, peanut cake, wheat straw, rice chaff, and alfalfa were applied under pot and field conditions in order to examine their efficacy in inhibiting the incidence of *Verticillium dahliae* Kleb. causing Verticillium wilt in cotton crops (Huang et al., 2006). The application of the amendments under both pot conditions in which pathogen was inoculated and field condition in which pathogen-infected naturally, positive results of disease suppression were observed. The most efficient treatment out of these were that of crab shell (72%), followed by soybean stalk (60%) and alfalfa (56%) in inhibiting the growth of the fungi in vascular tissues. The moderate effect was observed for the treatment with rice chaff while less efficiency was depicted by peanut cake, PM, and wheat straw treatment with 28%, 21% and 11% growth inhibition of the pathogen, respectively in vascular tissues. The organic amendment also contributed in increasing the population of beneficial microorganisms like actinomycetes, fungi, and bacteria in the rhizosphere of the cotton crops. The microbial population size varied according to the age of the crop. The crab shell treatments stimulated the maximum proliferation of antagonists, while the contribution of wheat straw, peanut cake, and PM in increasing the percentage of antagonists were comparatively less. The latter made only minute changes in the population of soil microbiota. Thus, the addition of organic amendment to the soil can cause an increase in the total amount of beneficial microbes and in inhibiting various soilborne phytopathogens.

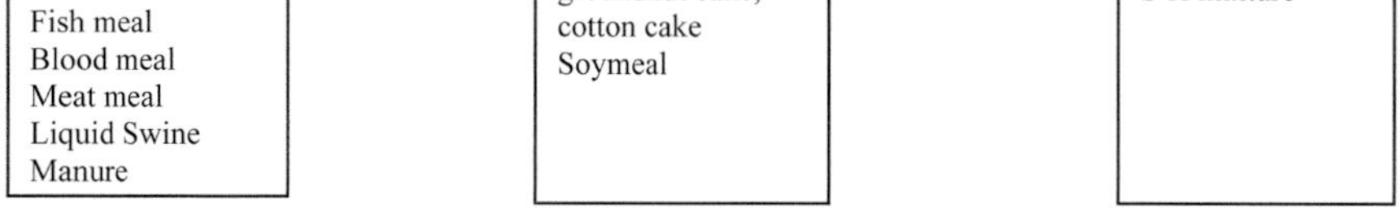

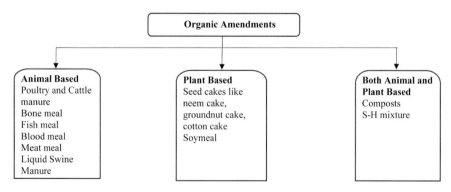

FIGURE 8.1 Different types of organic matter.

8.2.1 PLANT SEED-BASED CAKES

Efficacy of several plant seed-based cakes alone and in combination with biocontrol agents were observed in controlling many plant pathogens. Inhibition of growth of phytopathogens of sunflower crop by the application or amendment of soil using neem seed cake, cottonseed cake, *Datura fastuosa* and *Steochospermum marginatum* and each of them in combination with *Pseudomonas aeruginosa* were examined (Syed et al., 1998). All four of the above plant-based treatments successfully suppressed the population of *Fusarium solani*, cottonseed cake, and neem seed cake suppressed the growth of *Rhizoctonia solani* and *Macrophomina phaseolina* while *Datura fastuosa* and *Steochospermum marginatum* suppressed the population of *Rhizoctonia solani*. The beneficial microbe, *Pseudomonas aeruginosa* was capable of reducing the incidence of *Fusarium solani*, *Rhizoctonia solani* and *Macrophomina phaseolina*. Also, its combination with the plant-based treatments gave significant results in which, increased plant height and fresh weight of shoot was observed in treatment combining *Pseudomonas aeruginosa* with neem seed cake and *Steochospermum marginatum*.

8.2.2 SLURRY

Anaerobically digested slurry (ADS) when applied to the soil delivered a significant reduction in the mycelial growth of *Phytophthora capsici,* causing Phytophthora root rot disease in chili pepper (Cao et al., 2014). It also restricted zoospore germination of the pathogen. The decrease in disease incidence and zoospore germination was observed more with anaerobically digested pig slurry (ADP) treatment than with anaerobically digested dairy (ADD) slurry treatment. Isolation of humic substances (HS) and exogenously applied ammonium from ADS was performed, in which a higher percentage of inhibition of hyphal growth and zoospore germination was observed for HS derived from ADP than that from ADD. High ammonium content and specific structure of HS extracted from ADP was the root cause for its ability to carry out better control of the disease. Further, the efficiency of pig slurry in eliminating *Ralstonia solanacearum* biovar 2 strain 1609 infecting potato crops was studied, which showed a decrease in the population of infected and diseased crops according to the soil suppressiveness tests (Gorissen et al., 2004).

8.2.3 LIQUID SWINE MANURE (LSM)

Liquid swine manure (LSM) amendment reduced the common scab, wilt diseases, and plant-parasitic nematodes in potato field for 3 years after its first application at 55 $hLha^{-1}$ concentrations apart from the increasing occurrence of common scab disease due to application of animal manures

(Conn and Lazarovits, 1999). The LSM, which is incorporated into the soil, destroyed the microsclerotia of the Verticillium wilt causing pathogen, *Verticillium dahliae* directly at 5 pH whereas at a higher pH of 6.5 their activity was not appreciated. From this we can incur that the effectiveness of amendment is depended upon the soil pH. Similarly, soil temperature, soil moisture and volatile fatty acid (VFA) mixtures are also responsible for the activity of the organic amendment against phytopathogens such as enhancement in toxicity against common scab disease due to rise in soil temperature, reduction in toxicity due to increase in soil moisture causing dilution of the active component and immediate fatality of the microsclerotia due to VFA mixtures. The functioning of microsclerotia was ceased in the presence of all the VFA compounds like acetic acid, valeric acid, butyric acid, isobutyric acid, and propionic acid, whereas it failed with one or few of the VFA compounds (Tenuta et al., 2002). VFAs produced by LSM are of no use at a higher pH above 8 while ammonia takes place the role of causing detrimental effects to the pathogen at this pH. VFA (pKa 4.75) and nitrous acid play the essential role for the antifungal activity at low pH, showing the necessity of acidification of the organic amendments before applying in the soil (Tenuta et al., 2002). There was an increase in the amount of beneficial soil microbiota like *Trichoderma* when LSM acting as a stimulating agent for the microflora, was added to kill *Verticillium dahliae* (Conn and Lazarovits, 1999).

8.2.4 S-H MIXTURE

Positive results were obtained using S-H mixture, a formulated soil amendment, which included many industrial and agricultural wastes. The S-H mixture under greenhouse conditions was capable of controlling *Pseudomonas solanacearum*, which caused bacterial wilt disease in tomato, when applied at 0.5–1.0% (w/w) with the soil, one week prior to transplanting (Chang and Hsu, 1988). Further efficacy of the mixture by removing bagasse, rice husk powder, calcium superphosphate, potassium nitrate and oyster shell powder was evaluated, which depicted no reduction in the efficiency of S-H mixture in the soil, whereas the mixture devoid of urea or mineral ash gave less efficiency. And by applying the components of the mixture individually to the soil at the rate equivalent to the amount of each component present in 0.75% (w/w) of S-H mixture, urea alone was found to cause a reduction in the disease severity but was less efficient than the mixture comprising all of them. Moreover, disease suppression was par equivalent with the efficiency of S-H mixture, when urea was combined with mineral ash to amend the soil but not with each of any other constituents of the S-H mixture. The efficacy of urea amended soil was enhanced by increasing the period of treatment before planting the tomato seedlings, giving best results by planting after 2 weeks of amendment of the soil with urea.

Another bacteria infecting tomato crop was also inhibited by the S-H mixture in combination with plant growth-promoting rhizobacteria (PGPR). *Ralstonia solanacearum* (race 1, biovar 1) causing bacterial wilt disease in tomatoes were suppressed using soil amended with S-H mixture, acibenzolar-S-methyl (Actigard) and PGPR; *Bacillus pumilus* SE 34, *Pseudomonas putida* 89B61, BioYield, and Equity (Anith et al., 2004). The S-H mixture constituted of those components discussed before, i.e., urea, mineral ash, bagasse, potassium nitrate, rice husk, calcium superphosphate and oyster shell powder. Bio-Yield (Gustafson, LLC, Dallas, TX) and Equity (Naturize Inc., Jacksonville, FL) are the formulations in which Bio-Yield includes 2 strains of *Bacillus* spp.; *B. amyloliquefaciens* IN937a and *B. subtilis* GB03 and Equity includes 40 different microbial strains. S-H mixture alone was capable to lessen the disease incidence when applied 14 days before the transplantation of the tomato seedlings in the bacteria-infested soil. On application of *Pseudomonas putida* 89B61, the PGPR to the transplants at the seeding time and one week before the inoculation of the bacteria, prevented the disease incidence. The formulated PGPRs viz. Equity was not capable of inhibiting the bacteria and BioYield was efficient in reducing the disease incidence in three experiments. Actigard treatment did not exhibit much result in inhibiting the bacteria at inoculum densities of 1×10^5 and 1×10^6 CFU/ml. The amendment of soil with Actigard and S-H mixture efficiently inhibited the incidence of *Ralstonia solanacearum* in the tomato crops. The combination of PGPR; *P. putida* 89B61 or BioYield with Actigard significantly reduced the disease incidence compared to the control.

Mediterranean countries are leading in olive oil production. But the olive oil mill wastewater (Kotsou et al., 2004) and dry olive residues (DOR) (Sampedro et al., 2004) have caused a threat to the environment due to its high organic load, phytotoxic, and antimicrobial properties. Studies show that these waste particles render huge amount of fertilizers when applied to the soil and can play the role of organic amendment with high nutrients and organic matter (Paredes et al., 1999). Despite of its agronomic value, DOR is also harboring antimicrobial properties (Capasso et al., 1995). The effects of DOR were experimented on the growth of four crop species, on the saprophytic growth of four fungi causing plant disease and the impact of soil amendment with the DOR on five systems of plant and pathogen, in which three systems were plant with soil enduring pathogen while two were plant with foliar pathogen (Bonanomi et al., 2006). The four crop species on which the DOR effect tested were *Triticum aestivum*, *Lepidium sativum*, *Lactuca sativa* and *Lycopersicon esculentum*. The four fungi whose saprophytic growth was examined were *Botrytis cinerea* (BC), *Fusarium culmorum* (FC), *Fusarium oxysporum* f. sp. *lycopersici* (FOL) and *Sclerotinia minor* (SM), the plant-soil borne pathogen systems were *T. aestivum*-FC, *L. esculentum*-FOL and *L. sativa*-SM and plant-foliar pathogen systems were *L. sativa*-BC and *L. esculentum*-BC. Phytotoxic effects were shown by the residues on all plant species tested both under laboratory and greenhouse conditions in which the most sensitive ones were *L. sativum* and *L. sativa* which were followed by *L. esculentum* and *T. aestivum*. Favorable results were received in the tests against phytopathogenic fungi. Radial growth and hyphae density of the fungi were affected positively by the residues. In the plant pathogen systems under greenhouse conditions, no affirmative effects were observed for DOR amended soil. Thus, the report concludes that undecomposed DOR under controlled conditions cause phytotoxic effect on crop species, antimicrobial effect on plant pathogenic fungi and increased disease incidence with plant-fungus interaction systems.

8.2.5 COMPOSTS

The application of composts in the greenhouse has resulted in diminishing the soilborne diseases occurring in the plants (Hadar et al., 1992; Hoitink et al., 1993; Hoitink and Boehm, 1999). In the field conditions, these composts are also equipped with biocontrol agents that have a vital role in eliminating phytopathogens through parasitism (Keener et al., 2000). The population density of certain soilborne pathogens like *Pythium* and *Rhizoctonia* causing damping of disease and SM causing rotting in lettuce crops were declined by the application of composts due to elevated soil microbial activity (Lumsden et al., 1983, 1986). Due to the variability in the results of the composted products, the recommendation of these to the farmers remains as a major challenge. The outcome varies on the basis of different factors like season of application and the number of times they are applied. According to a report out of the 17 organic compounds derived from the household wastes, only 9 were mildly effective while the others had no role in decreasing the *Pythium* infection (Erhart et al., 1999). The efficacy of bark composts in suppressing disease-causing pathogens was apprehensive, but its activity was unable to relate to the presence of phenol in its extracts and the total biological activity.

Application of composts derived from municipal biosolids, sugar mill filter press cake, cottonwood mixed with pine bark or hardwood and cotton gin trash were found to reduce *Pythium arrhenomanes* causing stubble decline in sugarcane (Dissanayake and Hoy, 1999). Fresh compost was equipped with a high concentration of chemical components of VFA like propionic acid, isovaleric acid, acetic acid, butyric acid and isobutyric acid, which can contribute to reduce disease incidence (DeVleeschauwer and Assche, 1981). Few of these components like butyric and propionic acid was found inhibitory to seed germination and growth of the plant. The production of acetic acid was a major cause of phytotoxicity following anaerobic decomposition in wheat straw (Lynch, 1977, 1978). Such anaerobic conditions paved the way in reducing the inoculum of fungal pathogens like *Rhizoctonia solani*, *Verticillium dahliae* and *Fusarium oxysporum* f. sp. *asparagi* (Blok et al., 2000). Some of the effectiveness of the composts as fungicides is due to the biological control agents harboring them. *Pseudomonas fluorescens* a biocontrol agent produces antibiotic 2,4-diacetylphloroglucinol (DAPG) that can reduce the incidence of take-all disease of wheat (Raaijamakers and Weller, 1988). Application of organic amendment cannot be definitely linked with the biological control agent in soil. Application of organic amendments due to which the hydrolysis of fluorescein diacetate resulting high microbial activity has caused a decrease in fungal pathogen *Pythium* causing damping of diseases (Keener et al., 2000). This relation between organic matter and soil microbial activity has managed to reduce the incidence of Corky root rot disease, *Phytophthora parasitica* causing root infections and *Verticillium* wilt disease (van Bruggen, 1995; Davis et al., 1996, 2001). Cellulolytic and hemicellulolytic actinomycetes are the other contributors in developing antifungal activity in the organically amended soils (van Bruggen, 1995).

A study has been carried out by taking into consideration the efficacy of 101 microorganisms from compost prepared from urban organic and yard wastes, from which 28 microbes appeared to be antagonists against *Fusarium oxysporum* f. sp. *radicis-lycopersici* causing tomato wilt (Pugliese et al., 2008). The test was carried out under laboratory conditions in petri plates constituting perlite medium where tomato seedlings were grown. Further under greenhouse, the compost was tested against three pathogens infecting three different crops, viz. *Phytophthora nicotianae* on tomato, *Fusarium oxysporum* f. sp. *basilici* on basil and *Rhizoctonia solani* on bean. A *Fusarium* strain K5 was capable of controlling the fungi infecting basil, with 69% inhibition and 32% increase in the biomass production of the crop compared to the control. A bacterial strain B17 controlled the fungi *Phytophthora nicotianae* infecting tomato crop by 82% with 216% biomass production. *Rhizoctonia solani* causing stem rot and root rot in bean crop was controlled by E19 and P11 antagonists with 163% increase in the biomass production. Moreover, among these 28 microbes, best control at various dosages was imparted by two *Trichoderma* isolates; E28 and E36 and three *Fusarium* isolates; K7, K9 and K11, from which two isolates of *Fusarium*, i.e., K7 and K9 were regarded as commercial bio-control agents with its antagonistic activity against *F. oxysporum* f. sp. *basilica*. And no microbe could control all the soilborne phytopathogens. Composted grape marc and composted separated cattle manure can be regarded as a biological control against *Rhizoctonia solani* and *Pythium aphanidermatum* and were tested under laboratory conditions by amending the media with composts (Hadar and Mandelbaum, 1992). Microbial activity during the process of composting is responsible for the action of disease suppression. A unique experiment was conducted using heavily infested crop residues which were incorporated into 2.5–4.6 m^3 area of static compost heaps subjected to various temperature ranges (Bollen et al., 1989). On the basis of the type of crop residues, its location in the heap as well as the temperature to which it is subjected to, is changed. The temperature was brought up to 50–70°C in 6 days, which remained so for 2–3 weeks and later reduced below 40°C for 5 months using a.c. at maturation phase. Almost all the pathogens were eliminated after the test, while *Olpidium brassicae* and one formae speciales out of the four *Fusarium oxysporum* pathogen survived but were also less in inoculum. According to the study conducted, the elimination of phyto-pathogens from the crop residues was due to the evolution of heat during the process of composting. A short heat phase, *Fusarium oxysporum* f. sp. *melonis*, *Plasmodiophora brassicae* and *Olpidium brassicae* were successfully eliminated, while during the long-lasting maturation phase, they survived without losing their viability.

In order to eliminate the soil enduring pathogen, *Fusarium oxysporum* causing wilt in tomato, two residues from industrial and agricultural wastes were composted and evaluated (Borrero et al., 2004). Cork compost (CC) from cork (*Quercus suber* L.) production and grape marc compost (GMC) from alcohol producing industry (grape skins, seed, and stems) were compared with Sphagnum light peat (P) (Klasmann, Valinex, Palleter, Spain) and expanded vermiculite (V) (Vermiculitay derivadas, Gijón) and CC and GMC were used in its heated (60°C) form as CC-60 and GMC-60. Precisely 6.26 to 7.97 were the pH range maintained in the growth media. All the treatments were carried out in the growth media, and their effects were tabulated under AUDPC (area under disease progress curve) and RLSBX (relative length of stem with brown xylem). On the basis of AUDPC and RLSBX, the most effective treatment was of GMC, which showed 92% reduction compared to vermiculate (V) or peat (P) treatments. P and V were the most conducive ones while CC, CC-60 and GMC-60 gave intermediate results. CC and GMC had increased β-glucosidase activity and with this variation in the disease severity.

Beneficial microorganisms like *Bacillus amyloliquefaciens* strains QL-5 and QL-18 were combined with organic fertilizers to inhibit the growth of bacterial wilt pathogen, *Ralstonia solanacearum* in tomato crops under greenhouse conditions and field (Wei et al., 2011). The two strains of *Bacillus* sp. were isolated from the rhizosphere soil of the tomato crop. The most effective results were obtained when both the strains QL-5 and QL-18 were mixed with the fertilizer to obtain fortified bio-organic fertilizer (BOF). The field results however relied on season and climatic conditions. Less effect of the BOF was noticed in autumn crop season in 2008 and 2009 while the disease incidence was successfully reduced in the spring crop season in 2009 and 2010 significantly. The pathogen population in soil before transplantation of the tomato seedlings were found to be three times lower in the spring than autumn season. Significant and favorable results were obtained under field conditions when air temperature was correlated with the treatment while no correlations were observed with the relative humidity.

Green wastes are used as composts in order to cause mycelial inhibition (100%) of FC causing foot rot, *Gaeumannomyces graminis* f. sp. *tritici* causing take all disease and *Pseudocercosporella herpotrichoides* causing eyespot in winter wheat, inhibition (100%) of germination of pycniospores of *Phoma medicaginis* f. sp. *pinodella* causing black stem of garden pea, inhibition (100%) of spores of *Plasmodiophora brassicae* causing clubroot of Chinese cabbage (Tilston et al., 2002). Also, green wastes are meant to protect agricultural crops by inducing mycelial suppression (20%) of *Rhizoctonia solani* causing damping-off in radish and sugarbeet, *Pythium ultimum* causing damping-off in cucumber (Ryckeboer, 2001), *Phytophthora cinnamomi* causing root rot in lupin (Tuitert and Bollen, 1996), inhibiting mycelium by 33% of *Pythium ultimum* causing damping-off in cress (Fuchs, 2002), inhibiting conidia by 10% of FOL causing wilt of tomato (Cotxarrera et al., 2002), inhibiting spore production by 30% of *Mycosphaerella pinodes* causing foot rot of pea (Schuler et al., 1993) and inhibiting zoospore production by 100% of *Phytophthora fragariae* f. sp. *fragariae* causing red core in strawberry (Pitt et al., 1997).

8.3 MODE OF ACTION OF COMPOSTS ON PHYTOPATHOGENS

Composts comprising of various bio-control agents as discussed before conquer the pathogenic community through various mechanisms (Figure 8.2). Competition, physicochemical properties of the composts, antibiosis, ineffective pathogen proliferation, systemic acquired resistance (SAR), induced systemic resistance (ISR) and hyper-parasitism are the major factors to be considered (Mehta et al., 2014). These mechanisms are discovered, studied, and described by different researchers in order to control plant diseases and to maintain balance in the diversity of soil microbiota.

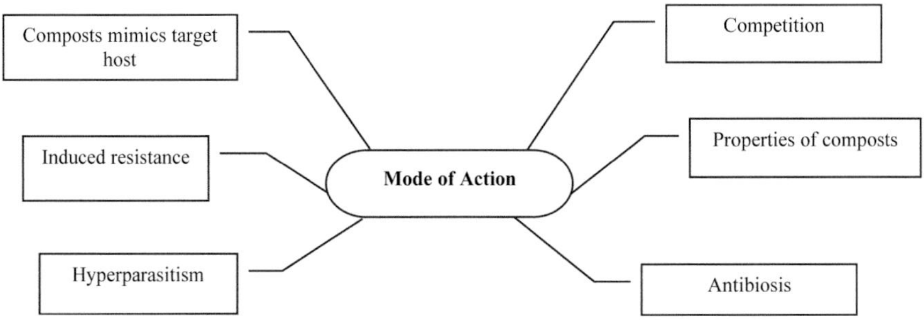

FIGURE 8.2 Mode of action of the organic matter.

8.3.1 COMPETITION OF THE PATHOGEN WITH THE BIO-CONTROL AGENT

The beneficial microbes become a competitor for the pathogens for nutrients (Chen et al., 1988) and space (Serra-Wittling et al., 1996). Lack of nutrients, infection sites, and efficiency in colonizing roots can lead to disease suppression, especially in the case of pathogens like *Pythium* and *Phytophthora,* which are nutrient-dependent (Diánez et al., 2004). As mentioned previously, many pathogens infecting tomato crops like *Rhizoctonia solani*, *Fusarium oxysporum* f. sp. *radicis-lycopersici*, *Pythium ultimum*, *Pyrenochaeta lycopersici* were eliminated in the soil amended by composts. The fact that composts are able to control these pathogens is due to the percentage increase in the siderophore production by the bio-control agents (de Brito-Alvarez et al., 1995). Siderophore are iron-chelating agents whose presence depletes the amount of iron required for the germination of the pathogen as well as its penetration in the host. *Pseudomonas fluorescens* is one such bacteria which is capable of producing siderophore inhibit the germination of chlamydospore, the resting structure of *Fusarium* spp. (Elad and Baker, 1985). According to a recent report, *Pseudomonas* spp. also help in the growth of the plant without causing any direct effects and remain as a competitor for the phytopathogens (Kyselková and Moënne-Loccoz, 2012).

8.3.2 PROPERTIES OF COMPOSTS

Physicochemical properties of the compost are also contributors to carry out protection of the crops from various disease-causing pathogens. Composts are rich in nutrients and organic compounds like phenolic, humic, or bioactive compounds (Hoitink et al., 1997; Siddiqui et al., 2008; Spatafora and Tringali, 2012) which are responsible for providing ISR, toxicity to pathogens and advanced nutritional status for self-defense. Mass spectrometry (MS) helps in determining the presence of antioxidant and antimicrobial compounds along with their chemical composition (Shahat et al., 2011), and the presence of volatile organic compounds (Font et al., 2011).

8.3.3 ANTIBIOSIS TO INHIBIT PATHOGEN GROWTH

Antibiosis is a process of production of specific or non-specific metabolites, volatile compounds, lytic agents, toxic substances, and enzymes by the bio-control agent to inhibit the development of pathogenic microbes (Jackson, 1965; Fravel, 1988). *Pseudomonas* and *Bacillus* are the two important microbe genus that carry out antibiosis by producing antibiotics to destroy the phytopathogens. *Pseudomonas* spp. cause disease control of potato scab, take-all disease in wheat, apple replant disease, Fusarium wilts (Weller et al., 2002) and also many pathogens infecting roots (Haas and Défago, 2005), while *Bacillus cereus* UW85 control diseases caused by oomycetes fungi like *Phytophthora*, *Pythium*, etc., by producing zwittermicin A and kanosamine (Silo-Suh et al., 1994; Milner et al., 1996). Two other bio-control agents from the composts are *Trichoderma* and *Gliocladium* (Howell et al., 1993) and another recently reported *Zygosporium masonii* which suppressed the anthracnose disease-causing *Colletotrichum capsici* in bell pepper (Ajith and Lakshmidevi, 2012).

8.3.4 COMPOSTS USED TO MIMIC TARGET HOST

At some instances, pathogens are deceived to proliferate prior to the appearance of the target host by the chemicals signals developed from the shoot or root exudates (Chen et al., 1988) of the composts. Usually, most of the pathogens are not capable of multiplication without any host other than the saprophytes, which even requires a dead medium to survive (Lockwood, 1990). Composts applied as seed cover or plug substitutes in tomato crops gave positive results in suppressing the Fusarium root rot and crown rot pathogens (Cheuk et al., 2005). Thus, compost act as a stimulator that trigger the pathogen to germinate and proliferate before the growth of the host crop due to which crops can escape from the soilborne diseases (Cheuk et al., 2005) and cause inactivation of the pathogens (Yogev et al., 2006).

8.3.5 INDUCED RESISTANCE BY BIO-AGENTS

Induced resistance is of many forms in which SAR and ISR are already developed in the plants before the infection of the pathogen (Vallad and Goodman, 2004). The plants having this induced resistance are stated to possess improved defensive capacity on stimulation by the presence of a pathogen or PGPR or any other bio-control agent (Bakker et al., 2003). Generally, SAR takes place in the presence of beneficial microbe, pathogens, or chemicals (Maurhofer et al., 1994; Pieterse et al., 1996; De Meyer and Höfte, 1997). Substrates amended by the composts possess an immense variety of bioagents that can induce systemic resistance in crops (Wei et al., 1991; Liu et al., 1995). Many plants growth-promoting bacterial (PGPB) and fungal isolates are found capable to induce ISR in crops (van Loon et al., 1998).

8.3.6 HYPER-PARASITISM CAUSING MYCELIAL LYSIS

Hyper-parasitism is a process through which a bio-control agent directly parasitizes on the pathogen and suffocates them to death (Heydari and Pessarakli, 2010). Generally, the agents imparting pathogenic control through hyper-parasitism are mainly classified into four categories, i.e., predators, obligate bacterial pathogens, facultative parasites and hypoviruses (Mehta et al., 2014). *Trichoderma* spp. are one of the most important microbes using this mechanism to kill pathogens in which, *Trichoderma harzianum* is commonly used to control the mycelial growth of *Rhizoctonia solani*

(Chet and Baker, 1980). Favorably, this bio-control agent often occurs in composts, as reported by the experiments conducted to suppress *R. solani* (Kuter et al., 1983). Many soilborne pathogens like *Verticillium*, *Pythium*, *Rhizoctonia*, *Sclerotium*, *Phytophthora*, and *Sclerotinia* are hyper parasitized by the bio-control agents through parasitization of their hyphae, mycelium, sclerotia or oospores (Diánez et al., 2005). In some cases, multiple hyper-parasitism also takes place to give a complete control of the pathogens. One such example is the control of powdery mildew pathogen by *Ampelomyces quisqualis*, *Acremonium alternatum*, *Cladosporium oxysporum*, *Gliocladium virens* and *Acrodontium crateriform* (Kiss, 2003). Few actinomycetes and fungi like *Verticillium chlamydosporium*, *Acremonium* spp. and *Humicola fuscoatra* are found capable of parasitizing oospores of *Phytophthora apsici* which infects crops of Solanaceae, Cucurbitaceae, and Fabaceae families (Sutherland and Papavizas, 2008).

8.4 ENHANCED EFFECT OF ORGANIC AMENDMENTS IN COMBINATION WITH OTHER AGRICULTURAL PRACTICES

The organic amendments have shown their efficacy in preventing disease incidence through various physical, chemical, and biological activities, but their outcome is slower compared to that of biofumigants and fungicides. Apart from this, these organic matters have a very advantageous factor that they give cumulative effects and are long-lasting. Merging of organic amendments with various agricultural practices like soil solarization, soil flooding, etc., is also being significantly used in the existing farming community. Organic matter along with tarping is another treatment conducted to check the increase or decrease in the anti-fungal activity of them (Blok et al., 2000). There was a significant inactivation in the inoculums of *Rhizoctonia solani*, *Verticillium dahliae* and *Fusarium oxysporum* f. sp. *asparagi* in the plots amended with organic matter and covered with plastic sheets (Blok et al., 2000). Here the amended materials were broccoli and grass, which contributed in inactivating the survival structures of the pathogen along with the rise in soil temperature due to the plastic sheet covering. The covered plots also facilitated anaerobic conditions for the soil microflora to activate their efficiency in parasitizing phytopathogens.

However, the temperature increases in the soil in covered plots treatment amended with plant material, have well contributed to pathogen inactivation by devitalizing the structures of pathogen meant for perpetuation. It was also noticed that the pathogen propagules were devitalized when many of the phytopathogens were exposed to sub-lethal temperatures (DeVay and Katan, 1991). The covered plots developed anaerobic conditions readily, regardless of incorporation of organic amendment. Diffusion of oxygen from adjacent no covered soil, several oxidative processes and consumption of these by soil microbiota contributed in increasing the concentration of the latter in soil atmosphere. The glucosinolates present in the cruciferous crops are the volatile product that is responsible for the inactivation of the survival structures of the fungus (Lewis and Papavizas, 1971; Muehlchen et al., 1990; Kirkegaard et al., 1996; Smolinska et al., 1997; Subbarao et al., 1999). These volatiles can be trapped by the plastic sheets under the covered plots, thus contributing to the anti-fungal activity (Blok et al., 2000). The role of fermentation products has also reported in inactivating the pathogen propagules. There was a reduction in the concentration of propagules of *Fusarium oxysporum* f. sp. *conglutinans* (Mitchell and Alexander, 1962), *Pyrenochaet aterrestris* (Watson, 1964), *Verticillium dahliae* (Menzies, 1962; Watson, 1964) and *Fusarium* spp. (Watson, 1964) when easily metabolizable organic materials were incorporated into flooded soil. The accumulation of several compounds like carbon dioxide, methane, hydrogen, ethylene, alcohol, aldehydes, and organic acids temporarily was examined in reducing soil conditions and found that these products are having fungi toxic effect which helps to inactivate the phytopathogens (Ponnamperuma, 1972). Few antibiotic and toxin producers like *Bacillus* spp. and *Clostridium* spp. are the anaerobic microorganisms which can also contribute in pathogen inactivation.

For controlling soil enduring pathogens, researchers have adopted many other methods keeping in consideration, anaerobic soil disinfestation (ASD) as their basic strategy. This approach has been carried out in different soil types, regions, and crop species along with various inputs considering the target pathogen (Rosskopf et al., 2015). It includes agricultural practices like application of carbon source, irrigating the field according to its optimum level and providing plastic tarp covering for the soil (Strauss and Kluepfel, 2015). These practices can bring about changes in the total number of beneficial organisms residing in the soil and their composition, cause availability of volatile organic compounds and create anaerobic conditions that can bring in lethal effect. It causes disease suppression through biological control and oxygen depletion, suffocating the phytopathogenic microbes (Rosskopf et al., 2015). The carbon sources, plastic tarps, increased soil temperature and anaerobic conditions can create a favorable environment for controlling diseases and suppressing phytopathogens. Soil solarization, though not essential, is found to enhance the efficiency of ASD. With effect to environmental conditions, carbon sources like rice bran (dry formulation) and ethanol (liquid formulation) are found extensively used. Application of ASD with rice bran, gave successful results in controlling the incidence of *Macrophomina phaseolina* causing charcoal rot in California strawberry (Muramoto et al., 2016). The experiment was also carried out with mustard seed meal (MSM) alone and with MSM along with ASD-rice bran, in which the latter gave intermediate effect while MSM failed in suppressing the pathogen. Further, Brassica seed meal amended soil has also shown suppression of *Macrophomina phaseolina* in the orchard and also prevented the recurrence of soilborne pathogens (Mazzola et al., 2016). It also resisted the re-infestation of *Pratylenchus penetrans*, a plant-parasitic nematode. It is more convincing to use these ASD strategies in areas where soil solarization is not possible and fumigant buffer zones, facilitating the enrichment of organic matter in soil (Rosskopf et al., 2015). Despite of providing enriched soil microbial community and disease suppression, the ASD technique shows its effect even after the first cropping system, preventing the recurrence of soilborne pathogens and nematodes (Rosskopf et al., 2015).

The outcome of applying solarization process and pig slurry to the soil microorganisms and field plots of potato crop was examined to control the amount of inoculums of *Ralstonia solanacearum* biovar 2 strain 1609 (Gorissen et al., 2004). Initially, the untreated soil was analyzed based on the suppressiveness assays, which depicted a slow reduction in the inoculum of the pathogen in about 9 weeks duration from 10^6–10^7 to 10^3–10^4 CFU (g dry soil)$^{-1}$. According to immunofluorescence colony staining and polymerase chain reaction (PCR), wilting, and sign of the pathogen was recognized in the lower part of the stem of most of the potato crops grown in untreated soil. The application of pig slurry and solarization of the soil were carried out separately in which both the techniques brought about sufficient reduction in the amount of the pathogen inoculum, but solarization alone appeared to be incapable of improving the soil suppressiveness. But by combining both methods, enhanced results were obtained. The latent existence of the pathogen in the lower parts of the stem was confirmed in the healthy-looking crops which were subjected to pig slurry and solarization treatments. Moreover, pig slurry is efficient in changing the bacterial community structure in the treated soil ensuring disease suppressiveness.

8.5 CONCLUSION

In viewing the various environmental issues and consequences which are facing in this era, we must realize that the unlimited and unconditional utilization of the synthetic products can only contribute to the immediate results. As discussed in this chapter as well as in many previous reports, the organic matters and other eco-friendly agricultural practices have the efficacy in reducing diseases and pest incidence and maintaining soil health and growth of the plant for long term results though they show a slower rate of action. With further research and innovations, the exceptional ability of these organic materials will be revealed, and their use for management of soilborne diseases will expand. Thus, it is the responsibility of every living being that, they ensure the protection and conservation of nature and its products and make sustainable utilization with respect to the future generation.

KEYWORDS

- **bio-control**
- **composts**
- **hyper parasitism**
- **induced resistance**
- **manure**
- **organic matter**
- **phytopathogenic**
- **proliferation**
- **seed cakes**
- **s-h mixture**
- **slurry**
- **suppression**

REFERENCES

Abawi, G. S., & Widmer, T. L., (2000). Impact of soil health management practices on soilborne pathogens, nematodes and root diseases of vegetable crops. *Applied Soil Ecology, 15*(1), 37–47.

Ajith, P. S., & Lakshmidevi, N., (2012). *Zygosporium masonii*: A new fungal antagonist against *Colletotrichum capsici* incitant of anthracnose on bellpeppers. *Journal of Agricultural Technology, 8*(3), 931–939.

Akhtar, M., & Malik, A., (2000). Roles of organic soil amendments and soil organisms in the biological control of plant-parasitic nematodes: A review. *Bioresource Technology, 74*(1), 35–47.

Anith, K. N., Momol, M. T., Kloepper, J. W., Marois, J. J., Olson, S. M., & Jones, J. B., (2004). Efficacy of plant growth-promoting rhizobacteria, acibenzolar-S-methyl, and soil amendment for integrated management of bacterial wilt on tomato. *Plant Disease, 88*(6), 669–673.

Bailey, K. L., & Lazarovits, G., (2003). Suppressing soilborne diseases with residue management and organic amendments. *Soil and Tillage Research, 72*(2), 169–180.

Bakker, P. A. H. M., Ran, L. X., Pieterse, C. M. J., & Van, L. L. C., (2003). Understanding the involvement of rhizobacteria-mediated induction of systemic resistance in biocontrol of plant diseases. *Canadian Journal of Plant Pathology, 25*(1), 5–9.

Blok, W. J., Lamers, J. G., Termorshuizen, A. J., & Bollen, G. J., (2000). Control of soilborne plant pathogens by incorporating fresh organic amendments followed by tarping. *Phytopathology, 90*(3), 253–259.

Bollen, G. J., Volker, D., & Wijnen, A. P., (1989). Inactivation of soilborne plant pathogens during small-scale composting of crop residues. *Netherlands Journal of Plant Pathology, 95*(1), 19–30.

Bonanomi, G., Giorgi, V., Neri, D., & Scala, F., (2006). Olive mill residues affect saprophytic growth and disease incidence of foliar and soilborne plant fungal pathogens. *Agriculture, Ecosystems and Environment, 115*(1–4), 194–200.

Borrero, C., Trillas, M. I., Ordovás, J., Tello, J. C., & Avilés, M., (2004). Predictive factors for the suppression of *Fusarium* wilt of tomato in plant growth media. *Phytopathology, 94*(10), 1094–1101.

Cao, Y., Chang, Z., Wang, J., Ma, Y., Yang, H., & Fu, G., (2014). Potential use of anaerobically digested manure slurry to suppress *Phytophthora* root rot of chilli pepper. *Scientia Horticulturae, 168*, 124–131.

Capasso, R., Evidente, A., Schivo, L., Orru, G. E. R. M. A. N. O., Marcialis, M. A., & Cristinzio, G., (1995). Antibacterial polyphenols from olive oil mill wastewaters. *Journal of Applied Bacteriology, 79*(4), 393–398.

Chang, M. L., & Hsu, S. T., (1988). Suppression of bacterial wilt of tomato by soil amendments. *Plant Prot. Bull. 30*, 349–359.

Chen, W., Hoitink, H. A. J., & Madden, L. V., (1988). Microbial activity and biomass in container media for predicting suppressiveness to damping-off caused by *Pythium ultimum*. *Phytopathology, 78*(11), 1447–1450.

Chet, I., & Baker, R., (1980). Induction of suppressiveness to *Rhizoctonia solani* in soil. *Phytopathology, 70*(10), 994–998.

Cheuk, W., Lo, K. V., Copeman, R., Joliffe, P., & Fraser, B. S., (2005). Disease suppression on greenhouse tomatoes using plant waste compost. *Journal of Environmental Science and Health, Part B, 40*(3), 449–461.

Conn, K. L., & Lazarovits, G., (1999). Impact of animal manures on verticillium wilt, potato scab, and soil microbial populations. *Canadian Journal of Plant Pathology, 21*(1), 81–92.

Conn, K. L., & Lazarovits, G., (2000). Soil factors influencing the efficacy of liquid swine manure added to soil to kill *Verticillium dahliae*. *Canadian Journal of Plant Pathology, 22*(4), 400–406.

Cook, R. J., (1986). Plant health and the sustainability of agriculture, with special reference to disease control by beneficial microorganisms. *Biological Agriculture and Horticulture, 3*(2, 3), 211–232.

Cook, R. J., (1990). Twenty-five years of progress towards biological control. *Biological Control of Soilborne Plant Pathogens*, 1–14.

Cotxarrera, L., Trillas-Gay, M. I., Steinberg, C., & Alabouvette, C., (2002). Use of sewage sludge compost and *Trichoderma asperellum* isolates to suppress *Fusarium* wilt of tomato. *Soil Biology and Biochemistry, 34*(4), 467–476.

Davey, C. B., (1997). *Nursery Soil Management-Organic Amendments* (pp. 6–18). United States Department of Agriculture Forest Service General Technical Report, PNW.

Davis, J. R., Huisman, O. C., Everson, D. O., & Schneider, A. T., (2001). *Verticillium* wilt of potato: A model of key factors related to disease severity and tuber yield in southeastern Idaho. *American Journal of Potato Research, 78*(4), 291.

Davis, J. R., Huisman, O. C., Westermann, D. T., Hafez, S. L., Everson, D. O., Sorensen, L. H., & Schneider, A. T., (1996). Effects of green manures on *Verticillium* wilt of potato. *Phytopathology, 86*(5), 444–453.

De Brito, A. M., Gagne, S., & Antoun, H., (1995). Effect of compost on rhizosphere microflora of the tomato and on the incidence of plant growth-promoting rhizobacteria. *Applied and Environmental Microbiology, 61*(1), 194–199.

De Meyer, G., & Höfte, M., (1997). Salicylic acid produced by the rhizobacterium *Pseudomonas aeruginosa* 7NSK2 induces resistance to leaf infection by *Botrytis cinerea* on bean. *Phytopathology, 87*(6), 588–593.

DeVay, J. E., & Katan, J., (1991). Mechanisms of pathogen control in solarized soils. *Soil Solarization*, 87–102.

DeVleeschauwer, D., Verdonck, O., & Van Assche, P., (1981) Phytotoxicity of refuse compost. *Biocycle, 22*, 44.

Diánez, F., Santos, M., & Tello, J. C., (2004). Suppression of soilborne pathogens by compost: Suppressive effects of grape marc compost on phytopathogenic oomycetes. In: *International Symposium on Soilless Culture and Hydroponics* (Vol. 697, pp. 441–460).

Dissanayake, N., & Hoy, J. W., (1999). Organic material soil amendment effects on root rot and sugarcane growth and characterization of the materials. *Plant Disease, 83*(11), 1039–1046.

Edwards, J. H., & Someshwar, A. V., (2000). Chemical, physical, and biological characteristics of agricultural and forest by-products for land application. *Land Application of Agricultural, Industrial and Municipal By-products, 6*, 1–62.

Erhart, E., Burian, K., Hartl, W., & Stich, K., (1999). Suppression of *Pythium ultimum* by biowaste composts in relation to compost microbial biomass, activity and content of phenolic compounds. *Journal of Phytopathology, 147*(5), 299–305.

Font, X., Artola, A., & Sánchez, A., (2011). Detection, composition and treatment of volatile organic compounds from waste treatment plants. *Sensors, 11*(4), 4043–4059.

Fravel, D. R., (1988). Role of antibiosis in the biocontrol of plant diseases. *Annual Review of Phytopathology, 26*(1), 75–91.

Fuchs, J. G., (2002). Practical use of quality compost for plant health and vitality improvement. In: *Microbiology of Composting* (pp. 435–444). Springer, Berlin, Heidelberg.

Gamliel, A., Austerweil, M., & Kritzman, G., (2000). Non-chemical approach to soilborne pest management-organic amendments. *Crop Protection, 19*(8–10), 847–853.

Gilpatrick, J. D., (1969). Role of ammonia in the control of avocado root rot with alfalfa meal soil amendment. *Phytopathology*, *59*, 973–978.

Gorissen, A., Van, O. L. S., & Van, E. J. D., (2004). Pig slurry reduces the survival of *Ralstonia solanacearum* biovar 2 in soil. *Canadian Journal of Microbiology, 50*(8), 587–593.

Haas, D., & Défago, G., (2005). Biological control of soilborne pathogens by fluorescent *Pseudomonas*. *Nature Reviews Microbiology, 3*(4), 307–319.

Hadar, Y., & Mandelbaum, R., (1992). Suppressive compost for biocontrol of soilborne plant pathogens. *Phytoparasitica, 20*(1), S113–S116.

Hadar, Y., Mandelbaum, R., & Gorodecki, B., (1992). Biological control of soilborne plant pathogens by suppressive compost. In: *Biological Control of Plant Diseases* (pp. 79–83). Springer, Boston, MA.

Hawke, M. A., & Lazarovits, G., (1994). Production and manipulation of individual microsclerotia of *Verticillium dahliae* for use in studies of survival. *Phytopathology*.

Heydari, A., & Pessarakli, M., (2010). A review on biological control of fungal plant pathogens using microbial antagonists. *Journal of Biological Sciences, 10*(4), 273–290.

Hoitink, H. A. J., & Boehm, M. J., (1999). Biocontrol within the context of soil microbial communities: A substrate-dependent phenomenon. *Annual Review of Phytopathology, 37*(1), 427–446.

Hoitink, H. A. J., (1993). Mechanisms of suppression of soilborne plant pathogens in compost amended substrates. *Science and Engineering of Composting: Design, Environmental, Microbiological and Utilization Aspects*, pp. 601–621.

Hoitink, H. A. J., Stone, A. G., & Han, D. Y., (1997). Suppression of plant diseases by composts. *HortScience, 32*(2), 184–187.

Howell, C. R., Stipanovic, R. D., & Lumsden, R. D., (1993). Antibiotic production by strains of *Gliocladium virens* and its relation to the biocontrol of cotton seedling diseases. *Biocontrol Science and Technology, 3*(4), 435–441.

Huang, H. C., & Huang, J. W., (1993). Prospects for control of soilborne plant pathogens by soil amendment. *Current Topics in Botanical Research, 1*, 223–235.

Huber, D. M., & Watson, R. D., (1970). Effect of organic amendment on soilborne plant pathogens. *Phytopathology, 60*(1), 22–26.

Jackson, R. M., (1965). Antibiosis and fungi stasis of soil microorganisms. In: Baker, K. F., & Snyder, W. C., (eds.), *Ecology of Soilborne Plant Pathogens*.

Keener, H. M., Dick, W. A., & Hoitink, H. A., (2000). Composting and beneficial utilization of composted by-product materials. *Land Application of Agricultural, Industrial and Municipal By-products, 6*, 315–341.

Kirkegaard, J. A., Wong, P. T. W., & Desmarchelier, J. M., (1996). In vitro suppression of fungal root pathogens of cereals by *Brassica* tissues. *Plant Pathology, 45*(3), 593–603.

Kiss, L., (2003). A review of fungal antagonists of powdery mildews and their potential as biocontrol agents. *Pest Management Science: Formerly Pesticide Science, 59*(4), 475–483.

Kuter, G. A., Nelson, E. B., Hoitink, H. A. J., & Madden, L. V., (1983). Fungal populations in container media amended with composted hardwood bark suppressive and conducive to *Rhizoctonia* damping-off. *Phytopathology, 73*(10), 1450–1456.

Kyselková, M., & Moënne-Loccoz, Y., (2012). *Pseudomonas* and other microbes in disease-suppressive soils. In: *Organic Fertilisation, Soil Quality and Human Health* (pp. 93–140). Springer, Dordrecht.

Lazarovits, G., Conn, K. L., & Potter, J., (1999). Reduction of potato scab, *Verticillium* wilt, and nematodes by soymeal and meat and bone meal in two Ontario potato fields. *Canadian Journal of Plant Pathology, 21*(4), 345–353.

Lazarovits, G., Conn, K. L., Abbasi, P. A., & Tenuta, M., (2005). Understanding the mode of action of organic soil amendments provides the way for improved management of soilborne plant pathogens. *Acta Horticulturae, 698*, 215.

Lazarovits, G., Tenuta, M., & Conn, K. L., (2001). Organic amendments as a disease control strategy for soilborne diseases of high-value agricultural crops. *Australasian Plant Pathology, 30*(2), 111–117.

Lewis, J. A., & Papavizas, G. C., (1971). Effect of sulfur-containing volatile compounds and vapors from cabbage decomposition on *Aphanomyces euteiches*. *Phytopathology, 61*, 208–214.

Liu, L., Kloepper, J. W., & Tuzun, S., (1995). Induction of systemic resistance in cucumber by plant growth-promoting rhizobacteria: Duration of protection and effect of host resistance on protection and root colonization. *Phytopathology, 85*(10), 1064–1068.

Lockwood, J. L., (1990). Relation of energy stress to behavior of soilborne plant pathogens and to disease development. *Biological Control of Soilborne Plant Pathogens*, 197–214.

Löffler, H. J. M., Cohen, E. B., Oolbekkink, G. T., & Schippers, B., (1986). Nitrite as a factor in the decline of *Fusarium oxysporum* f. sp. *dianthi* in soil supplemented with urea or ammonium chloride. *Netherlands Journal of Plant Pathology, 92*(4), 153–162.

Lumsden, R. D., Lewis, J. A., & Millner, P. D., (1983). Effect of composted sewage sludge on several soilborne pathogens and diseases. *Phytopathology, 73*(11), 1543–1548.

Lumsden, R. D., Millner, P. D., & Lewis, J. A., (1986). Suppression of lettuce drop caused by *Sclerotinia minor* with composted sewage sludge. *Plant Disease, 70*(3), 197–201.

Lynch, J. M., (1977). Phytotoxicity of acetic acid produced in the anaerobic decomposition of wheat straw. *Journal of Applied Bacteriology, 42*(1), 81–87.

Lynch, J. M., (1978). Production and phytotoxicity of acetic acid in anaerobic soils containing plant residues. *Soil Biology and Biochemistry, 10*(2), 131–135.

Mathre, D. E., Cook, R. J., & Callan, N. W., (1999). From discovery to use: Traversing the world of commercializing biocontrol agents for plant disease control. *Plant Disease, 83*(11), 972–983.

Maurhofer, M., Hase, C., Meuwly, P., Metraux, J. P., & Defago, G., (1994). Induction of systemic resistance of tobacco-to-tobacco necrosis virus by the root-colonizing *Pseudomonas fluorescens* strain CHA0: Influence of the gacA gene and of pyoverdine production. *Phytopathology, 84*, 139–146.

Mazzola, M., Hewavitharana, S. S., Strauss, S. L., Shennan, C., & Muramoto, J., (2016). Anaerobic soil disinfestation and Brassica seed meal amendment alter soil microbiology and system resistance. *International Journal of Fruit Science, 16*(S1), 47–58.

Menzies, J. D., (1962). Effect of anaerobic fermentation in soil on survival of sclerotia of *Verticillium dahliae*. *Phytopathology, 52*(8), 743.

Mian, I. H., & Rodriguez-Kabana, R., (1982). Survey of the nematicidal properties of some organic materials available in Alabama as amendments to soil for control of *Meloidogyne arenaria*. *Nematropica*, 235–246.

Michel, V. V., & Mew, T. W., (1998). Effect of a soil amendment on the survival of *Ralstonia solanacearum* in different soils. *Phytopathology, 88*(4), 300–305.

Milner, J. L., Silo-Suh, L., Lee, J. C., He, H., Clardy, J., & Handelsman, J. O., (1996). Production of kanosamine by *Bacillus cereus* UW85. *Applied and Environmental Microbiology, 62*(8), 3061–3065.

Mitchell, R., & Alexander, M., (1962). Microbiological changes in flooded soils. *Soil Science, 93*(6), 413–419.

Muehlchen, A. M., Rand, R. E., & Parke, J. L., (1990). Evaluation of crucifer green manures for controlling *Aphanomyces* root rot of peas. *Plant Disease, 74*(9), 651–654.

Muramoto, J., Shennan, C., Zavatta, M., Baird, G., Toyama, L., & Mazzola, M., (2016). Effect of anaerobic soil disinfestation and mustard seed meal for control of charcoal rot in California strawberries. *International Journal of Fruit Science, 16*(S1), 59–70.

Paredes, C., Cegarra, J., Roig, A., Sánchez-Monedero, M. A., & Bernal, M. P., (1999). Characterization of olive mill wastewater (alpechin) and its sludge for agricultural purposes. *Bioresource Technology, 67*(2), 111–115.

Pieterse, C. M., Van, W. S. C., Hoffland, E., Van, P. J. A., & Van, L. L. C., (1996). Systemic resistance in *Arabidopsis* induced by biocontrol bacteria is independent of salicylic acid accumulation and pathogenesis-related gene expression. *The Plant Cell, 8*(8), 1225–1237.

Pitt, D., Tilston, E. L., & Groenhof, A. C., (1997). Recycled organic materials (ROM) in the control of plant disease. *International Symposium on Composting and Use of Composted Material in Horticulture, 469*, 391–404.

Ponnamperuma, F. N., (1972). The chemistry of submerged soils. *Advances in Agronomy, 24*, 29–96.

Pugliese, M., Liu, B. P., Gullino, M. L., & Garibaldi, A., (2008). Selection of antagonists from compost to control soilborne pathogens. *Journal of Plant Diseases and Protection, 115*(5), 220–228.

Raaijmakers, J. M., & Weller, D. M., (1998). Natural plant protection by 2,4-diacetylphloroglucinol-producing *Pseudomonas* spp. in take-all decline soils. *Molecular Plant-Microbe Interactions, 11*(2), 144–152.

Rodriguez-Kabana, R., Boube, D., & Young, R. W., (1989). Chitinous materials from blue crab for control of root-knot nematode. I. Effect of urea and enzymatic studies. *Nematropica*, 53–74.

Rosskopf, E. N., Serrano-Pérez, P., Hong, J., Shrestha, U., Rodríguez-Molina, M. D. C., Martin, K., & Butler, D., (2015). Anaerobic soil disinfestation and soilborne pest management. In: *Organic Amendments and Soil Suppressiveness in Plant Disease Management* (pp. 277–305). Springer, Cham.

Ryckeboer, J., (2001). *Biowaste and Yard Waste Composts: Microbiological and Hygienic Aspects-Suppressiveness to Plant Diseases* (p. 278). Leuven, Belgium: Katholieke Universiteit Leuven (Doctoral Dissertation).

Schüler, C., Pikny, J., Nasir, M., & Vogtmann, H., (1993). Effects of composted organic kitchen and garden waste on *Mycosphaerella pinodes* (Berk, et Blox) vestergr., causal organism of footrot on peas (*Pisum sativum* L.). *Biological Agriculture and Horticulture, 9*(4), 353–360.

Serra-Wittling, C., Houot, S., & Alabouvette, C., (1996). Increased soil suppressiveness to *Fusarium* wilt of flax after addition of municipal solid waste compost. *Soil Biology and Biochemistry, 28*(9), 1207–1214.

Shahat, A. A., Ibrahim, A. Y., Hendawy, S. F., Omer, E. A., Hammouda, F. M., Abdel-Rahman, F. H., & Saleh, M. A., (2011). Chemical composition, antimicrobial and antioxidant activities of essential oils from organically cultivated fennel cultivars. *Molecules, 16*(2), 1366–1377.

Shiau, F. L., Chung, W. C., Huang, J. W., & Huang, H. C., (1999). Organic amendment of commercial culture media for improving control of *Rhizoctonia* damping-off of cabbage. *Canadian Journal of Plant Pathology, 21*(4), 368–374.

Siddiqui, Y., Meon, S., Ismail, M. R., & Ali, A., (2008). *Trichoderma*-fortified compost extracts for the control of *Choanephora* wet rot in okra production. *Crop Protection, 27*(3–5), 385–390.

Silo-Suh, L. A., Lethbridge, B. J., Raffel, S. J., He, H., Clardy, J., & Handelsman, J., (1994). Biological activities of two fungistatic antibiotics produced by *Bacillus cereus* UW85. *Applied and Environmental Microbiology, 60*(6), 2023–2030.

Smolinska, U., Morra, M. J., Knudsen, G. R., & Brown, P. D., (1997). Toxicity of glucosinolate degradation products from *Brassica napus* seed meal toward *Aphanomyces euteiches* f. sp. *pisi*. *Phytopathology, 87*(1), 77–82.

Spatafora, C., & Tringali, C., (2012). Valorization of vegetable waste: Identification of bioactive compounds and their chemo-enzymatic optimization. *The Open Agriculture Journal, 6*(1).

Strauss, S. L., & Kluepfel, D. A., (2015). Anaerobic soil disinfestation: A chemical-independent approach to pre-plant control of plant pathogens. *Journal of Integrative Agriculture, 14*(11), 2309–2318.

Subbarao, K. V., Hubbard, J. C., & Koike, S. T., (1999). Evaluation of broccoli residue incorporation into field soil for *Verticillium* wilt control in cauliflower. *Plant Disease, 83*(2), 124–129.

Sutherland, E. D., & Papavizas, G. C., (1991). Evaluation of oospore hyperparasites for the control of *Phytophthora* crown rot of pepper. *Journal of Phytopathology, 131*(1), 33–39.

Tenuta, M., & Lazarovits, G., (2002). Ammonia and nitrous acid from nitrogenous amendments kill the microsclerotia of *Verticillium dahliae*. *Phytopathology, 92*(3), 255–264.

Tenuta, M., Conn, K. L., & Lazarovits, G., (2002). Volatile fatty acids in liquid swine manure can kill microsclerotia of *Verticillium dahliae*. *Phytopathology, 92*(5), 548–552.

Tilston, E. L., Pitt, D., & Groenhof, A. C., (2002). Composted recycled organic matter suppresses soilborne diseases of field crops. *New Phytologist, 154*(3), 731–740.

Tsao, P. H., & Oster, J. J., (1981). Relation of ammonia and nitrous acid to suppression of *Phytophthora* in soils amended with nitrogenous organic substances. *Phytopathology, 71*(1), 53–59.

Tuitert, G., & Bollen, G. J., (1996). The effect of composted vegetable, fruit and garden waste on the incidence of soilborne plant diseases. In: *The Science of Composting* (pp. 1365–1369). Springer, Dordrecht.

Vallad, G. E., & Goodman, R. M., (2004). Systemic acquired resistance and induced systemic resistance in conventional agriculture. *Crop Science, 44*(6), 1920–1934.

Van, B. A. H., (1995). Plant disease severity in high-input compared to reduced-input and organic farming systems. *Plant Disease, 79*(10), 976–984.

Van, L. L. C., Bakker, P. A. H. M., & Pieterse, C. M. J., (1998). Systemic resistance induced by rhizosphere bacteria. *Annual Review of Phytopathology, 36*(1), 453–483.

Watson, R. D., (1964). Eradication of soil fungi by a combination of crop residue flooding and anaerobic fermentation. *Phytopathology, 54*(12), 1437.

Wei, G., Kloepper, J. W., & Tuzun, S., (1991). Induction of systemic resistance of cucumber to *Colletotrichum orbiculare* by select strains of plant growth-promoting rhizobacteria. *Phytopathology, 81*(11), 1508–1512.

Wei, Z., Yang, X., Yin, S., Shen, Q., Ran, W., & Xu, Y., (2011). Efficacy of *Bacillus*-fortified organic fertilizer in controlling bacterial wilt of tomato in the field. *Applied Soil Ecology, 48*(2), 152–159.

Weller, D. M., Raaijmakers, J. M., Gardener, B. B. M., & Thomashow, L. S., (2002). Microbial populations responsible for specific soil suppressiveness to plant pathogens. *Annual Review of Phytopathology, 40*(1), 309–348.

Wilhelm, S., (1951). Effect of various soil amendments on the inoculum potential of the *Verticillium* wilt fungus. *Phytopathology, 41*(8), 684–690.

Yogev, A., Raviv, M., Hadar, Y., Cohen, R., & Katan, J., (2006). Plant waste-based composts suppressive to diseases caused by pathogenic *Fusarium oxysporum*. *European Journal of Plant Pathology, 116*(4), 267–278.

CHAPTER 9

ORGANIC FARMING IMPROVES SOIL HEALTH SUSTAINABILITY AND CROP PRODUCTIVITY

ABDEL RAHMAN M. AL-TAWAHA,[1] ELIF GÜNAL,[2] İSMAIL ÇELIK,[3] HIKMET GÜNAL,[2] ABDULKADIR SÜRÜCÜ,[4] ABDEL RAZZAQ M. AL-TAWAHA,[5] ALLA ALEKSANYAN,[6] DEVARAJAN THANGADURAI,[7] and JEYABALAN SANGEETHA[8]

[1]*Department of Biological Sciences, Al-Hussein Bin Talal University, P.O. Box 20, Maan, Jordan*

[2]*Gaziosmanpaşa University, Faculty of Agriculture, Department of Soil Science and Plant Nutrition, Tokat, Turkey*

[3]*Çukurova University, Faculty of Agriculture, Department of Soil Science and Plant Nutrition, Adana, Turkey*

[4]*Harran University, Faculty of Agriculture, Department of Soil Science and Plant Nutrition, Şanliurfa, Turkey*

[5]*Department of Crop Science, Faculty of Agriculture, University Putra Malaysia, Serdang–43400, Selangor, Malaysia*

[6]*Institute of Botany aft. A.L. Takhtajyan NAS RA/Department of Geobotany and Plant Eco-Physiology, Yerevan, Armenia*

[7]*Department of Botany, Karnatak University, Dharwad–580003, Karnataka, India*

[8]*Department of Environmental Science, Central University of Kerala, Kasaragod–671316, Kerala, India*

ABSTRACT

Organic farming increases soil sustainability and crop production. Demands for healthy food and a sustainable environment have increased interest in alternative conservative production systems and led to the introduction of the concept of organic farming. Plants are continually dealing with increasingly evolving and potentially harmful external environmental factors. Organic farming is very important to protect the environment, minimize soil degradation and erosion, reduce pollution, optimize biological productivity, and promote a healthy state of health. Soil quality and fertility are a concern today to boost the sustainability of our agricultural system. Soil fertility is more important than the supply of macro and micronutrients to plants in organic farming systems (OFSs). Effective management of fertility takes into account plant, soil organic matter (SOM), and soil biology. This chapter focuses on: (i) impact of organic farming on physical properties of soils; (ii) status of organic farming; (iii) impact of organic farming on plant growth and yield.

9.1 INTRODUCTION

The growth of many plants is affected by many environmental factors that affect in more than one way the physiological and morphological changes of the plant (Hayyawi et al., 2018; Amanullah et al., 2019; Hani et al., 2019; Aludatt et al., 2019). Organic farming is a plant and animal production process that requires far more than opting not to use pesticides, fertilizers, GMOs, antibiotics, and growth hormones. In organic farming systems (OFSs), soil fertility means more than providing macro and micronutrients to plants. In organic farming, providing mineral nutrients are considered very key factor that limits plant biomass and productivity in many ecosystems (Turk et al., 2003a–c; Nikus et al., 2004a, b; Tawaha and Turk, 2004; Turk and Tawaha, 2004). Also, the effective fertility management considers plant, soil organic matter (SOM) and soil biology. Organic production is a comprehensive method designed to maximize the efficiency and health within the agroecosystem of different populations, including soil species, plants, livestock, and humans. The key aim of organic production is the creation of sustainable enterprises in accordance with the climate. Many researchers (Al-Kiyyam et al., 2008; Al-Ajlouni et al., 2009; Al-Tawaha et al., 2010) reported about general principles for organic production:

- Environmental protection, reduces soil degradation, reduces pollution, improves biological productivity, and promotes good health of the soil;
- Maintaining long term soil fertility by enhancing soil biological activity;
- Preserving the system's biodiversity;
- Provide care that promotes health and meets livestock behavioral needs; and
- Preparation of organic products, stressing preparation procedures and thorough handling to protect the biological integrity and essential characteristics of products at all stages of production.

9.2 STATUS OF ORGANIC FARMING

Soil health and fertility is a matter of concern these days to make our agricultural system more sustainable (Al-Tawaha et al., 2005; Lee et al., 2005; Sulpanjan et al., 2005). On the other hand, adequate plant nutrition is essential for commercial crop production (Amanullah et al., 2019a, b; Hani et al., 2019; Rajendran et al., 2019). For OFSs, soil fertility requires more than supplying macro and micronutrients to plants. Good fertility management takes plant, SOM, and soil biology into consideration. Ideally OFSs are planned to increase soil fertility so as to achieve multiple goals such as preserving and, if necessary, enhancing the soil's physical condition such that the soil maintains healthy plants and living organisms in the soil and has the capacity to withstand and recover from stress such as flooding or violent tillage; maintaining the soil's buffering capacity to reduce the deterioration of the environment caused by soil loss or the inability of the OFSs are designed to keep nutrients in organic reservoirs or bioavailable forms of minerals, rather than simply supplying nutrients through frequent fertilizer additions. This is accomplished as nutrients flow through organic reservoirs. Soil fertility is increased by organic matter control and not by input substitution (Tawaha et al., 2005; Yang et al., 2005; Assaf et al., 2006; Turk et al., 2006). Regardless of the most human food requirement the agriculture plays a vital role in human life. Traditional agriculture's main objective was to achieve maximum productivity from a unit of agricultural land, and these farming methods are distinguished by a high degree of crop specialization and intensive land, labor-work, and capital contributions per unit of soil. Extensive use of natural resources in traditional agriculture has caused problems such as soil, water, and air contamination, food residues of agricultural chemicals, increased depletion of natural resources and, consequently, increased social cost of production. The transition to organic farming is described as one of the best strategies for achieving sustainable management of the agri-environment at national and international level. Organic agriculture is getting more and more relevant for agriculture and agro-industry in many developed and developing countries, due to public knowledge of the environment and health, and the negative effects of conventional agriculture (Rehber and Turhan, 2002). Organic farming is a modern method of processing in which agricultural pesticides, concentrated feed and additives are avoided or eliminated from the farm completely. Organic activities rely on crop rotation, crop residues, compost, legumes, green manures (GMs), organic off-farm waste, mechanical agriculture, mineral rocks, and techniques for pest control (Lampkin and Padel, 1994; Hameed et al., 2008; Al-Tawaha et al., 2010, 2017; Ananthi et al., 2017; Abu Obaid et al., 2018).

Organic farming has the potential to provide advantages in terms of protecting the environment, preserving non-renewable resources, improving nutritional quality, reducing the production of surplus products and reorienting agriculture toward evolving market demand (Lampkin and Padel, 1994). Demand for organic food, and hence trade in organic food and organic farming, is increasingly important worldwide, particularly in developed countries. Also reinforcing the development of organic farming is the new conservation of rural concepts, improving understanding of environmental issues and encouraging sustainable agricultural development.

9.3 IMPACT OF ORGANIC FARMING ON SOIL HEALTH

Demands on healthy food and sustainable environment increased the interest in alternative conservative production systems and gave rise to the introduction of organic agriculture concept (Sudjatmiko et al., 2018). The sustainability of agricultural ecosystems to produce health food can be obtained by improving and maintaining soil health (Tully and McAskill, 2019). Higher amounts of antioxidants and less or negligible quantity of pesticides in organic foods attract the attention of consumers (Barański et al., 2014). Utilization of carbon-based soil amendments, diversifying the crop rotations, including cover crops during the fallow period, absence of synthetic agrochemicals, using organic mulches, weed control without herbicides, and integrating crop and livestock production are the main strategies of organic farming to improve the functioning potential of soils (Schonbeck et al., 2017). Therefore, organic cropping systems are expected to have an accumulation of organic C compared to conventionally agricultural production systems. Soil nutrient sources in organic farming are met by organic sources, including solid or liquid animal manure, GMs, crop residues, yard debris, and waste from food processing industries (Tully and McAskill, 2019). OFS plainly deals with the management of SOM, which in turn has a strong influence on the physical, chemical, and biological properties of soil (Stockdale et al., 2001).

Despite the concerns on the low yield in organic agriculture (Wittwer et al., 2017), organic production land is continuously increasing due to the detrimental impacts, such as erosion, degradation of soil structure, soil compaction, losses of SOMs, and nutrient depletion, of conventional farming systems (Abebe et al., 2005; Abera et al., 2005; Al-Tawaha et al., 2005, 2010; Elser et al., 2007; Al-Kiyyam et al., 2008; Bünemann et al., 2018). The area under organic farming was estimated at 57.8 million hectares worldwide in 2016, including in conversion areas. Total area of organic farming corresponds to 1.2% of the agricultural land in the world (Willer et al., 2018). The decrease in crop yield is the most common disappointment of the producers when converted from conventional to organic agriculture. The meta-analysis conducted by Niggly (2015) revealed that average crop yield in organic agriculture is lower between 0.75 and 0.80 of conventional agriculture. However, most studies reported a gradual increase in crop yield overtime along with improved soil quality (Reeve and Drost, 2012; Suja and Sreekumar, 2014) associated mainly with increased soil biological activity (Knapp and van der Heijden, 2018).

Organic farming enhances the delivery of several ecosystem services by improving the soil health, quality of surface and ground waters. Some of the important ecosystem services are increasing the efficiency nutrients use, preventing nutrient, buffering nutrients according to their removal, reducing the erosion by increasing the stability of aggregates and increasing the water holding capacity (Peigné et al., 2014; Abbott and Manning, 2015) (Figure 9.1). Organic amendments in organic farming significantly improve soil moisture retention, total organic carbon and nitrogen contents, population, and diversity of microorganisms, and enzyme activities, which indicates significant enhancement on soil health (Nautiyal et al., 2010).

Conventional tillage using plows are frequently used by organic farmers to prepare seedbed and control weeds prior to planting, bury intermediate crops, and incorporate organic fertilizers and amendments (Peigne et al., 2018). However, continuous, and intensive soil tillage by plows may neutralize positive effects to soil aggregates and functions due to loss of organic carbon, compaction, and crust formation (Crittenden et al., 2015), negative impact on infiltration, water retention capacity and nutrient cycling (Williams and Hedlund, 2013). Therefore, some studies indicated the positive effect of combining conservation tillage and organic farming on functioning potential soils. Introduction of reduced tillage without plowing in OFSs improves soil physical health and helps to preserve long-term soil fertility compared to intensive tillage in organic farming (Cooper et al., 2016; Seitz et al., 2019) by decreasing erosion rate (Seitz et al., 2019) and promoting aggregation of microaggregates into macroaggregates (Puerta et al., 2018). In contrast, higher losses of microaggregates have been reported in tilled organic fields by Green et al. (2005). Peigne et al. (2018) indicated that 10 years of very superficial tillage at 5 to 7 cm and superficial tillage at 15 cm increased organic carbon, total nitrogen, Olsen P in the 0 to 15 cm of soil surface compared to moldboard plowing treatments.

FIGURE 9.1 Soil aggregates under organically managed field and conventionally managed adjacent field from central Anatolia, Turkey.

Sustainable production in OFS relies on organic matter management to improve soil health and provide nutrients for growing plants (Turk and Tawaha, 2001, 2002a, b; Tawaha and Turk, 2002a, b; Tawaha et al., 2003). The farmers in organic agriculture simulate the natural cycle to add essential nutrients to maintain and improve the soil health (Biswas et al., 2014). Productivity function of soils in organic farming is preserved or improved by the addition of GMs and organic amendments such as animal manure, compost, biochar, etc. More than 70 years of a conventional and an organic farming and permanent pasture without tillage resulted in 15, 24, and 46 g kg^{-1} organic matter content at 0–20 cm depth, respectively (Pulleman et al., 2003).

Soil health is defined as the functioning ability or capacity of soils as an essential component of the living ecosystem that is vital to sustaining the lives of plants, animals, and humans (Moebius-Clune et al., 2016). Therefore, soil health management is a crucial issue to sustain the agricultural productivity and environmental quality (Reeves, 1997).

The performance of soils for all its functions cannot be assessed by determining only crop yield, water quality, or any single soil property (USDA, 2020). The application of organic amendments increases the concentration of essential nutrients in soils. Organic material in soil improves aggregation of mineral particles, increases the stability against destructive forces, thus; reduces soil compaction and erosion, increases porosity and consequently the water holding capacity (Kaje et al., 2018).

9.3.1 IMPACT ON PHYSICAL PROPERTIES OF SOILS

Physical quality of soil is closely related to the maintenance of the stable aggregates and an improved storing and transmitting potential of water, air, and nutrients to support crops and to reduce degradation (Reynolds et al., 2007). The results conducted to assess the impacts of organic farming on the sustainability of agricultural production have shown the positive effects of animal manure and diversified crop rotations on soil quality (Schjonning et al., 2002). However, contrasting reports on the impacts of organic farming have been published for different soil physical properties. In general, soil physical environment in terms of soil pore volume and stability of aggregates favors sustainable use of organically managed arable lands (Papadopoulos et al., 2014). The incorporation of organic matter to soil activates biological processes that help to supply plant nutrients along with an increase in the strength and stability of soil structure for better plant growth environment (Abbott and Manning, 2015). Long-term conservative practices in organic farming are reported to result in lower bulk density and higher available water retention capacity compared to conventional farms. Improved soil physical health by the introduction of organic farming indicates a highly resistant soil environment to wind and water erosion, to the crusting and creating compaction, in addition, reduces plant stress related to the long-term drought (Liebig and Doran, 1999).

9.3.1.1 AGGREGATE STABILITY

Soil organic C is the most important agent in the formation and stability of soil aggregates (Franzleubbers et al., 2000). Addition of greater amount of plant and animal biomass in OFSs result in a stable soil surface cover and increase proportion of water-stable macro aggregation (Pulleman et al., 2003; Mikha and Rice, 2004; Zhang et al., 2014), and aeration (Shukla et al., 2006) by stabilizing soil aggregates (Erhart and Hartl, 2009) and changing soil pore characteristics (Papadopoulos et al., 2014). However, the positive effects of organic farming on aggregate stability and soil porosity may not be constant over time. Aggregate formation is primarily related to the stabilization of organic matter and stability of long-term soil mass (Williams and Petticrew, 2009), therefore, aggregate stability is positively correlated with the increase in soil organic C content (Green et al., 2005). Papadopoulos et al. (2014) reported significantly higher aggregate stability for short term (5 years) organically managed sandy loam soils (mean weight diameter, 1.24 mm) compared to conventionally managed soils (mean weight diameter, 0.64 mm), while mean weight diameter in longer-term (>10 years) organically and conventionally managed silty clay soils was similar to each other (Table 9.1). In addition to the commonly stated effects of organic carbon on the stability of aggregates, Lori et al. (2017) indicated the importance of bacteria and fungi on the improvement and stability of soil aggregates.

9.3.1.2 SOIL WATER RETENTION

Soil structure, pore-size distributions, SOM content, and soil texture that is influential on retention at higher matric potential, are the major determinants of water holding capacity and water infiltration rate of a soil (Hillel, 1998). Increasing the organic C content of soils under OFSs has a beneficial effect on soil hydraulic properties by improving the important soil physical properties. Soils under organic farming are reported to have higher soil water retention capacity, therefore, higher potential water-limited crop yields compared to soils under conventional farming (Crittenden et al., 2015; Crittenden and Goede, 2016). Similarly, Nautiyel et al. (2010) also indicated that the addition of organic manure and crop residues in organic farming increased the moisture retention of sandy loam soils. Reducing soil surface evaporation by organic mulching increases the moisture content of soils (Bayer et al., 2006).

Mechanical weed control in organic farming and intensive tractor passes in the field may cause a compaction, which decreases the volume of macropores, increases the volume of smaller pores. The average amount of macropores under organic farming (0.65%) was significantly lower than the macropores (0.82%) under conventional farming. The differences in pore size distribution between organic and conventional farming systems

TABLE 9.1 Mean Weight Diameter (MWD), Aggregate Stability (AS) and Bulk Density Values for Organic (O) versus Conventional (C) Farming Systems

Duration Year	Organic Practice	Soil Type	Depth (cm)	MWD (mm)		AS (g g^{-1})		BD (g cm^{-3})		References
				O	C	O	C	O	C	
37-year	Green manure	Silt Loam	0–10	–	–	–	–	0.98	0.95	Reganold et al. (1987)
46-year	Forage crop, animal manure	Sandy loam	6–13	–	–	0.51	0.36	1.54	1.44	Schjonning et al. (2002)
47-year						0.87	0.68	1.35	1.35	
40-year						0.64	0.87	1.36	1.49	
>70-year	5.1 and 1.8 tDMha±1 yr±1 crop residue and animal manure	Loam	0–10	5.3[&]	0.6	–	–	–	–	Pulleman et al. (2003)
			10–20	3.7	0.5					
7-year	Vetch cover	–	0–5	0.40	0.40	0.31*	0.35*	1.53	1.53	Green et al. (2005)
	5.1 Mg/ha poultry litter									
–	–	Silty	0–20	1.24	0.64	–	–	1.06	0.93	Papdopoulos et al. (2014)
		Silty Clay		0.44	0.33			1.27	1.45	
		Sandy Loam		2.41	3.09			1.05	1.03	
7-year	20–40 Mg/ha/yr (fresh weight) slurry and/or solid cow manure	Clay Loam	0–10	0.64	0.42	–	–	1.42	–	Crittenden et al. (2015)
			10–20	0.50	0.45			1.42		
8-year			0–10	0.57		–	–	1.34	–	
			10–20	0.56				1.42		
6-year	Potato	Silt Loam	0–40	–	–	–	–	1.52	1.50	Glab et al. (2016)
	Wheat							1.59	1.57	
	Oats/vetch							1.57	1.57	
	Spelt							1.53	1.49	

TABLE 9.1 *(Continued)*

Duration Year	Organic Practice	Soil Type	Depth (cm)	MWD (mm)		AS (g g^{-1})		BD (g cm^{-3})		References
				O	C	O	C	O	C	
20-year	Chipped pruned branches, weeds, composted sheep manure (10 Mg ha^{-1})	Sandy Loam	0–10	–	–	–	–	0.99	1.12	Hondebrink et al. (2017)
10-year		Sandy loam	0–10	–	–	–	–	1.67	1.67	
15-year	–	Sandy loam	–	–	–	–	–	1.22	1.32	Sihi et al. (2017)
3-year	Horse manure, green cover crop (mustard)	Silt	0–15	1.46	0.97	82.0%	76.9%	1.39	1.29	Morwan et al. (2018)

Macro aggregates (g/g); &: Wet sieving.

were more evident for micropores (50–100 μm). The decrease in volume of macropores and increase in micropores have significant impacts on water retention characteristics of soils, especially at high matric potential resulting in a higher amount of plant-available water content (Glab et al., 2016).

9.3.1.3 SOIL COMPACTION

Pleasing positive effects of organic management on bulk density and hydraulic conductivity have been reported (Oquist et al., 2006). The increase in organic matter content of a silty clay soil under more than 10 years of organic farming led to a significantly lower bulk density (1.27 g cm^{-3}) compared to companion conventionally managed soils (1.45 g cm^{-3}). However, the use of heavy machinery and conventional tillage methods may cause severe subsoil below plowing depth, even under OFSs (Schjonning et al., 2002).

Intensive tractor traffic for mechanical weed protection under OFS resulted in higher bulk density value (1.54 g cm^{-3}) compared to the conventional farming system (1.51 g cm^{-3}) (Glab et al., 2016). Moreover, Pulleman et al. (2003) raised their concerns on a higher risk of soil compaction under organic farming applied with reduced or no-till practices, if not properly managed. Crittenden et al. (2015) also reported significantly higher mean weight diameter at 0–10 cm depth in conventional farming system compared to organic farming. Therefore, shallow mechanical disturbance was recommended in the organically managed fields to break up the aggregates in the very surface layer.

Penetration resistance is a commonly used indicator of soil physical health and considered a representation of roots penetration through soil. The penetration resistance of a soil mainly depends on texture, water content, and dry bulk density of a soil (Crittenden et al., 2015). Reganold (1995) reported greater soil porosity and lower penetration resistance under organic compared to conventional management system. Compaction in organic farming has been reported to be a major constrain for plant growth, and cause of the yield losses. Around 10 years of shallow and very shallow tillage in organic agriculture increased the soil compaction, especially between 15 and 30 cm depth (Peigne et al., 2018). In contrast, positive effects of short-term organic farming (5 years) on microporosity have been reported by Papadopoulos et al. (2014), however, microporosity of silty clay soil was reduced in the longer term (more than 10 years) organically managed land (18.4%) in comparison to the conventionally managed land (25.0%).

9.3.1.4 SOIL EROSION

Soil erosion is an important agent of land degradation and causes severe environmental problems with detrimental impacts on the provisioning of ecosystem services (Smith et al., 2016). Sediment losses in organic farming including cover crop is significantly lower compared to conventional farming (Green et al., 2005). The greater amount of improved water-stable macroaggregates due to the increased organic matter content in organic farming has been correlated with the decrease in soil erosion (Lado et al., 2004). Therefore, Morwan et al. (2018) explained the runoff and soil loss differences by lower stability of soil aggregates under conventional farming compared to organic farming. Seitz et al. (2019) reported 30% less sediment loss (0.54 t ha^{-1} h^{-1}) in organic farming than the conventional farming. Repeated losses of sediments from surface layers decrease the thickness of the surface horizon. Despite the great sediment losses every year, farmers may not realize the severity of the problem in the short term due to the mixing with subsurface horizon by plowing. But in the long term (37 years), the thickness of the most productive layer (topsoil) has been substantially decreased by water and tillage erosion (Reganold et al., 1987). Average soil loss from organically managed fields was reported to be 8.3 tons ha^{-1} $year^{-1}$, while it was fourfold higher (32.4 tons ha^{-1} $year^{-1}$) on conventionally managed fields (Table 9.2). Reganold et al. (1987) related the difference in erosion rates between organically and conventionally managed fields to the differences in crop rotations and tillage intensities of the production systems.

Crop residues or other organic amendments applied to soil surface are incorporated and buried with conventional tillage practices, which accelerate the decomposition of biomass. Therefore, sediment loss in organic farming can be further decreased by the introduction of reduced tillage due to increased soil surface cover. Otherwise, conventional tillage using moldboard plow may nullify some of the benefits on physical properties expected from the addition of animal manure and plant residues in organic farming. Soil loss under conventional tillage and organic farming, in which moldboard plow and disk used before planting, and rotary hoeing for weed control, was estimated 7.5 and 5 times greater relative to no-tillage (Green et al., 2005). Therefore, unless conservational tillage practices are applied, the risk for nutrient loss through sediment transport in soils under organic farming are much higher than the non-organic reduced or no-till soils.

TABLE 9.2 Macro Porosity (MP), Runoff, and Soil Loss Values for Organic (O) versus Conventional (C) Farming Systems

Duration	Organic Practice	Soil Type	Depth	MP (%)		Runoff (mm/year)		Soil Loss (Mg/ha/yr)		References
				O	C	O	C	O	C	
46	Forage crop, animal manure	Sandy loam	6–13 cm	11.2	15.9	–	–	–	–	Schjonning et al. (2002)
47				14.1	14.9					
40				15.9	11.0					
7-year	Vetch cover 5.1 Mg/ha poultry litter	–	0–5	–	–	108	116	43	64	Green et al. (2005)
–	–	Silty	0–20	23.0*	16.7*	–	–	–	–	Papdopoulos et al. (2014)
		Silty clay		20.6	15.7					
		Sandy loam		21.1	12.2					
6-year	Potato	Silt Loam	0–20	0.428[&]	0.435	–	–	–	–	Glab et al. (2016)
	Wheat			0.402	0.408					
	Oats/vetch			0.406	0.408					
	Spelt			0.423	0.436					
37-year	Green manure	Silt Loam	0–10	–	–	–	–	8.3	32.4	Reganold et al. (1987)
13-year	Bean	25 different fields, predominantly sandy loam, loamy sand	0–30	–	–	–	–	32.5	32.8	Arnhold et al. (2014)
	Potato							38.2	30.6	
	Radish							45.0	54.8	
	Cabbage							34.7	34.7	

[&]: *Total porosity ($cm^3\ cm^{-3}$).*

Soil loss under OFSs is closely related to the agricultural practices needed to grow crops. In general, OFSs have been reported to decrease the sediment loss in production of row crops due to the development of weeds in the furrows. Similarly, soil loss in radish growth was also decreased due to higher weed coverage. Conversely, potato cultivation has been reported to increase the sediment loss (Arnhold et al., 2014).

9.3.2 IMPACT ON CHEMICAL PROPERTIES OF SOIL

Long-term manure, compost or crop residue amendments in organic farming significantly influence the soil physical and chemical environments (Abbott and Manning, 2015). The extend of changes in the chemical properties of the soils are closely related to the nature of organic amendments (Mäder et al., 2002) and agricultural practices such as tillage (Peigne et al., 2018). Conservation tillage treatments are reported to have favorable effects on soil chemical components in the surface layer and contributed to the improvement of the productivity function of soils (Peigne et al., 2018).

9.3.2.1 SOIL PH

Organic farming generally leads soil pH closer to neutral, higher macro and micronutrients, higher microbial biomass C and N as compared with conventional farms (Liebig and Doran, 1999) (Table 9.3). Di Prima et al. (2018) compared the impacts of organic and conventional management practices on soil characteristics of citrus orchards in eastern Spain. Although an increase in soil pH with long-term organic fertilization was found in several studies, addition of chipped pruned branches, weeds, and manure from sheep and goats (annually at 8 Mg ha^{-1}) caused a lower soil pH over a decade and developed a 5 mm litter layer and a 3 mm organic layer. In contrast to lower soil pH with the incorporation of manure reported by Di Prima et al. (2018), higher soil pH was found for an organically managed sandy loam soil (Nautiyal et al., 2010). Similar to the findings of Nautiyal et al. (2010); and Nobile et al. (2020) also determined an increase in soil pH with 10 years of compost and slurry at high dose compared to the control soil by up to 0.7 and 1.2 units on average, respectively. The increase in pH under OFSs is mainly attributed to the alkaline nature of the applied organic amendments (von Arb et al., 2020).

TABLE 9.3 Soil pH and Total Phosphorus (TP) Values for Organic (O) versus Conventional (C) Farming Systems

Duration	Organic Practice	Soil Type	Depth	pH		TP (mg/kg)		References
				O	C	O	C	
>70-year	5.1 and 1.8	Loam	0–10	8.2	8.4	1.06 kg/m³	1.07 kg/m³	Pulleman et al. (2003)
	tDMha±1 yr±1 crop residue and animal manure		10–20	8.3	8.3			
6-year	Cultivation of *Asparagus officinalis* L.	Silty clay loam	0–10	7.50	7.99	–	–	Monokrousos et al. (2006)
5-year				7.83				
3-year				7.89				
2-year				7.94				
9.4-year	Animal manure and cover crop	Silt Loam	0–20	7.32	7.39	899.8	868.6	van Diepeningen et al. (2006)
10.3-year		Loamy fine sand		5.38	5.46	781.3	784.7	
5-year	25–35 tons/ha in 2 years, mixture of sheep and hen manure	Silty clay loam	0–10	7.8	7.9	–	–	Garcia-Ruiz et al. (2009)
4-year		Clay loam		7.6	7.6			
5-year		silt loam		7.8	7.8			
14-year	20 ton of composted Cow manure per hectare each year	Sandy loam	0–15	7.9	6.5*	85.5 kg/ha	94.5 kg/ha*	Nautiyal et al. (2010)
20-year	Chipped pruned Branches, weeds, composted sheep manure (10 Mg ha^{-1})	Sandy loam	0–10	7.48	7.65	–	–	Hondebrink et al. (2017)
10-year		Sandy loam	0–10	7.75	7.52	–	–	
3-year	Horse manure, green cover crop (mustard)	Silt	0–15	7.0	7.2	–	–	Morwan et al. (2018)

Fallow grassland.

TABLE 9.4 Soil Organic Matter (SOM), Total Organic Carbon (TOC), and Total Organic Nitrogen (TON) Values for Organic (O) versus Conventional (C) Farming Systems

Duration	Organic Practice	Soil Type	Depth	SOM (g/kg)		TOC (mg/kg)		TON (mg/kg)		References
				O	C	O	C	O	C	
>70-year	5.1 and 1.8 tDMha±1 yr±1 crop residue and animal manure	Loam	0–10 10–20	24.9 23.2	16.2 14.6	–	–	–	–	Pulleman et al. (2003)
6-year	Grass clover, cereal crops	Sandy loam	0–15	–	–	12.26	11.47	0.99	0.79	Bending et al. (2004)
7-year	Vetch cover 5.1 Mg/ha poultry litter	–	0–5	–	–	25.2 kg/m^3	25.6 kg/m^3	2.3 kg/m^3	2.3 kg/m^3	Green et al. (2005)
1993-09/2000 04/2001 11/2001	Poultry composted manure (8 t/ha), green manuring of cover crops residues	Sandy Clay Loam	5–20	–	–	0.90% 0.93% 0.99%	0.87% 0.95% 0.93%	0.20% 0.21% 0.12%	0.17% 0.13% 0.08%	Marinari et al. (2006)
			20–35	–	–	0.95% 0.81% 0.83%	0.88% 0.80% 0.81%	0.18% 0.22% 0.12%	0.08% 0.13% 0.07%	
5-year 4-year 5-year	25–35 tons/ha in 2 years, mixture of sheep and hen manure	Silty clay loam Clay loam Silt Loam	0–10	–	–	8.2% 5.6% 8.9%	7.9% 4.6% 7.8%	0.33% 0.22% 0.21%	0.26% 0.15% 0.13%	Garcia-Ruiz et al. (2009)
14-year	20 ton of composted cow manure per hectare each year	Sandy loam	0–15	–	–	10.20%	0.81%	2.55%	0.08%	Nautiyal et al. (2010)

TABLE 9.4 *(Continued)*

Duration	Organic Practice	Soil Type	Depth	SOM (g/kg)		TOC (mg/kg)		TON (mg/kg)		References
				O	C	O	C	O	C	
–	–	Silty	0–20	5.0%	3.5%	–	–	–	–	Papdopoulos et al. (2014)
		Silty clay		2.5%	1.7%					
		Sandy loam		8.2%	7.9%					
7-year	20–40 Mg/ha/yr (fresh weight) slurry and/or solid cow manure	Clay loam	0–10	37.07*	32.33	–	–	–	–	Crittenden et al. (2015)
			10–20	33.95	31.00					
8-year			0–10	41.11*		–	–	–	–	
			10–20	35.94						
20-year	Chipped pruned branches, weeds, composted sheep manure (10 Mg ha^{-1})	Sandy loam	0–10	10.57%	3.50%	–	–	–	–	Hondebrink et al. (2017)
10-year		Sandy loam	0–10	3.05%	2.03%	–	–	–	–	
15-year	Manure (5 t/ha), decorticated neem cake (125 kg/ha), green manure	Sandy loam	0–15	1.37%	0.88%	0.80%	0.51%	–	–	Sihi et al. (2017)
3-year	Horse manure, cover crop (mustard)	Silt	0–15	–	–	13.3 g/kg	10.2 g/kg	–	–	Morwan et al. (2018)

**Tillage systems were moldboard plowing.*

9.3.2.2 TOTAL ORGANIC CARBON AND NITROGEN

Soil organic C (SOC) increases under steady high doses of animal manure or plant residue additions. The changes in SOC content of a soil at a given site rely on the duration of addition, soil texture, mineralogy, and structural stability, among other factors (Frossard et al., 2016). In contrast to some of the research findings which stated that organic farms significantly accumulated organic matter in the soil, Marinari et al. (2006) indicated that 7 years of consistent poultry manure (PM) (8 t ha^{-1}) and cover crops residues additions were not sufficient to increase the SOC content under organic farming relative to the conventional farming (Table 9.4). The results of van Diepeningen et al. (2006) also found no significant increase in SOC content under organic farming, despite the repeated application of animal and GMs. Non-significant difference in SOC content between organic and conventional farming was attributed to the priming effect due to the high ratio of labile C species in manure and mechanical weed control, which accelerates the decomposition of organic matter. In addition, enzymatic activities as the reliable indicators of microbial activity are significantly higher in organically managed soil compared to conventional. Therefore, mineralization of organic matter is boosted with the increased enzymatic activities in the organically managed soil (Marinari et al., 2006).

Low doses of organic matter amendments may not be sufficient to compensate the losses of SOM due to the tillage and mechanical weeding during the crop growth period (von Arb et al., 2020). Addition of manure, compost, and other organic amendments and using cover crops supply essential macro- and micronutrients to meet crop needs (Reeve et al., 2016), though the carbon to nitrogen (C/N) ratio of organic amendments is the main determinant on the extent of impact on soil health. Expected benefit may not be obtained by the addition of organic materials with very low or very high C/N (Cates et al., 2015), organic amendments with C/N less than 20 decompose rapidly and become available to crops in a short time (Hadas et al., 2004). Total organic carbon and nitrogen contents, microbial biomass and enzymatic activities under 14-year of organically cultivated field were higher compared to adjacent fallow grassland soil. Total organic C content in organically managed field with additions of composted cow manure, crop residues and no-tillage for the 14 years was about 12.5 times compared to the fallow grassland soil, thus C/N ratio of organically managed soils was higher than the fallow grassland soils (Nautiyal et al., 2010).

9.3.2.3 MICRO AND MACRO PLANT NUTRIENTS

SOM enhances the availability of micronutrients for crops (Li et al., 2007). The concentrations of nutrients in soils that were under long term OFSs in different parts of the world are not similar to each other (Maltais-Landry et al., 2015). Differences in soil nutrient concentrations between the studies can be associated with different rates of nutrient load of organic amendments, the extents of nutrient uptake by crops and losses such as leaching (Nobile et al., 2020), biomass removal, burning, etc. Long term repeated application of organic amendments, especially in high input system, progressively increases the levels of nutrients in organically managed farms. Therefore, nutrients concentrations usually exceed crop requirements, which cause a decrease in nutrient-use efficiency and potentially contribute to non-point source pollution through losses to the environment. In contrast, anaerobic N concentration in organic farms is reported higher than crop N requirement during the growing season. Therefore, the potential for nonpoint source pollution due to the low level of NO_3-N on organic farms is lower than the conventional farms (Liebig and Doran, 1999). Continues cow manure and crop residue incorporation in organic farming resulted in significantly higher Zn, Mn, and S, while lower P, K, Fe, and Cu concentrations as compared with fallow grassland (Nautiyal et al., 2010). Contrarily, application of organic fertilizers in organic farming led to a significant increase in total P concentration of soils. Nobile et al. (2020) reported that total P concentration was 4.3 g kg^{-1} in soil receiving 10 years of compost, likewise 10-year slurry addition caused P accumulation (4.2 g kg^{-1}).

Long-term compost and slurry applications decreased the soil inorganic P sorption capacity of Andolsols, principally due to the increasing soil pH level (Nobile et al., 2020). The decrease in inorganic P sorption was ascribed by the relationship between electric potential and pH level of soils (Antelo et al., 2005). The electric potential of colloidal surfaces is lower at high pH levels, which causes an electrostatic repulsion between the charged surface and inorganic P and thus decrease in inorganic P sorption (Antelo et al., 2005).

9.3.3 ORGANIC FARMING AND BIODIVERSITY

Usually any agricultural activity as negative factor effects on biodiversity. First of all, agricultural activity leads exclusion of natural ecosystems and their transformation into semi-natural or full agroecosystems. The result is a change in the composition of biodiversity, extension of plant and animal species and the emergence of new ones:

- Agriculture has reduced habitat for wild species due to a 500% expansion in the extent of cropland and pasture worldwide in the last 300 years;
- Agriculture has expanded into sensitive ecosystems and had far-reaching effects on biodiversity, carbon storage and important environmental services;
- Clearing tropical forests for agriculture results in the loss of about 5–10 million hectares of forest annually;
- Habitat loss is now identified as the main threat to 85–90% of all species described by IUCN as threatened or endangered and is the most commonly recorded reason for species extinction during the last 20 years.

In the last 30 years, there has been a big change in use of agricultural lands and the same time an active development of alternative types of agriculture due to the deterioration of soil resources, decrease in the quality of food, the loss of biodiversity and the degradation of natural ecosystems as a result of their intensive use in agricultural production. The intensification and expansion of modern agriculture is amongst the greatest current threats to worldwide biodiversity. Over the last quarter of the 20th century, dramatic declines in both range and abundance of many species associated with farmland have been reported in Europe, leading to growing concern over the sustainability of current intensive farming practices (Hole et al., 2005).

In Europe, loss of biodiversity is primarily reflected in the decline of many species of plants and in the disappearing of local and old plant varieties. In 2011, the European Parliament adopted the European Union (EU) Biodiversity Strategy to 2020 with the aim to preventing biodiversity loss and degradation of ecosystems (Bavec and Bavec, 2015). One of the options to reduce the negative impact of agriculture on biodiversity while maintaining productivity and product quality and the most common type of agriculture is organic agriculture.

Organic agriculture is a complete system of production, reduced subsistence and viability of agro-ecosystems, maintaining, and restoring soil fertility; conservation of water resources, high biodiversity, production of quality products using technologies that do not harm the environment and human health. Organic agriculture must rely on all vital processes in ecosystems (EU Council Regulation No. 834/2007). At the same time, biodiversity, and soil fertility are the basis of the sustainable functionality of organic agriculture. In the process of formation of organic agroecosystem, it is necessary to use different agricultural techniques, contributing to conservation and increase of biodiversity of different living organisms, which will feed capacity and enhanced biodegradation of living organism and SOM. According to the results of investigations of last years, organic agriculture increased species richness by about 30% and had a greater effect on biodiversity, as the percentage of the landscape consisting of arable fields increased. It was found that organic fields had up to five times higher plant species richness compared to conventional fields, the same impact was recorded for different animal groups (Bavec and Bavec, 2015). However, very often, the increase in biodiversity both in the agricultural fields and on pastures occurs due to the increase of a number of weeds, and for insects due to the increase of agricultural pests. But on the other hand, the abundance of weeds and the absence of the use of chemical pesticides improve the conditions for pollinating insects, which can increase the yield of entomophilous agricultural plants. Biodiversity is one of the most important ecosystem services of organic agriculture is connected to biocontrol and pollination services. Organic farming might decrease the biomass of the crop by 25%, but increases the diversity of most functional species groups. The abundance of cereal aphids was five-times lower in organic fields, while predator abundances were 20 times higher in organic fields, indicating significantly higher potential for biological pest control in organic fields. Organic fields had 20 times higher pollinator species richness compared to conventional fields. Pollinators and predator abundance was higher at field edges compared to field centers, highlighting the importance of field edges for ecosystem services. Edges provide important nesting, feeding, and sheltering sites for birds in agricultural areas (Krauss et al., 2011).

The importance of organic agriculture in the maintenance and conservation of biodiversity, natural resources is reflected in the international agreements and conventions, as well as the national laws and regulations. For example Convention on Biological diversity encourages the development of agricultural technologies and methods that not only increase the productivity, but also prevent degradation, as well as contribute to the restoration, rehabilitation, and enhancement of biological diversity, including organic agriculture (UN Convention on Biological Diversity). In addition, the CBD includes the global strategy for plant conservation (GSPC) (2010–2020), which considers organic agriculture as one of the sustainable uses of plant diversity.

Organic fields usually accommodate a greater variety of plants, animals, and microorganisms. The organic agroecosystem is thus more resistant to stress and disturbance. The structure and species composition of plant communities in organic agricultural lands are very close to the natural ecosystems, in contrast to traditional farms, which helps to preserve the natural appearance of the landscape and genetic diversity. In the surroundings of organic farms can occur localities of rare and endangered species. In total biodiversity of organic farms is 6 times higher than in traditional farms. With Organic agriculture is possible to create suitable conditions for beneficial insects, birds, other animals, soil mycoflora, micro, and macrofauna, which promotes the increase of plant biodiversity.

According to different investigations for ecological balance of ecosystem, the optimal percentage ratio between the natural and transformed (agro) ecosystems in the same area should be 60:40 (Chernikov et al., 2000). Organic agriculture has a smaller negative impact on natural biodiversity, and has many advantages in this respect compared to modern intensive agriculture, including big potential for conservation of current biodiversity of agricultural landscapes and areas.

As conclusion we should mention that in rural areas the transition from traditional to organic agriculture should not only help to reduce the negative impacts on natural biodiversity, but also to contribute the increase biodiversity. On the other hand, still there are not enough comparative scientific studies, which can prove that: organic agriculture approach provides 100% greater benefits to biodiversity than carefully targeted prescriptions applied to relatively small conventional agriculture habitats (agri-environmental schemes). This issue is especially important for mountainous countries, where a significant part of agriculture is built on the basis of small-sized fields and pastures. With organic agriculture there are:

- Limitations of quantitative studies and analysis depends on methodological limitations
- Big knowledge gap on long term impacts of organic agriculture in specific habitats and climatic conditions, sustainability under the climate change, etc.
- Limitations and gaps in national and international regulations and strategies on organic agriculture and relations with biodiversity conservation

9.3.4 ORGANIC FARMING PLANT GROWTH AND YIELD

Organic manure can help improve soil conditions such as pH rises in acid soils, soil water retaining efficiency, hydraulic conductivity and infiltration levels and the reduction in soil bulk density. Mahmoodabadi et al. (2010) reported that manure also offers sufficient plant nutrients and improves soil structure. Applying cow manure may enhance SOM. SOM enhances soil quality, thus increasing the supply of nutrient sources. Organic matter also contributes to crop growth and yields directly by providing nutrients and indirectly by modifying soil physical properties such as aggregate stability and porosity to increase root growth, rhizosphere, and stimulate plant growth. In addition, manure application also affects the accumulation of macro-protected carbon. Silvester and Sujalu (2013) reported that the spread of cow manure may provide nutrients for the formation of leaves. On the other hand, organic fertilizer nutrients support the rapid root development (Baldi et al., 2010), which may have increased the growth of the leaves towards the end of plant life. Ghosh et al. (2004) reported that the total content of chlorophyll was increased by an increase in the inorganic fertilizer dose complemented by organic fertilizer.

When applied correctly, organic manure supports plants such as maize and increases in general size, height, and leaf count (Asiegbu and Uzo, 1984). If well supplied with organic materials, maize can tolerate sandy soil. On the heavier soil, the output is generally large and longer than the lighter soil.

The use of organic manure (poultry dropping) improves the soil's chemical and physical properties, thus increasing the growth and yield of maize, according to Okoroafor et al. (2013). Organic manure increases plant height, number of maize leaves, stem circumference, cob numbers and weight of fresh maize during harvest. Jeptoo et al. (2013) reported that the use of *Tithonia diversifolia* manure led to a higher fresh weight, dry root, and biomass and volume of the root compared to the control system. In season 1 and 2, the total yield of carrots subjected to 3.0 t/ha increased by 33% and 18%, respectively, compared with the control (Jeptoo et al., 2013). At the highest tithonia level the sweetness of carrot was influenced (Jeptoo et al., 2013).

9.4 CONCLUSION

Many environmental factors affect the growth of many plants, which influence the physiological and morphological changes of the plant in more than one way. Organic farming is a method of growing plants and animals that needs far more than deciding not to use pesticide, fertilizers, GMOs, antibiotics, and growth hormones. Soil fertility means more than supplying macro and micronutrients to plants in OFSs. In organic farming that supplies mineral nutrients, plant biomass and productivity are limited in many ecosystems. On the other hand, organic farming has been planned to optimize the productivity and health of various communities, including soil, plants, livestock, and people, within the agroecosystem.

KEYWORDS

- **agroecosystem**
- **environment**
- **farming systems**
- **global strategy for plant conservation**
- **micronutrients**
- **mineral nutrients**
- **organic farming**

REFERENCES

Abbott, L. K., & Manning, D. A., (2015). Soil health and related ecosystem services in organic agriculture. *Sustain Agriculture Res., 4*(3), 116–125.

Abebe, G., Hattar, B., & Al-Tawah, A. M., (2005). Nutrient availability as affected by manure application to cowpea (*Vigna unguiculata* L. Walp.) on calcareous soils. *Journal of Agriculture and Social Sciences, 1*(1), 1–6.

Abera, T., Feyisa, D., Yusuf, H., Nikus, O., & Al-Tawaha, A. R., (2005). Grain yield of maize as affected by biogas slurry and NP fertilizer rate at Bako, Western Oromiya, Ethiopia. *Bioscience Research, 2*, 31–38.

Abu, O. A. M., Melnyk, A. V., Onychko, V. I., Usmael, F. M., Abdullah, M. J., Rifaee, M. K., & Tawaha, A. M., (2018). Evaluation of six sunflower cultivars for forage productivity under salinity conditions. *Advances in Environmental Biology, 12*(7), 13–15.

Al Zoubi, O. M., & Al-Tawaha, A. R., (2019). Allelopathic effect of beetroot (*Beta vulgaris* L.) on germination and growth *Zea mays* and *Vigna umbellata*. *International Journal of Botany Studies, 4*(4), 47–51.

Al-Ajlouni, M. M., Al-Ghzawi, A. L. A., & Al-Tawaha, A. R., (2010). Crop rotation and fertilization effect on barley yield grown in arid conditions. *Journal of Food, Agriculture and Environment, 88*(3), 869–872.

Al-Juthery, H. W., Habeeb, K. H., Jawad, F., & Altaee, K., AL-Taey, D. K. A., & Al-Tawaha, A. R. M., (2018). Effect of foliar application of different sources of nano-fertilizers on growth and yield of wheat. *Bioscience Research, 15*(4), 3988–3997.

Al-Kiyyam, M. A., Turk, M., Al-Mahmoud, M., & Al-Tawaha, A. R., (2008). Effect of plant density and nitrogen rate on herbage yields of marjoram under Mediterranean conditions. *American-Eurasian Journal of Agricultural and Environmental Science, 3*, 153–158.

Al-Tawaha, A. M., & Turk, M. A., (2002a). Lentil (*Lens culinaris* Medic.) productivity as influenced by rate and method of phosphate placement in a Mediterranean environment. *Acta Agronomica Hungarica, 50*(2), 197–201.

Al-Tawaha, A. M., & Turk, M. A., (2004). Field pea seeding management for semi-arid Mediterranean conditions. *Journal of Agronomy and Crop Science, 190*, 86–92.

Al-Tawaha, A. M., Singh, V. P., & Turk, M. A., (2003). A review on growth, yield components and yield of barley as influenced by genotypes, herbicides, and fertilizer application. *Research on Crop, 4*(1), 1–9.

Al-Tawaha, A. M., Turk, M., Hameed, K., Assaf, T., Tahhan, R., & Al-Jamali, A. F., (2010). Effects of application of olive mill by-products on lentil yield and their symbiosis with mycorrhizal fungi under arid conditions. *Crop Research, 40*(3), 66–75.

AL-Tawaha, A. R. M., & Nidal, O. D. A. T., (2010). Use of sorghum and maize allelopathic properties to inhibit germination and growth of wild barley (*Hordeum spontaneum*). *Notulae Botanicae Horti Agrobotanici Cluj-Napoca, 38*(3), 124–127.

Al-Tawaha, A. R. M., Turk, M. A., Lee, K. D., Zheng, W. J., Ababneh, M., Abebe, G., & Musallam, I. W., (2005). Impact of fertilizer and herbicide application on performance of ten barley genotypes grown in northeastern part of Jordan. *International Journal of Agriculture and Biology, 7*(2), 162–166.

Al-Tawaha, A. R., Al-Karaki, G., Al-Tawaha, A. R., Sirajuddin, S. N., Makhadmeh, I., Wahab, P. E. M., Youssef, R. A., et al., (2018a). Effect of water flow rate on quantity and quality of lettuce (*Lactuca sativa* L.) in nutrient film technique (NFT) under hydroponics conditions. *Bulgarian Journal of Agricultural Science, 245*, 791–798.

Al-Tawaha, A. R., Al-Tawaha, A. R., Alu'datt, M., Al-Ghzawi, A. L., Wedyan, M., Al-Obaidy, S. D. A., & Al-Ramamneh, E. A. D., (2018b). Effects of soil type and rainwater harvesting treatments in the growth, productivity, and morphological trains of barley plants cultivated in semi-arid environment. *Australian Journal of Crop Science, 12*(6), 975–979.

Al-Tawaha, A. R., McNeil, D. L., Yadav, S. S., Turk, M., Ajlouni, M., Abu-Darwish, M. S., Al-Ghzawi, A. L. A., et al., (2010). Integrated legume crops production and management technology. In: *Climate Change and Management of Cool Season Grain Legume Crops* (pp. 325–349). Springer, Dordrecht.

Al-Tawaha, A. R., Turk, M. A., Abu-Zaitoon, Y. M., Aladaileh, S. H., Al-Rawashdeh, I. M., Alnaimat, S., Al-Tawaha, A. R. M., et al., (2017). Plants adaptation to drought environment. *Bulgarian Journal of Agricultural Science, 23*(3), 381–388.

Aludatt, M. H., Rababah, T., Alhamad, M. N., Al-Tawaha, A., Al-Tawaha, A. R., Gammoh, S., Ereifej, K. I., et al., (2019). Herbal yield, nutritive composition, phenolic contents and antioxidant activity of purslane (*Portulaca oleracea* L.) grown in different soilless media in a closed system. *Industrial Crops and Products, 141*, 111746.

Ananthi, T., Amanullah, M. M., & Al-Tawaha, A. R. M. S., (2017). A review on maize-legume intercropping for enhancing the productivity and soil fertility for sustainable agriculture in India. *Advances in Environmental Biology, 11*(5), 49–63.

Antelo, J., Avena, M., Fiol, S., López, R., & Arce, F., (2005). Effects of pH and ionic strength on the adsorption of phosphate and arsenate at the goethite-water interface. *J Colloid Interface Sci., 285*(2), 476–486.

Arnhold, S., Lindner, S., Lee, B., Martin, E., Kettering, J., Nguyen, T. T., & Hume, B., (2014). Conventional and organic farming: Soil erosion and conservation potential for row crop cultivation. *Geoderma, 219*, 89–105.

Assaf, T. A., Hameed, K. M., Turk, M. A., & Al-Tawaha, A. M., (2006a). Effect of soil amendment with olive mill by-products under soil solarization on growth and productivity of faba bean and their symbiosis with mycorrhizal fungi. *World Journal of Agricultural Sciences, 2*(1), 21.

Assaf, T. A., Hameed, K. M., Turk, M. A., & Al-Tawaha, A. M., (2006b). Effect of soil amendment with olive mill by-products under soil solarization on growth and productivity of faba bean and their symbiosis with mycorrhizal fungi. *International Journal of Plant Production, 2*(4), 341–352.

Baldi, E., & Toselli, M., (2013). Root growth and survivorship in cow manure and compost amended soils. *Plant Soil Environ., 59*, 221–226.

Barański, M., Średnicka-Tober, D., Volakakis, N., Sanderson, R., Stewart, G. B., Benbrook, C., Biavati, B., et al., (2014). Higher antioxidant and lower cadmium concentrations and lower incidence of pesticide residues in organically grown crops: A systematic literature review and meta-analyses. *British J. Nutr., 112*, 794–811.

Bavec, M., & Bavec, F., (2015). Impact of organic farming on biodiversity. In: Lo, Y. H., & Blanco, J. A., (eds.), *Biodiversity in Ecosystems-Linking Structure and Function* (pp. 1–18). doi: 10.5772/58974.

Bayer, C., Martin-Neto, L., Mielniczuk, J., Pavinato, A., & Dieckow, J., (2006). Carbon sequestration in two Brazilian cerrado soils under no-till. *Soil Till. Res., 86*, 237–245.

Bending, G. D., Turner, M. K., Rayns, F., Marx, M. C., & Wood, M., (2004). Microbial and biochemical soil quality indicators and the potential for differentiating areas under contrasting agricultural management regimes. *Soil Biol. Biochem., 36*(11), 1785–1792.

Biswas, M. S., Ali, N., Goswami, R., & Chakraborty, S., (2014). Soil health sustainability and organic farming: A review. *J. Food Agric. Environ., 12*(3, 4), 237–243.

Bünemann, E. K., Bongiorno, G., Bai, Z., Creamer, R. E., De Deyn, G., De Goede, R., & Pullman, M., (2018). Soil quality: A critical review. *Soil Biol. Biochem., 120*, 105–125.

Cates, A. M., Ruark, M. D., Hedtke, J. L., & Posner, J. L., (2015). Long term tillage, rotation and perennialization effects on particulate and aggregate organic matter. *Soil and Till Research*, 371–380.

Chernikov, V., Aleksakhin, R., & Golubyev, A., (2000). *Agroecology, Textbook for HEIs* (p. 536). Moskow.

Cooper, J., Baranski, M., Stewart, G., Nobel-de, L. M., Bàrberi, P., Fließbach, A., & Crowly, O., (2016). Shallow non-inversion tillage in organic farming maintains crop yields and increases soil C stocks: A meta-analysis. *Agron. Sustain. Dev., 36*(1), 22.

Crittenden, S. J., & De Geode, R. G. M., (2016). Integrating soil physical and biological properties in contrasting tillage systems in organic and conventional farming. *Euro J. Soil Biol., 77*, 26–33.

Crittenden, S. J., Poot, N., Heinen, M., Van, B. D. J. M., & Pulleman, M. M., (2015). Soil physical quality in contrasting tillage systems in organic and conventional farming. *Soil Till. Res., 154*, 136–144.

Di Prima, S., Rodrigo-Comino, J., Novara, A., Iovina, M., Pirastru, M., Keessta, S., & Cerda, A., (2018). Soil physical quality of citrus orchards under tillage, herbicide and organic managements. *Pedosphere, 28*(3), 463–477.

Erhart, E., & Hartl, W., (2009). Soil protection through organic farming: A review. In: *Organic Farming, Pest Control and Remediation of Soil Pollutants* (pp. 203–226). Springer, Dordrecht.

European Union Council, (2007). Regulation No. 834/2007 of 28 June 2007 on organic production and labelling of organic products and repealing Regulation No. 2092/91. *Official Journal of the European Union, L189*, 1–23.

Franzleubbers, A. J., Wright, S. F., & Stuedemann, J. A., (2000). Soil aggregation and glomalin under pastures in the southern piedmont, USA. *Soil Sci Soc Am J., 64*, 1018–1026.

Frick, (2000). *Organic Farming Enhances Soil Fertility and Biodiversity: Results from a 21 Year old Field Trial* (pp. 1–96). Research Institute of Organic Framing, Switzerland.

Frossard, E., Buchman, N., Bünemann, E. K., Kiba, D. I., Lompo, F., Oberson, A., & Traore, O. Y., (2016). Soil properties and not inputs control carbon: Nitrogen: Phosphorous ratios in cropped soils in the long term. *Soil, 2*(1), 83–99.

García-Ruiz, R., Ochoa, V., Vingela, B., Hinojosa, M. B., Pena-Santiago, R., Liébanas, G., & Carreira, J. A., (2009). Soil enzymes, nematode community and selected physic-chemical properties as soil quality indicators in organic and conventional Olive oil farming: Influence of seasonality and site features. *Appl. Soil Ecol., 41*(3), 305–314.

Ghosh, P. K., Ajay, B. K. K., Manna, M. C., Mandal, K. G., Misra, A. K., Hati, K. M., et al., (1984). *Evaluation Principle in Fertilizer*, 50.

Ghosh, P. K., Ajay, B. K. K., Manna, M. C., Mandal, K. G., Misra, A. K., & Hati, K. M., (2004). Comparative effectiveness of cattle manure, poultry manure, phosphocompost and fertilizer-NPK on three cropping systems in vertisols of semi-arid tropics. II. Dry matter yield, nodulation, chlorophyll content and enzyme activity. *Bioresource Technology, 95*(1), 85–93.

Głąb, T., Pużyńska, K., Pużyński, S., Palmowska, J., & Kowalik, K., (2016). Effect of organic farming on a Stagnic Luvisol soil physical quality. *Geoderma, 282*, 16–25.

Green, V. S., Cavigelli, M. A., Dao, T. H., & Flanagan, D. C., (2005). Soil physical properties and aggregate-associated C, N and P distributions in organic and conventional cropping systems. *Soil Sci., 170*(10), 822–831.

Hadas, A., Kautsky, L., Goek, M., & Erman, K. E., (2004). Rates of decomposition of plant residues and available nitrogen in soil, related to residue composition through the stimulation of carbon and nitrogen turnover. *Soil Biol. Biochem., 36*, 255–266.

Hani, N. B., Al-Ramamneh, E. A. D., Haddad, M., Al-Tawaha, A. R., & Al Satari, Y., (2019). The impact of cattle manure on the content of major minerals and nitrogen uptake from ^{15}N isotope-labeled ammonium sulphate fertilizer in maize (*Zea mays* L.) plants. *Pakistan Journal of Botany, 51*(1), 185–189.

Hillel, D., (1998). Irrigation and water-use efficiency. In: *Environmental Soil Physics* (pp. 617–646). Academic Press, London.

Hole, D., Perkins, A., Wilson, J., Alexander, I., Grice, P., & Evans, A., (2005). Does organic farming benefit biodiversity? *Biological Conservation, 122*(1), 113–130.

Hondebrink, M. A., Cammeraat, L. H., & Cedra, A., (2017). The impact of agricultural management on selected soil properties in citrus orchards in Eastern Spain: A comparison between conventional and organic citrus orchards with drip and flood irrigation. *Sci Total Environ., 581*, 153–160.

Jeptoo, A., Aguyoh, J. N., & Saidi, M., (2013). Tithonia manure improves carrot yield and quality. *GJBAHS, 2*(4), 136–142.

Kaje, V. V., Sharma, D. K., Shivay, Y. S., Jat, S. L., Bhatia, A., Purakayastha, T. J., & Bhattacharyya, A. N., (2018). Long-term impact of organic and conventional farming on soil physical properties under rice (*Oryza sativa)* – wheat (*Triticum aestivum*) cropping system in northwestern Indo-Gangetic plains. *Indian J. Agric. Sci., 88*(1), 107–113.

Khan, N., Khan, M. I., Khalid, S., Iqbal, A., & Al-Tawaha, A. R., (2019). Wheat biomass and harvest index increases with integrated use of phosphorus, zinc, and beneficial microbes under semiarid climates. *Journal of Microbiology, Biotechnology and Food Sciences*, 242–247.

Knapp, V. V., & Van, D. H. M. G., (2018). A global meta-analysis of yield stability in organic and conservation agriculture. *Nat Commun., 9*(1), 3632.

Krauss, J., Gallenberger, I., & Steffan-Dewenter, I., (2011). Decreased functional diversity and biological pest control in conventional compared to organic crop fields. *PLoS One, 6*(5), 1–9.

Lado, M., Paz, A., & Ben-Hur, M., (2004). Organic matter and aggregate size interactions in infiltration, seal formation and soil loss. *Soil Sci. Soc. Am. J., 68*, 935–942.

Lampkin, N., & Padel, S., (1994). *The Economics of Organic Agriculture: An International Perspective*. CAB International, Wallingford, UK.

Leibig, M. A., & Doran, J. W., (1999). Impact of organic production practices on soil quality indicators. *J. Environ. Qual., 28*(5), 1601–1609.

Li, B. Y., Zhou, D. M., Cang, L., Zhand, H. L., Fan, X. H., & Qin, S. W., (2007). Soil micronutrient availability to crops as affected by long-term inorganic and organic fertilizer applications. *Soil Till. Res., 96*, 166–173.

Lori, M., Symnaczik, S., Mäder, P., De Deyn, G., & Gattinger, A., (2017). Organic farming enhances soil microbial abundance and activity: A meta-analysis and meta-regression. *PLoS One, 12*(7), 1–25, e0180442.

Mäder, P., Fliessbach, A., Dubois, D., Gunst, L., Fried, P., & Niggli, U., (2002). Soil fertility and biodiversity in organic farming. *Science, 296*(5573), 1694–1697.

Mahmoodabadi, M., Amini, R. S., & Khazaeepour, K., (2010). Using animal manure for improving soil chemical properties under different leaching conditions. *Middle-East J. Sci. Res., 5* (4), 214–217.

Maltais-Landry, G., Scow, K., Brennan, E., & Vitousek, P., (2015). Long-term effect of compost and cover crops on soil phosphorous in two California agroecosystems. *Soi Sci Soc Am J., 79*(2), 688–697.

Marinari, M. M., Marinari, R., Campiglia, E., & Grego, S., (2006). Chemical and biological indicators of soil quality in organic and conventional farming systems in Central Italy. *Ecolog. Indic., 6*(4), 701–711.

Mikha, M. M., & Rice, C. W., (2004). Tillage and manure effects on soil and aggregate-associated carbon and nitrogen. *Soil Sci. Am. J., 68*, 809–816.

Moebius-Clune, B. N., Moebius-Clune, D. J., Gugino, B. K., Idowu, O. J., Schindelbeck, R. R., Ristow, A. J., Van, E. H. M., et al., (2016). *Comprehensive Assessment of Soil Health. The Cornell Framework, Edition 3.1*. (p. 123). Cornell University, NY.

Monokrousos, N., Papatheodorou, E. M., Diamantopoulos, J. D., & Stamou, G. P., (2006). Soil quality variables in organically and conventionally cultivated field sites. *Soil Biol. Biochem., 38*(6), 1282–1289.

Moravan, X., Verbeke, L., Laratte, S., & Schneider, A. R., (2018). Impact of recent conversion to organic farming on physical properties and their consequences on runoff, erosion and crusting in a silty soil. *Cetena, 165*, 398–407.

Nautiyal, C. S., Chauhan, P. S., & Bhatia, C. R., (2010). Changes in soil physicochemical properties and microbial functional diversity due to 14 years of conversion of grassland to organic agriculture in semi-arid agroecosystem. *Soil Till. Res., 109*(2), 55–60.

Niggli, U., (2015). Sustainability of organic food production: Challenges and innovations. *Proc. Nutr. Soc., 74*(1), 83–88.

Nikus, O., Al-Tawaha, A. M., & Turk, M. A., (2004a). Effect of manure supplemented with phosphate fertilizer on the fodder yield and quality of two sorghum cultivars (*Sorghum bicolor* L.). *Bioscience Research, 1*, 1–7.

Nikus, O., Al-Tawaha, A. M., & Turk, M. A., (2004b). Yield response of Sorghum (*Sorghum bicolor* L.) to manufacture supplemented with phosphate fertilizer under semi-arid Mediterranean conditions. *International Journal of Agriculture and Biology, 6*, 889–893.

Nobile, C. M., Bravin, M. N., Becquer, T., & Paillat, J. M., (2020). Phosphorous sorption and availability in an andosol after a decade of organic or mineral fertilizer application: Importance of pH and organic carbon modifications in soil as compared to phosphorous accumulation. *Chemosphere, 239*, 124709.

Okoroafor, I. B., Okelola, E. O., Edeh, O., Nemehute, V. C., Onu, C. N., Nwaneri, T. C., & Chinaka, G. I., (2013). Effect of organic manure on the growth and yield performance of maize in Ishiagu, Ebonyi State, Nigeria. *IOSR Journal of Agriculture and Veterinary Science, 5*(4), 28–31.

Oquist, K. A., Strock, J. S., & Mulla, D. J., (2006). Influence of alternative and conventional management practices on soil physical and hydraulic properties. *Vadose Zone, 5*, 356–364.

Papadopoulos, A., Bird, N. R. A., Whitmore, A. P., & Mooney, S. J., (2014). Does organic management leads to enhanced soil physical quantity. *Geoderma, 213*, 435–443.

Peigné, J., Vian, J. F., Payet, V., & Saby, N. P., (2018). Soil fertility after 10 years of conservation tillage in organic farming, *Soil Till. Res., 175*, 194–204.

Puerta, V. L., Pereira, E. I. P., Wittwer, R., Van, D. H. M., & Six, J., (2018). Improvement of soil structure through organic crop management, conservation tillage and grass-clover ley. *Soil Till. Res., 180*, 1–9.

Pulleman, M., Jongmans, A., Marinissen, J., & Bouma, J., (2003). Effect of organic versus conventional arable farming on soil structure and organic matter dynamics in a marine loam in Netherlands. *Soil Use Manag., 19*(2), 157–165.

Reeves, J. R., & Drost, D., (2012). Yields and soil quality under transitional organic high tunnel tomatoes. *HortScience, 47*(1), 38–44.

Reeves, J. R., Hoagland, L. A., Villalba, J. J., Carr, P. M., Atucha, A., Cambardella, C., Davis, D. R., & Delate, K., (2016). Organic farming, soil health, and food quality: Considering possible links. In: *Advances in Agronomy* (pp. 319–367). Academic Press, London.

Reeves, W. D., (1997). The role of soil organic matter in maintaining soil quality in continuous cropping systems. *Soil Till. Res., 43*, 131–167.

Reganold, J. P., (1995). Soil quality and profitability of biodynamic and conventional farming systems: A review. *Am. J. Alter. Agric., 10*(1), 36–45.

Reganold, J. P., Elliott, L. F., & Unger, Y. L., (1987). Long-term effects of organic and conventional farming on soil erosion. *Nature, 330*(6146), 370.

Rehber, E., & Turhan, S., (2002). Prospects and challenges for developing countries in trade and production of organic food and fibers-The case of Turkey. *British Food Journal, 104*, 371–390.

Reynolds, W. D., Drury, C. F., Yang, X. M., Fox, C. A., Tan, C. S., & Zhang, T. Q., (2007). Land management effects on the near-surface physical quality of a clay loam soil. *Soil Till. Res., 96*, 316–330.

Schjønning, P., Elmholt, S., Munkholm, L. J., & Debosz, K., (2002). Soil quality aspects of humid sandy loams as influenced by organic and conventional long-term management. *Agric. Ecosys. Environ., 88*(3), 195–214.

Schonbeck, B. M., Jerkins, D., & Ory, J., (2017). *Soil Health and Organic Farming* (pp. 1–44). Santa Cruz, USA.

Seitz, S., Goebes, P., Puerta, V. L., Pereira, E. I. P., Wittwer, R., Six, J., & Scholten, T., (2019). Conservation tillage and organic farming reduce soil erosion. *Agron Sustain Dev., 39*(1), 4.

Shukla, M. K., Lal, R., & Ebinger, M., (2006). Determination of soil quality indicators by factor analysis. *Soil Till. Res., 87*, 194–204.

Sihi, D., Dari, B., Sharma, D. K., Pathak, H., Nain, L., & Sharma, O. P., (2017). Evaluation of soil health in organic vs. conventional farming of basmati rice in North India. *J. Plant Nutr. Soil. Sci., 180*(3), 389–406.

Silvester, M. N., & Sujalu, A. P., (2013). Effects of organic and inorganic fertilizers on growth, activity of nitrate reductase and chlorophyll contents of peanuts (*Arachis hypogaea*). *J. Agrofor., 7*, 206–211.

Smith, P., House, J. I., Bustamante, M., Sobocká, J., Harper, R., Pan, G., West, P. C., et al., (2016). Global change pressures on soils from land use and management. *Glob. Chang. Biol., 22*, 1008–1028.

Stockdale, E. A., Lampkin, N. H., Hovi, M., Keatinge, R., Lennartsson, E. K. M., MacDonald, D. W., Padel, S., et al., (2001). Agronomic and environmental implications of organic farming systems. *Adv. Agron., 70*, 261–327.

Sudjatmiko, S., Muktamar, Z., Chozin, M., Setyowati, N., & Fahrurrozi, F., (2018). Changes in chemical properties of soil in an organic agriculture system. In: *IOP Conference Series: Earth and Environmental Science, 215*(1), 012016.

Suja, G., & Sreekumar, J., (2014). Implications of organic management on yield, tuber quality and soil health in yams in humid tropics. *International Journal of Plant Production, 8*(3), 291–310.

Tully, K. L., & Mc Askill, C., (2019). Promoting soil health in organically managed systems: A review. *Org. Agr., 1*–20.

Turk, M. A., & Tawaha, A. M., (2001). Common vetch (*Vicia sativa* L.) productivity as influenced by rate and method of phosphate fertilization in a Mediterranean environment. *Agricultura Mediterranea, 131*, 108–111.

Turk, M. A., & Tawaha, A. M., (2002a). Onion (*Allium cepa* L.) as influenced by rate and method of phosphorus placement. *Crop Research, 23*(1), 105–107.

Turk, M. A., & Tawaha, A. M., (2002b). Impact of seeding rate, seeding date and method of phosphorous application in faba bean (*Vicia faba* L. Minor) in the absence of moisture stress. *Biotechnology, Agronomy, Society and Environment, 6*(3), 171–178.

Turk, M. A., & Tawaha, A. M., (2003a). Allelopathic effect of black mustard (*Brassica nigra* L.) on germination and growth of wild oat (*Avena fatua* L.). *Crop Protection, 22*(4), 673.

Turk, M. A., & Tawaha, A. M., (2003b). Allelopathic effect of black mustard (*Brassica nigra* L.) on germination and growth of wild barley (*Hordeum spontaneum*). *Journal of Agronomy and Crop Science, 189*(5), 298–303.

Turk, M. A., & Tawaha, A. M., (2004). Effect of variable sowing ratios and sowing rates of bitter vetch on the herbage yield of oat-bitter vetch mixed cropping. In: Peltonen-Sainio, P., & Topi-Hulmi, M., (eds.), *Proceedings 7th International Oat Conference*. MTT.

Turk, M. A., Hmaeed, K. M., Aqeel, A. M., & Tawaha, A. M., (2003c). Nutritional status of durum wheat grown in soil supplemented with olive mill by-products. *Agrochimica, 47*(5, 6), 209–219.

Turk, M. A., Lee, K. D., & Tawaha, A. M., (2005). Inhibitory effects of aqueous extracts of black mustard on germination and growth of radish. *Research Journal of Agriculture and Biological Sciences, 1*(3), 227–231.

Turk, M. A., Tawaha, A. M., & Samara, N., (2003b). Effects of seedling rate and date, and phosphorous application on growth and yield of carbon vetch (*Vicia narbonensis*). *Agronomie, 23*, 1–4.

Turk, M. A., Tawaha, A. M., & Shatnawi, M., (2003a). Lentil (*Lens culinaris* Medik) response to plant density, sowing date, phosporous fertilization and Ethephon application in the absence of moisture stress. *Journal of Agronomy and Crop Science, 189*(1), 1–6.

UN Convention on Biological Diversity, (1996). *Conference of Parties Decision III/11 on Conservation and Sustainable Use of Agricultural Biological Diversity*. Buenos Aires, Argentina.

USDA, (2020) *Natural Resources Conservation Service*. Soil health assessment. https://www.nrcs.usda.gov/wps/portal/nrcs/main/soils/health/assessment/ (accessed on 12 July 2021).

Van, D. A. D., De Vos, O. J., Korthals, G. W., & Van, B. A. H., (2006). Effects of organic versus conventional management on chemical and biological parameters in agricultural soils. *App. Soil Ecol., 31*(1, 2), 120–135.

Von, A. C., Bünemann, E. K., Schmalz, H., Portmann, M., Adamtey, N., Musyoka, M. W., & Fliessbach, A., (2020). Soil quality and phosphorus status after nine years of organic and conventional farming at two input levels in the central highlands of Kenya. *Geoderma, 362*, 114112.

Willer, H., Lernoud, J., & Kemper, L., (2018). The world of organic culture 2018: Summary. In: *The World of Organic Agriculture*: *Statistics and Emerging Trends* (pp. 22–31). Research Institute of Organic Agriculture, FiBL and IFOAM-Organics International.

Williams, A., & Hedlund, K., (2013). Indicators of soil ecosystem service in conventional and organic arable fields along a gradient of landscape heterogeneity in southern Sweden. *Appl. Soil Ecol., 65*, 1–7.

Williams, N. D., & Petticrew, E. L., (2009). Aggregate stability in organically and conventionally farmed soils. *Soil Use Manag., 25*(3), 284–292.

Wittwer, R. A., Dorn, B., Jossi, W., & Van, D. H. M. G., (2017). Cover crops support ecological intensification of arable cropping systems. *Scientific Reports, 7*, 41911.

Zhang, X., Wu, X., Zhang, S., Xing, Y., Wang, R., & Liang, W., (2014). Organic amendment effects on aggregate-associated organic C, microbial biomass C and glomalin in agricultural soils. *Catena, 123*, 188–194.

CHAPTER 10

USE OF BIOCHAR IN AGRICULTURE: AN INSPIRING WAY IN EXISTING SCENARIO

IMRAN,[1] AMANULLAH,[1] ABDEL RAHMAN M. AL-TAWAHA,[2] ABDEL RAZZAQ M. AL-TAWAHA,[3] SAMIA KHANUM,[4] DEVARAJAN THANGADURAI,[5] JEYABALAN SANGEETHA,[6] HIBA ALATRASH,[7] PALANI SARANRAJ,[8] NIDAL ODAT,[9] MAZEN A. ATEYYA,[10] MUNIR TURK,[11] ARUN KARNWAL,[12] SAMEENA LONE,[13] and KHURSHEED HUSSAIN[13]

[1]*Department of Agronomy, The University of Agriculture, Peshawar, Pakistan*

[2]*Department of Biological Sciences, Al-Hussein Bin Talal University, P.O. Box 20, Maan, Jordan*

[3]*Department of Crop Science, Faculty of Agriculture, University Putra Malaysia, Serdang–43400, Selangor, Malaysia*

[4]*Department of Botany, University of the Punjab, Lahore, Pakistan*

[5]*Department of Botany, Karnatak University, Dharwad–580003, Karnataka, India*

[6]*Department of Environmental Science, Central University of Kerala, Kasaragod–671316, Kerala, India*

[7]*General Commission for Scientific Agricultural Research, Syria*

[8]*Department of Microbiology, Sacred Heart College (Autonomous), Tirupattur–635601, Tamil Nadu, India*

[9]*Department of Medical Laboratories, Al-Balqa Applied University, Al-Salt–19117, Jordan*

[10]*Faculty of Agricultural Technology, Al Balqa Applied University, Al-Salt–19117, Jordan*

[11]*Department of Plant Production, Jordan University of Science and Technology, Irbid, Jordan*

[12]*Department of Microbiology, School of Bioengineering and Biosciences, Lovely Professional University, Phagwara, Punjab, India*

[13]*Division of Vegetable Science, SKUAST-Kashmir, Jammu and Kashmir, India*

ABSTRACT

Numerous factors restrain food crop production in semi-arid and Mediterranean regions. Soil fertility and plant nutrition have had a significant effect on crop production in the 20th century. Biochar has been utilized in agriculture to increase soil health, improve soil fertility, and increase crop productivity. Disturbed edaphic properties like high levels of rocks and a low pH level, high radiation levels, rapid drainage during the germination process, compaction of the soil, contamination, and other adverse effects, can be improved with biochar deposits. There is little understanding of the specific mechanisms that underlie biochar's contribution to plant response. This chapter focuses on: (i) biochar application and phosphate starvation in plants; (ii) soil microbiota and biochar as a symbiosis for plant life; (iii) biochar act as a drought-resistant tool; (iv) reclamation of soil salinity through biochar amendments; (v) biochar and soil fertility; and (vi) biochar amendments and plant growth.

10.1 INTRODUCTION

Biochar application in agriculture is getting much attention in changing climatic scenario. A large number of environmental and agricultural factors limit the production of major food crops in semi-arid and Mediterranean regions (Turk and Tawaha, 2002b; Abebe et al., 2005a, b; Lee et al., 2005d; Nikus et al., 2005a). Biochar application paly an important role in soil organic carbon building and soil fertility status (Turk and Tawaha, 2001,

2002a; Tawaha and Turk, 2002). Biochar is utilized in agriculture for the last decades, to enhance soil health, soil fertility, and crop production (Bos et al., 2005; Chimenti et al., 2006; Brussaard et al., 2007; Ping et al., 2007; Zhan et al., 2011; Akhtar et al., 2015b; Khan et al., 2016). It is stated that soil biochar and nutrients stress is most harmful in combinations with other abiotic stress and could be treated with biochar application. These findings were concerning those of Roelofs et al. (2008) who revealed that abiotic stress reduce the grain yield of cereal crops. The production of major food crops is restricted by many environmental and agricultural factors. The integration of biochar and especially of biochar is an antique farming practice aimed at enhancing soil carbon, nutrient, soil porosity, soil quality, soil fertility, and water infiltration and water retention (Palta et al., 2006; Ping et al., 2007; Peng et al., 2008; Palm and Volkenburgh, 2012; Paz et al., 2014; Pant et al., 2015). Biochar treatment with chemical and biological fertilizers may help in improved crop production and soil quality through successful management. Integrated nutrients management is essential to improve crop growth, seed yield, nodulation, and crop quality (Tawaha et al., 2005a, b; Yang et al., 2005; Zheng et al., 2005; Assaf et al., 2006). In this chapter, we have focused on biochar and it can be the right way to soil sustainability. Biochar application plays a key role in providing nutrients, improving soil functions, and reducing soil degradation (Turk and Tawaha, 2002b; Turk et al., 2003a–c). Biochar application improves soil fertility and increases crop yield. When chemical fertilizer without organic amendments is continually applied to the soil, it decreases soil bulk density and increases the porosity and buffering capacity of the soil (Nikus et al., 2005b; Mesfine et al., 2005; Sulpanjani et al., 2005a). Adding organic carbon and biochar have agricultural and economical importance due to enhancing soil fertility, improving soil health and properties by providing important macro and micronutrients, as well as sustaining soil productivity such as, mobilization of nutrients and sorption capacity which, resulting in increasing crop growth and production. (Abebe et al., 2005c, d; Abera et al., 2005; Assefa et al., 2005; Lee et al., 2005a, b). The use of organic sources along with the application of chemical fertilizers has the potential to extend the assembly of soybean and maize. Frequent use of biochar with chemical fertilizers enhances crop production. Biochar amendments may help in improving soil properties by obtaining a high yield as compared to inorganic fertilizers. Adequate agricultural practices with the recycling of organic sources (plant biomass) and the addition of biochar into the soil help decrease soil fertility degradation, improve soil quality (chemical, physical, and biological characters), nutrients supply, and crop production, as well as enhance farmer's income. Low soil quality and poor supply of nutrients are a major cause of low crop production. (Sulpanjani et al., 2005b, c; Tawaha et al., 2005a, b; Yang et al., 2005; Zheng et al., 2005; Assaf et al., 2006). Low soil quality and poor supply of nutrients are a major cause of low crop production. Regardless of soil types and crop seasons, organic applications are usually more effective to increase soil C and biochar status as compared to mineral fertilizers (Sulpanjani et al., 2005b, c; Tawaha et al., 2005a, b; Yang et al., 2005; Zheng et al., 2005; Assaf et al., 2006).

10.2 BIOCHAR APPLICATION AND PHOSPHATE STARVATION IN PLANTS

World soil is lacking frequent availability of important plant nutrients of phosphorus plant (Prado et al., 2000; Palta et al., 2006; Ping et al., 2007; Peng et al., 2008; Paz et al., 2014; Pant et al., 2015). Manure, poultry manure (PM) and different sources of organic carbon and biochar provide organic material to the soil. Maize production is improved by adding PM which will provide essential nutrients to the soil which improves soil fertility. After the application of PM, the crop yield and yield component of maize were significantly increased. Some scientists have shown that growth attributes of rice were significantly increased after the application of biochar at 12.5 t ha^{-1} combined with Azospirillum (2 kg ha^{-1}). On the other hand, biofertilizer application resulted in 6.5% in 1st year and 8.16% in 2nd year in rice due to biofertilizer application (Bos et al., 2005; Chimenti et al., 2006; Brussaard et al., 2007). Soil having maximum biochar and carbon is considered healthy and biologically diverse, resulting in more chances of the survival of the plant (Duan et al., 2007; Farahbakhsh, 2012; Imran and Khan, 2015, 2017; Imran, 2018). Biochar act synergistically to improve crop performance under harmful condition (Gonçalves-Alvim et al., 2001; Flowers, 2004; Gasco et al., 2012; González-Villagra et al., 2018). Several local conditions including climate, soil chemistry, and soil condition all influence biochar agronomic benefits (Lehmann et al., 2011; Imran et al., 2015–2018). Microbial biomass increased in biochar amended soils act as a biocontrol agent and mitigate abiotic stress induce damages (Heans, 1984; Imran and Khan, 2015; Imran et al., 2020). It has been shown that biochar application is a master regulator of P-starvation response in plants (González-Villagra et al., 2018). Biochar amended soil has been shown to regulate extensive secondary metabolites during phosphorus limitation stress (Bos et al., 2005; Brussaard et al., 2007).

10.3 SOIL MICROBIOTA AND BIOCHAR A SYMBIOSIS FOR PLANT LIFE

Biochar has positive impacts on soil microbes. Lehmann et al. (2011) stated microbial biomass increased in biochar-amended soils (Palta et al., 2006; Ping et al., 2007; Peng et al., 2008; Palm and Volkenburgh, 2012; Paz et al., 2014; Pant et al., 2015). Also, biochar application causes significant changes in microbial community composition and enzyme activities in both bulk soil and the rhizosphere (Lu and Vonshak, 2002; Lehmann et al., 2011; Lei et al., 2011; Imran et al., 2020).

10.4 BIOCHAR ACT AS A DROUGHT RESISTANT TOOL

Water is needed in so many essential processes in plant growth (Figure 10.1) and its scarcity can cause tremendous loss of economic yield (Van et al., 2001; Wang and Jin, 2007; Wolfe, 2007; Sidari et al., 2008; Tiwari et al., 2011; Zargar et al., 2017; Tombesi et al., 2018; Karim and Imran, 2019). Plants have come up with many strategies to deal with drought stress (Figure 10.2). One of the theses strategies is to increase the rate of water infiltration and soil porosity (Yeo et al., 1990; Xu et al., 2008; Zakaria et al., 2012; Younis et al., 2015; Zargar et al., 2017). Biochar application can enhance soil porosity and increasing water infiltration in various types of soil. It has been reported that biochar application regulating hormone secretion that maintains stomatal opening and closing. The hormone which is regulating by biochar application is abscisic acid (ABA). Current studies showed that levels of ABA increased in drought-stressed plants (Van et al., 2001; Wang and Jin, 2007; Wang and Xu, 2013). They reported that exposing plants to drought stress increases the levels of ABA in order to save whatever water they had in their leaves (Rubio et al., 2001; Shamsi and Kobraee, 2013; Savvides et al., 2015; Imran et al., 2020). Additionally, biochar application can regulate another important factor that improves drought tolerance and water uptake aquaporins (AQPs).

10.5 RECLAMATION OF SOIL SALINITY THROUGH BIOCHAR AMENDMENTS

Soil salinity which is one of the most serious abiotic stress that can be affect not only on the plants growth but also on the on the plant's production

(Martinez-Beltran and Manzur, 2005; Lehmann and Joseph, 2009; Khan et al., 2010; Kammann et al., 2011; Lashari et al., 2013; Imran et al., 2015, 2019; Imran and Amanullah, 2018), this problem is most significant in both dry and semi-arid area where there is a dearth of water resources (Martinez-Beltran and Manzur, 2005; Maestre et al., 2007; Lei et al., 2011; Imran et al., 2015, 2017; Imran, 2017, 2018).

FIGURE 10.1 Biochar could mitigate drought stress and enhance the growth and productivity of soybean.

FIGURE 10.2 Comparison of soybean crop under drought stress with and without biochar application.

All stages of plants can be negatively affected because of the salinity of soil from the germination of seeds through to the vegetative growth stage and reproductive development. All these harmful effects lead to decrease in the plant production soil. Soil salinity causes ion poisoning, osmotic, and oxidative stresses, and nutrient deficiency (Bano and Fatima, 2009). This reduces the plant's absorption of soil water which leads to inhibition of photosynthesis (Rubio et al., 2001; Lu and Vonshak, 2002; Lehmann et al., 2011; Lei et al., 2011). One of the most important biochar applications to improve plant tolerance to salt stress is the disruption of the Na^+/K^+ ratio in the cytoplasm of the plant cell (Joseph and Jini, 2011; Jagatheeswari and Ranganathan, 2012; Kamara et al., 2014).

Increasing concentrations of Na^+ in the soil can cause a decrease in the plant's ability to utilize soil water; here application of organic carbon and biochar improved the physical, chemical, and biological properties of soil. Many studied showed the benefit of using biochar as fertilizers to improve the fertility of soil which can be resulted in the increase in production of plants (Bos et al., 2005; Chimenti et al., 2006; Brussaard et al., 2007) mentioned that using biochar increased the yield for the rice and in the same way increase in nutrient use efficiency.

10.6 BIOCHAR AND SOIL FERTILITY

Biochar application use is still very limited in the present agriculture scenario because of poor understanding of the mechanisms by which it improves soil fertility (Bruinsma, 1963; Capell and Doerffling, 1993; Bohnert and Jensen, 1996; Bos et al., 2005; Chimenti et al., 2006; Brussaard et al., 2007). Addition of Biochar in the soil results in change of the microbial populations in the rhizosphere leading to the beneficial microorganism populations ultimately promoting plant growth and resistance to biotic stresses. Although the mechanism, resulting in the alteration of the microbial populations in the rhizosphere is still unknown.

The development of systemic resistance towards several foliar pathogens in three crop systems has been shown earlier for biochar-induced plant protection against soil-borne diseases (Lehmann and Joseph, 2009; Khan et al., 2010; Lashari et al., 2013; Imran et al., 2015, 2019; Imran and Amanullah, 2018). Some studies pointed that biochar induces responses along with both systemic acquired resistance (SAR) and induced systemic resistance (ISR) pathways leading to a major controlling system (Lu and Vonshak, 2002; Lei et al., 2011). The result of biochar soil amendment on the different soil-plant-microbe interactions leading to a significant role in plant health is being studied in this review.

Attainment of resistance to different diseases by the plant can be one of the benefits gained from applying biochar to soil (Raghothama, 1999; Prado et al., 2000; Rengasamy, 2002; Rashid et al., 2003; Ping et al., 2007; Peng et al., 2008; Roelofs et al., 2008; Raza, 2012; Paz et al., 2014; Imran, 2015, 2018; Rajalakshmi et al., 2015; Iqbal et al., 2017). Application of biochar at 12.5 t ha^{-1} significantly increased the growth and yield attributes and yield of rice as reported earlier by several researchers.

Biochar is utilized in agriculture for the last decades, in improving soil quality, soil fertility, and crop productivity (Bos et al., 2005; Chimenti et al., 2006; Brussaard et al., 2007; Ping et al., 2007; Zhan et al., 2011; Akhtar et al., 2015b; Khan et al., 2016). Mittler (2006) stated that nutrients stress is most harmful in combinations with other abiotic stress and could be treated with biochar application. These findings were about those of Roelofs et al. (2008) who revealed that salinity and drought stress combined limits the grain yield of maize. The production of major food crops is restricted by many environmental and agricultural factors. The integration of biochar is an antique farming practice aimed at enhancing soil air, nutrient provision, soil porosity, soil quality, soil fertility, and water infiltration and water retention.

10.7 BIOCHAR AND PLANT PRODUCTION

Biochar application has been applied to improve tolerance to abiotic stresses in crop plants. Biochar application enhances enzyme activation and may increase a stress-inducing chemical in the plant; hence, it stimulates the plant's defense mechanism (Raghothama, 1999; Prado et al., 2000; Ping et al., 2007; Peng et al., 2008; Paz et al., 2014). So, when abiotic stress happens, the plant has already prepared defenses strategies that can be activated faster and improve tolerance (Lu and Vonshak, 2002). Over generations, many plants have mutated and built various strategies to reduce salinity effects (Rashid et al., 2003; Imran, 2015, 2018; Rajalakshmi et al., 2015; Iqbal et al., 2017). Biochar application has positive effects on regulating plant growth and development under stress conditions (Palta et al., 2006; Palm and Volkenburgh, 2012; Pant et al., 2015). With biochar application under stress conditions, plants respond very differently from one another, even from a plant living in the same area (Neto et al., 2004; Shamsi and Kobraee,

2013; Savvides et al., 2015). Several studies have stated that adding charcoal to the soil increases the yield of moong, soybean, and pea (Raghothama, 1999; Prado et al., 2000; Rengasamy, 2002; Rashid et al., 2003; Ping et al., 2007; Peng et al., 2008; Roelofs et al., 2008; Raza, 2012; Paz et al., 2014; Imran, 2015, 2018; Rajalakshmi et al., 2015; Iqbal et al., 2017), also applying charcoal to the soil increase the shoot and root biomass of birch and pine (Rashid et al., 2003; Imran, 2015, 2018; Iqbal et al., 2017). Likewise, Sugi trees (*Cryptomeria japonica*) biomass production increased due to the soil application of charcoal (Sidari et al., 2008; Karim and Imran, 2019). Several studies reported that applying biochar to Colombian savannah soils increased the maize yield by 28 to 140% compared to the unamended control (Prado et al., 2000; Ping et al., 2007; Peng et al., 2008).

10.8 BIOCHAR AMENDMENTS AND PLANT GROWTH

The capacity of biochar to increase crop production, crop response can be attributed to direct effects via biochar-supplied nutrients (Munns, 2002; Munns et al., 2006; Munns and Tester, 2008), or to many other indirect effects, including: increasing the retention of nutrient and water in soil (Rengasamy, 2002; Rashid et al., 2003; Roelofs et al., 2008; Raza, 2012; Imran, 2015, 2018; Rajalakshmi et al., 2015; Iqbal et al., 2017). Biochar resulted in higher crop production with low P availability and enhanced the response to N and NP chemical fertilizer treatments (Munns, 2002; Neto et al., 2004; Munns et al., 2006; Munns and Tester, 2008; Nazarli et al., 2011). Utilizing biochar (30 and 60 t ha^{-1}) in the Mediterranean basin improved the biomass and grain yield by up to 30% in durum wheat, and the effect lasted for two consecutive seasons (Neto et al., 2004; Nazarli et al., 2011).

Generally, several results confirm the efficiency of biochar applications in improving crops production, which could be either through the direct effects via biochar-supplied nutrients (Munns, 2002; Munns et al., 2006; Munns and Tester, 2008), or through many other indirect effects, such as: (i) increasing the retention of nutrients and water in soil; (ii) increasing the cation exchange capacity in soil; (iii) improving the physical properties of the soil; (iv) improving the transformations and turnover of P and S (Rengasamy, 2002; Rashid et al., 2003; Roelofs et al., 2008; Raza, 2012; Imran, 2015, 2018; Rajalakshmi et al., 2015; Iqbal et al., 2017); (v) promoting the mycorrhizal fungi (Rajalakshmi et al., 2015); (vi) neutralizing the phytotoxic compounds in the soil (Rengasamy, 2002; Roelofs et al., 2008; Raza, 2012; Rajalakshmi et al., 2015); (vii) making changes in soil microbial populations and functions (Rubio et al., 2001; Shamsi and Kobraee, 2013; Savvides et al., 2015). These effects are associated with each other and may act synergistically to enhance crop production (Rubio et al., 2001; Shamsi and Kobraee, 2013).

There are many factors that influence on the physical and chemical properties of biochar and its agronomical benefits, such as, climate, chemical properties of the soil, conditions of the soil, and variation of biomass feedstocks and pyrolysis conditions (Savvides et al., 2015). The production of major food crops is restricted by many environmental and agricultural factors. The integration of biochar is an antique farming practice aimed at enhancing soil air, nutrient provision, soil porosity, soil quality, soil fertility, and water infiltration and water retention. The combination of organic and mineral fertilizers with inoculation of beneficial microbes enhances mutual output. Organic, chemical, and biological interactions not only help plant production and soil quality through successful management but also optimize the use of chemical fertilizers in various crop systems. Integrated management of P is crucial to enhance crop growth, seed yield, nodulation, and human health. This chapter discusses how organic amendments can be the right way to soil sustainability.

Biochar application contains an essential role in providing nutrients, improving soil functions, and reducing soil degradation. The addition of organic carbon and biochar enhances soil fertility and increases crop productivity. Frequent use of fertilizers increases the soil porosity and decreases soil bulk density. Biochar is generally utilized in different forms. The use of organic sources along with the application of chemical fertilizers has the potential to extend the assembly of soybean and maize. Frequently use of biochar with chemical fertilizers enhances crop production. Biochar amendments may help in improving soil properties by obtaining high yield as compared with inorganic fertilizers.

10.9 CONCLUSION

Biochar is an activated carbon soil conditioner that can alleviate the negative impacts of salinity, drought, heat, and cold, and heavy metal stresses. Soil fertility is harshly effected by drought and salt stress which ultimately reduce crop growth and development. It could be concluded that biochar application may be helpful for plant growth, biomass, and yield under either drought or salt stress.

KEYWORDS

- **amendments**
- **biochar**
- **drought**
- **induced systemic resistance**
- **plant production**
- **soil fertility**

REFERENCES

Abebe, G., Assefa, T., Harrun, H., Mesfine, T., & Al-Tawaha, A. M., (2005d). Participatory selection of drought-tolerant maize varieties using mother and baby methodology: A case study in the semi-arid zones of the central rift valley of Ethiopia. *World Journal of Agricultural Sciences, 1*(1), 22–27.

Abebe, G., Hattar, B., & Al-Tawaha, A. M., (2005c). Nutrient availability as affected by manure application in cowpea [*Vigna unguiculata* (L.) Walp.] on calcareous soils. *Journal of Agriculture and Social Science, 1*(1), 1–6.

Abebe, G., Sahile, G., & Al-Tawaha, A. M., (2005a). Evaluation of potential trap crops on Orobanche soil seed bank and tomato yield in the central rift valley of Ethiopia. *World Journal of Agricultural Sciences, 1*(2), 148–151.

Abebe, G., Sahile, G., & Al-Tawaha, A. M., (2005b). Effect of soil solarization on Orobanche soil seed bank and tomato yield in the central rift valley of Ethiopia. *World Journal of Agricultural Sciences, 1*(2), 143–147.

Abera, T., Feyisa, D., Yusuf, H., Nikus, O., & Al-Tawaha, A. M., (2005). Grain yield of maize as affected by biogas slurry and N-P fertilizer rate at Bako, Western Oromiya, Ethiopia. *Bioscience Research, 2*(1), 31–38.

Akhtar, S. S., Andersen, M. N., & Liu, F., (2015a). Biochar mitigates salinity stress in potato. *Journal of Agronomy and Crop Sciences, 201*, 321–400.

Akhtar, S. S., Andersen, M. N., & Liu, F., (2015b). Residual effects of biochar on improving growth, physiology, and yield of wheat under salt stress. *Agricultural Water Management, 158*, 61–68.

Akhtar, S. S., Andersen, M. N., Naveed, M., Zahir, Z. A., & Liu, F., (2015). Interactive effect of biochar and plant growth-promoting bacterial endophytes on ameliorating salinity stress in maize. *Functional Plant Biology, 42*, 770–781.

Ali, I., Khan, A. A., Imran, Ullah, I., Khan, A., Asim, M., Ali, I., et al., (2019). Humic acids and nitrogen levels optimizing productivity of green gram (*Vigna radiata* L.). *Russian Agriculture Sciences, 45*(1), 43–47.

Al-Tawaha, A. M., & Turk, M. A., (2001). Effects of dates and rates of sowing on yield and yield components of carbon vetch under semi-arid conditions. *Acta Agronomica Hungarica, 49*(1), 103–105.

Al-Tawaha, A. M., & Turk, M. A., (2002). Lentil (*Lens culinaris* Medic.) productivity as influenced by rate and method of phosphate placement in a Mediterranean environment. *Acta Agronomica Hungarica, 50*(2), 197–201.

Al-Tawaha, A. M., Turk, M. A., Lee, K. D., Supanjani, Nikus, O., Al-Rifaee, M., & Sen, R., (2005a). Awnless barley response to crop management under Jordanian environment. *Bioscience Research, 2*(3), 125–129.

Al-Tawaha, A. R. M., Turk, M. A., & Lee, K. D., (2005b). Adaptation of chickpea to cultural practices in a Mediterranean type environment. *Research Journal of Agriculture and Biological Science, 1*(2), 152–157.

Amonette, J. E., & Joseph, S., (2009). Characteristics of biochar: Microchemical properties. In: Lehmann, J., & Joseph, S., (eds.), *Biochar for Environmental Management: Science and Technology* (pp. 33–52). Earthscan, London.

Anwar, S., Israeel, Iqbal, B., Khan, A. A., Imran, Shah, W. A., Islam, M., et al., (2016). Nitrogen and phosphorus fertilization of improved varieties for enhancing phenological traits of wheat. *Pure and Applied Biology, 5*(3), 511–519.

Armor, N. B., Hamed, K. B., Debez, A., & Grignonand, C. C., (2005). Physiological and antioxidant responses of perennial halophyte *Crithmum maritimum* to salinity. *Plant Science, 168*(4), 889–899.

Arshad, M., Shaharoona, B., & Mahmood, T., (2008). Inoculation with *Pseudomonas* spp. containing ACC-deaminase partially eliminates the effects of drought stress on growth, yield, and ripening of pea (*Pisum sativum* L.). *Pedosphere, 18*, 611–620.

Assaf, T. A., Hameed, K. M., Turk, M. A., & Al-Tawaha, A. M., (2006). Effect of soil amendment with olive mill by-products under soil solarization on growth and productivity of faba bean and their symbiosis with mycorrhizal fungi. *World Journal of Agricultural Sciences, 2*(1), 21–28.

Assefa, T., Abebe, G., Fininsa, C., Tesso, B., & Al-Tawaha, A. M., (2005). Participatory bean breeding with women and smallholder farmers in eastern Ethiopia. *World Journal of Agricultural Sciences, 1*(1), 28–35.

Ayaz, F. A., Kadioglu, A., & Turgut, T. R., (2000). Water stress effects on the content of low molecular weight carbohydrates and phenolic acids in *Cienanthe setosa*. *Candian Journal of Plant Sciences, 80*, 373–378.

Badridze, G., Weidner, A., Asch, F., & Borner, A., (2009). Variation in salt tolerance within Georgian wheat germplasm collection. *Genetic Resources and Crop Evolution, 56*, 1125–1130.

Bano, A., & Fatima, M., (2009). Salt tolerance in *Zea mays* (L.) following inoculation with *Rhizobium* and *Pseudomonas*. *Biol. Fertility Soils., 45*, 405–413.

Baqa, S., Khan, A. Z., Inamullah, Imran, Khan, A. A., Anwar, S., Iqbal, B., Khan, S., & Usman, A., (2015). Influence of biochar and phosphorus application on yield and yield components of wheat. *J. Pure Appl. Biol., 4*(4), 458–464.

Bates, L. S., Waldren, R. P., & Teare, I. D., (1973). Rapid determination of free proline for water-stress studies. *Plant and Soil, 39*, 205–207.

Bohnert, H. J., & Jensen. R. G., (1996). Strategies for engineering water-stress tolerance in plants. *Trends in Biotechnology, 14*, 89–97.

Bos, C., Juillet, B., Fouillet, H., Turlan, L., Dare, S., Luengo, C., Ntounda, R., et al., (2005). Postprandial metabolic utilization of wheat protein in humans. *The American Journal of Clinical Nutrition, 81*, 87–94.

Bruinsma, J., (1963). The quantitative analysis of chlorophyll a and b in plant extract. *Photochemistry and Photobiology, 2*, 241–249.

Brussaard, L., De Ruiter, P. C., & Brown, G. G., (2007). Soil biodiversity for agricultural sustainability. *Agriculture, Ecosystems and Environment, 121*(3), 233–244.
Capell, B., & Doerffling, K., (1993). Genotype specific difference in chilling tolerance of maize in relation to chilling induce changes in water status and abscisic acid accumulation. *Physiologia Plantarum, 88*, 638–646.
Chimenti, C. A., Marcantonio, M., & Hall, J. A., (2006). Divergent selection for osmotic adjustment results in improved drought tolerance in maize (*Zea mays* L.) in both early growth and flowering phases. *Field Crops Research, 95*, 305–315.
Conde, A., Chaves, M. M., & Gero´s, H., (2011). membrane transport, sensing and signaling in plant adaptation to environmental stress. *Plant and Cell Physiology, 52*(9), 1583–1602.
Cramer, G. R., Urano, K., Delrot, S., Pezzotti, M., & Shinozaki, K., (2011). Effects of abiotic stress on plants: A systems biology perspective. *BMC Plant Biology, 11*, 163.
Duan, D. Y., Li, W. Q., Liu, X. J., Ouyang, H. P., & An, P., (2007). Seed germination and seedling growth of *Suaeda salsa* under salt stress. *Annales Botanici Fennici, 44*, 161–169.
Dubois, M., Gilles, K., Hammiltron, J. K., Robers, P. A., & Smith, F., (1951). A colorimetric method for the determination of sugars. *Nature, 168*, 167–168.
Farahbakhsh, H., (2012). Germination and seedling growth in un-primed and primed seeds of fennel as affected by reduced water potential induced by NaCl. *International Research Journal of Applied and Basic Sciences, 3*, 737–744.
Flowers, T. J., (2004). Improving crop salt tolerance. *Journal of Experimental Botany, 55*, 307–319.
Gasco, G., Paz-Ferreiro, J., & Mendez, A., (2012). Thermal analysis of soil amended with sewage sludge and biochar from sewage sludge pyrolysis. *Journal of Thermal Analysis and Calorimetry, 108*, 769–775.
Giannopolitis, C. N., & Ries, S. K., (1977). Superoxide dismutase occurrence in higher plants. *Plant Physiology, 59*, 304–314.
Gonçalves-Alvim, S. J., & Fernandes, G. W., (2001). Biodiversity of galling insects: historical, community and habitat effects in four neotropical savannas. *Biodiversity and Conservation, 10*, 79–98.
González Villagra, J., Rodrigues Salvador, A., Nunes-Nesi, A., Cohen, J. D., & Reyes-Díaz, M. M., (2018). Age-related mechanism and its relationship with secondary metabolism and abscisic acid in *Aristotelia chilensis* plants subjected to drought stress. *Plant Physiology and Biochemistry, 124*, 136–145.
Heans, D. L., (1984). Determination of total organic carbon in soils by an improved chromic acid digestion and spectrophotometric procedure. *Communications in Soil Science and Plant Analysis, 15*, 1191–1213.
Imran, & Amanullah, (2018). Global impact of climate change on water, soil resources and threat towards food security: Evidence from Pakistan. *Adv. Plants Agric. Res., 8*(5), 350–355.
Imran, & Khan, A. A., (2015). Effect of transplanting dates on yield and yield components of various rice genotypes in hilly area cold climatic region of Khyber Pakhtunkhwa-Pakistan. *Journal of Biology, Agriculture and Healthcare, 5*(7), 1–9.
Imran, & Khan, A. A., (2017). Canola yield and quality enhanced with Sulphur fertilization. *Russian Agricultural Sciences, 43*(2), 113–119.
Imran, (2015). Effect of germination on proximate composition of two maize cultivars. *Journal of Biology, Agriculture and Healthcare, 3*(5), 123–128.
Imran, (2017). Climate change is a real fact confronting to agricultural productivity. *Int. J. Environ. Sci. Nat. Res., 3*(3), 555613.

Imran, (2018). Biochar amendments improve soil health, productivity, and profitability of maize and soybean. *Ann. Rev. Res., 1*(3), 555–564.

Imran, (2018). Phosphorous fertilization influenced weeds attributes and phenological characteristics of mung bean cultivars (*Vigna radiata* L.). *Russian Agricultural Sciences, 44*(3), 229–234.

Imran, Amanullah, Arif, M., Shah, Z., & Bari, A., (2020). Soil application of *Trichoderma* and Peach (*Prunus persica* L.) residues possesses biocontrol potential for weeds and enhances growth and profitability of Soybean (*Glycine max*). *Sarhad Journal of Agriculture, 36*(1), 10–20.

Imran, Amanullah, Bari, A., &Ali, R., (2018). Peach sources, phosphorous and beneficial microbes enhance productivity of soybean. *Soy. Res., 16*(2), 39–48.

Imran, Amanullah, Khan, A. A., Bari, A., Ud Din, R., Ali, R., Naveed, S., Ullah, I., Zada, H., & Khan, G. R., (2019). Production statistics and modern technology of maize cultivation in Khyber Pakhtunkhwa Pakistan. *Plant Sci., 2*(2) 1–12.

Imran, Bari, A., Ali, R., Ahmad, N., Ahmad, Z., Khattak, M. I., Ali, A., et al., (2017). Traditional rice farming accelerate CH_4 and N_2O emissions functioning as a stronger contributors of climate change. *Int. J. Environ. Sci. Nat. Res., 9*(3), 89–92.

Imran, Hussain, I., Khattak, I., Rehman, A., Ahmad, F., & Shah, S. T., (2015). Roots nodulation, yield and yield contributing parameters of mung bean cultivars as influenced by different phosphorous levels in Swat-Pakistan. *J. Pure Appl. Biol., 4*(4), 557–567.

Imran, Hussain, I., Naveed, S., Shah, S., Rehman, A., Zada, H., Ullah, I., et al., (2016). Growth and yield of maize hybrids as effected by different sowing dates in Swat Pakistan. *J. Pure App. Biol., 5*(1), 114–120.

Imran, Jamal, N., Alam, A., & Khan, A. A., (2017). Grain yield, yield attributes of wheat and soil physicochemical characteristics influenced by biochar, compost and inorganic fertilizer application. *Agri. Res. Tech., 10*(4), 555795.

Imran, Khan, A. A., Irfanullah, & Ahmad, F., (2014). Production potential of rapeseed (*Brassica napus* L.) as influenced by different nitrogen levels and decapitation stress under the rainfed agro-climatic condition of swat Pakistan, *J. Glob. Innov. Agric. Soc. Sci., 2*(3), 112–115.

Iqbal, A., Amanullah, Ali, A., Iqbal, M., Ikramullah, & Imran, (2017). Integrated use of phosphorus and biochar improve fodder yield of moth bean (*Vigna aconitifolia* Jacq.) under irrigated and dryland conditions of Pakistan. *Journal of AgriSearch, 4*(1), 10–15.

Iqbal, B., Ahmad, B., Ullah, I., Imran, Khan, A. A., Anwar, S., Ali, A., Shahzad, K., & Khan, S., (2016). Effect of phosphorus, Sulphur and different irrigation levels on phenological traits of triticale. *Pure and Applied Biology, 5*(2), 303–310.

Iqbal, B., Jan, M. T., Imran, Zar, M., Khan, A. A., Anwar, S., & Shahzad, K., (2015). Growth and phenology of maize as affected by integrated management of compost and fertilizers nitrogen. *Journal of Pure and Applied Biology*, 4.

Iqbal, B., Jan, M. T., Inamullah, Imran, Khan, A. A., Muhammad, Z., Anwar, S., et al., (2015). Integrated management of compost type and fertilizer in Maize. *J. Pure Appl Biol., 4*(4), 453–457.

Iqbal, B., Jan, M. T., Muhammad, Z., Khan, A. A., Anwar, S., Imran, & Shahzad, K., (2016). Phenological traits of maize influenced by integrated management of compost and fertilizer nitrogen. *J. Pure Appl Biol., 5*(1), 58–63.

Islam, M., Anwar, S., Bashir, S., Khattak, W. A., Imran, Ali, M., & Khan, N., (2015). Growth and yield components of wheat varieties as affected by dual-purpose practices. *Journal of Pure and Applied Biology, 4*(4), 491–496.

ISTA, (2005). *International Rules for Seed Testing*. Zurich, Switzerland: International Seed Testing Association.

Jackson, M., (1962). *Soil Chemical Analysis* (p. 496). Constable and Company Ltd, London.

Jagatheeswari, D., & Ranganathan, P., (2012). Influence of mercuric chloride on seeds germination, seedling growth and biochemical analysis of green gram (*Vigna radiata* (L.) Wilczek. var. Vamban-3). *International Journal of Pharmaceutical and Biological Archives, 3*, 291–295.

Joseph, B., & Jini, D., (2011). Development of salt stress-tolerant plants by gene manipulation of antioxidant enzymes. *Asian Journal of Agricultural Research, 5*, 17–27.

Kamara, A., Kamara, A., Mansaray, M. M., & Sawyerr, P. A., (2014). Effects of biochar derived from maize stover and rice straw on the germination of their seeds. *American Journal of Agriculture and Forestry, 2*, 246–249.

Kammann, C. I., Linsel, S., Gobling, J. W., & Koyro, H., (2011). Influence of biochar on drought tolerance of *Chenopodium quinoa* Willd and on soil-plant relation. *Plant and Soil, 345*, 195–210.

Karim, S. A., & Imran, (2019). Reduction in substrate moisture content reduce the final yield of wheat. *Biomed J. Sci. Tec. Res.*, 1–5.

Khan, A. A., Imran, Ali, F., Inamullah, S., Laiq, Z., Muhammad, N., Muhammad, N., & Khan, K., (2015). Phenological traits of rice as influenced by seedling age and number of seedlings per hill under temperate region. *Journal of Biology, Agriculture and Healthcare, 5*(3), 145–149.

Khan, A. A., Khan, M. N., Inamuallah, Shah, S., Arshad, I. R., Muhammad, I., Zeb, A., & Imran, (2015). Effect of potash application on growth, yield and yield components of spring maize hybrids. *Pure and Applied Biology, 4*(2), 195–203.

Khan, A. Z., Imran, Muhammad, A., Khalil, A., Gul, H., Akbar, H., & Wahab, S., (2016). Impact of fertilizer priming on seed germination behavior and vigor of maize. *Pure and Applied Biology, 5*(4), 744–751. http://dx.doi.org/10.19045/bspab.2016.50093.

Khan, I., (2018). Climate change is a threat toward agronomy (base of food, fiber system), and food security. *Food Nutr. J., 3*(10), 1–8.

Khan, M. A., Islam, E., Shirazi, M. U., Mumtaz, S., Mujtaba, S. M., Khan, M. A., Shereen, A., et al., (2010). Physiological responses of various wheat genotypes to salinity. *Pakistan Journal of Botany, 42*, 3497–505.

Lashari, M. S., Liu, Y., Li, W., Pan, J., Pan, G., Zheng, J., et al., (2013). Effect of amendment of biochar manure compost in conjunction with pyroligneous solution on soil quality and wheat yield of a salt-stressed cropland from central China great plain. *Food Crop Research, 144*, 113–118.

Lee, K. D., Tawaha A. R. M., & Supanjani, (2005d). Antioxidant status, stomatal resistance and mineral composition of hot pepper under salinity and boron stress. *Bioscience Research, 2*(3), 148–154.

Lee, K. D., Turk, M. A., & Tawaha, A. M., (2005b). Nitrogen fixation in rice based farming system. *Bioscience Research, 2*(3), 130–138.

Lee, K., Sulpanjani, D., Tawaha, A. M., & Min, Y. S., (2005a). Effect of Phosphorus application on yield, mineral contents and active components of *Chrysanthemum coronarium* L. *Bioscience Research, 2*(3), 118–124.

Lehmann, J., & Joseph, S., (2009). Biochar for environmental management: An introduction. In: Lehmann, J., & Joseph, S., (eds.), *Biochar for Environmental Management: Science and Technology* (pp. 1–12). London: Earthscan.

Lehmann, J., Rillig, M. C., Thies, J., Masiello, C. A., Hockaday, W. C., & Crowley, D., (2011). Biochar effects on soil biota. *Soil Biology and Biochemistry, 43*, 1812–1836.

Lei, G., Shen, M., Li, Z., Zhang, B., Duan, K., Wang, N., Cao, Y., Zhang, W., & Ma, B., (2011). EIN2 regulates salt stress response and interacts with a MA3 domain-containing protein ECIP1 in Arabidopsis. *Plant, Cell and Environment, 34*(10).

Lu, C., & Vonshak, A., (2002). Effects of salinity stress on photosystem II function in cyanobacterial *Spirulina platensis* cells. *Physiol. Plant, 114*, 405–413.

Maestre, F. T., Cortina, J., & Bautista, S., (2007). Mechanisms underlying the interaction between *Pinus halepensis* and the native late-successional shrub *Pistacia lentiscus* in a semi-arid plantation. *Ecography, 27*(6), 776–786.

Martinez-Beltran, J., & Manzur, C. L., (2005). Overview of salinity problems in the world and FAO strategies to address the problem. *Proceedings of the International Salinity Forum* (pp. 311–313). Riverside, California.

Mazhar, R., Ilyas, N., Saeed, M., Bibi, F., & Batool, N., (2016). Biocontrol and salinity tolerance potential of *Azospirillum lipoferum* and its inoculation effect in wheat crop. *International Journal of Agriculture and Biology, 3*, 494–500.

Mesfine, T., Abebe, G., & Al-Tawaha, A. M., (2005). Effect of reduced tillage and crop residue ground cover on yield and water use efficiency of sorghum (*Sorghum bicolor* (L.) Moench) under semi-arid conditions of Ethiopia. *World Journal of Agricultural Sciences, 1*(2), 152–160.

Ministry of Food, Agriculture and Livestock, (2007). *Agricultural Statistics of Pakistan*. Ministry of Food, Agriculture and Livestock, Government of Pakistan, Islamabad.

Miri, Y., & Mirjalili, S. A., (2013). Effect of salinity stress on seed germination and some physiological traits in primary stages of growth in purple coneflower (*Echinacea purpurea*). *International Journal of Agronomy and Plant Production, 4*, 142–146.

Mittler, R., (2006). Abiotic stress, the field environment and stress combination. *Trends in Plant Science, 11*(1), 15–19.

Munns, R. R., James, A., & Lauchli, A., (2006). Approaches to increasing the salt tolerance of wheat and other cereals. *Journal of Experimental Botany, 57*, 1025–1043.

Munns, R., & Tester, M., (2008). Mechanisms of salinity tolerance. *Annual Review of Plant Biology, 59*, 651–681.

Munns, R., (2002). Comparative physiology of salt and water stress. *Plant, Cell and Environment, 25*, 239–250.

Naveed, S., Ibrar, M., Khattak, I., Khan, I., Imran, Khan, M. I., & Khan, H., (2016). Anthelmintic, antilice, insecticidal, cytotoxic and phytotoxic potential of ethanolic extracts of two wild medicinal plants *Iphiona grantioides* and *Pluchea arguta*. *J. Woulfenia, 23*(11), 13–25.

Nazarli, H., Faraji, F., & Zardashti, M. R., (2011). Effect of drought stress and polymer on osmotic adjustment and photosynthetic pigments of sunflower. *Carcetriari Agronomica in Maldova, 1*, 45.

Neto, A. D. A., Prisco, J. T., Enéas-Filho, J., De Lacerda, C. F., Silva, J. V., De Costa, P. H. A., & Gomes-Filho, E., (2004). Effects of salt stress on plant growth, stomatal response and solute accumulation of different maize genotypes. *Brazilian Journal of Plant Physiology, 16*(1), 31–38.

Nikus, O., Abebe, G., Takele, A., Harrun, H., Chanyalew, S., Al Tawaha, A. M., & Mesfin, T., (2005b). Yield response of tef (*Eragrostis tef* (Zucc.) Trotter) to NP fertilization in the semi-arid zones of the central rift valley in Ethiopia. *European Journal of Scientific Research, 4*(4), 49–60.

Nikus, O., Nigussie, M., & Al Tawaha, A. M., (2005a). Agronomic performance of Maize varieties under irrigation in Awash Valley, Ethiopia. *Bioscience Research, 2*(1), 26–30.

Olsen, S. R., & Sommers, L. E., (1982). Phosphorous. In: Bunton, D., (ed.), *Methods of Soil Analysis: Chemical and Microbiological Properties* (pp. 403–427). Madison, WI: Soil science society of America.

Palm, E., Brady, K., & Van, V. E., (2012). Serpentine tolerance in *Mimulus guttatus* does not rely on exclusion of magnesium. *Functional Plant Biology, 39*(8), 679.

Palta, J. P. & Farag, K., (2006). *Methods for Enhancing Plant Health, Protecting Plants from Biotic and Abiotic Stress-Related Injuries and Enhancing the Recovery of Plants Injured as a Result of Such Stresses*. United States Patent, 7101828.

Pant, B. D., Burgos, A., Pant, P., Cuadros-Inostroza, A., Willmitzer, L., & Scheible, W., (2015). The transcription factor PHR1 regulates lipid remodeling and triacylglycerol accumulation in *Arabidopsis thaliana* during phosphorus starvation. *Journal of Experimental Botany, 66*(7), 1907–1918.

Pant, B. D., Pant, P., Erban, A., Huhman, D., Kopka, J., & Scheible, W., (2015). Identification of primary and secondary metabolites with phosphorus status-dependent abundance in *Arabidopsis*, and of the transcription factor PHR1 as a major regulator of metabolic changes during phosphorus limitation. *Plant, Cell and Environment, 38*(1), 172–187.

Paz-Ferreiro, J., Fu, S., Mendez, A., & Gasco, G., (2014). Interactive effects of biochar and the earthworm *Pontoscolex corethrurus* on plant productivity and soil enzymes activities. *Journal of Soils Sediments, 14*, 483–494.

Peng, Y., Gao, Z., Gao, Y., Liu, G., Sheng, L., & Wang, D., (2008). Eco-physiological characteristics of alfalfa seedlings in response to various mixed salt-alkaline stresses. *Journal of Integrative Plant Biology, 50*, 29–39.

Ping, J., Geo, K., & Ron, M., (2007). Understanding abiotic stress tolerance mechanisms: Recent studies on stress response in rice. *Journal of Integrative Plant Biology, 49* (6), 742–750.

Prado, F. E., Boero, C., Gallarodo, M., & Gonzalez, A. J., (2000). Effect of NaCl on germination, growth and soluble sugar content in *Chenopodium quinoa* Wil seeds. *Botanical Bulletin-Academia Sinica Taipei, 14*, 27–34.

Raghothama, K. G., (1999). Phosphate acquisition. *Annual Review of Plant Physiology and Plant Molecular Biology, 50*(1), 665–693.

Rajalakshmi, A. G., Kumar, S. K., Bharathi, C. D., Karthika, R., Visalakshi, K. D., Meera, R., & Mohanapriya, S., (2015). Effect of biochar in seed germination: *In-vitro* study. *International Journal of Biosciences and Nanosciences, 2*, 132–136.

Rashid, A., Saleem, Q., Nazir, A., & Kazım, H. S., (2003). Yield potential and stability of nine wheat varieties under water stress conditions. *International Journal of Agriculture and Biology, 5*, 7–9.

Raza, A., (2012). *Grain and Field Annual 2012 Gain Report* (pp. 1–3). Global Agricultural Information Network, USDA Foreign Agricultural Services.

Rengasamy, P., (2002). Transient salinity and subsoil constraints to dryland farming in Australian sodic soils: An overview. *Australian Journal of Experimental Agriculture, 42*, 351–361.

Roelofs, D., Jiwan, P., & Kwangy, I., (2008). Functional ecological genomics to demonstrate general and specific responses to abiotic stress. *Functional Ecology, 22*, 8–18.

Rubio, V., Linhares, F., Solano, R., Martín, A. C., Iglesias, J., Leyva, A., & Paz-Ares, J., (2001). A conserved MYB transcription factor involved in phosphate starvation signaling both in vascular plants and in unicellular algae. *Genes and Development, 15*(16), 2122–2133.

Samreen, U., Ibrar, M., Lalbadshah, Naveed, S., Imran, & Khatak, I., (2016). Ethnobotanical study of subtropical hills of Darazinda, Takht-e-Suleman range F.R D.I. Khan, Pakistan. *Journal of Pure and Applied Biology, 5*(1), 149–164.

Savvides, A., Ali, S., Tester, M., & Fotopoulos, V., (2015). chemical priming of plants against multiple abiotic stresses: Mission possible?. *Trends in Plant Science, 21*(4), 329–340.

Scholander, P. F., Hammel, H. T., Bradstreet, E. D., & Hemminogsen, E. A., (1965). Sap pressure in vascular plants: Negative hydrostatic pressure can be measured in plants. *Plant Sciences, 148*, 339–346.

Shamsi, K., & Kobraee, S., (2013). Biochemical and physiological responses of three wheat cultivars (*Triticum aestivum* L.) to salinity stress. *Annals of Biological Research, 4*, 180–185.

Sidari, M., Mallamaci, C., & Muscolo, A., (2008). Drought, salinity and heat differently affect seed germination of *Pinus pinea*. *Journal of Forest Research, 13*, 326–330.

Sulpanjani, A., Yang, M. S., Tawaha, A. R. M., & Lee, K. D., (2005a). Effect of magnesium application on yield, mineral contents and active components of *Chrysanthemum coronarium* L. under hydroponics conditions. *Bioscience Research, 2*(2), 73–79.

Supanjani, Tawaha, A. M., Min, Y. M. S., & Lee, K. D., (2005b). Role of calcium in yield and medicinal quality of *Chrysanthemum coronarium* L. *Journal of Agronomy, 4* (3), 188–192.

Supanjani, Tawaha, A. M., Yang, M. S., & Lee, Y. D., (2005c). Calcium effects on yield, mineral uptake and terpene components of hydroponic *Chrysanthemum coronarium* L. *Research Journal of Agriculture and Biological Science, 1*(1), 146–151.

Tiwari, S., Singh, P., Tiwari, R., Meena, K. K., Yandigeri, M., Singh, D. P., & Arora, D. K., (2011). Salt-tolerant rhizobacteria-mediated induced tolerance in wheat (*Triticum aestivum*) and chemical diversity in rhizosphere enhance plant growth. *Biology and Fertility of Soils, 47*, 907–916.

Tombesi, S., Frioni, T., Poni, S., & Palliotti, A., (2018). Effect of water stress "memory" on plant behavior during subsequent drought stress. *Environmental and Experimental Botany, 150*, 106–114.

Turk, A. M., & Tawaha, A. M., (2002a). Impact of seeding rate, seeding date, rate and method of phosphorus application in faba (*Vicia faba* L. Minor) in the absence of moisture stress. *Biotechnology, Agronomy, Society and Environment, 6*(3), 171–178.

Turk, A. M., & Tawaha, A. M., (2002b). Response of winter wheat to applied with or without ethrel spray under irrigation planted in semi-arid Environments. *Asian Journal of Plant Sciences, 1*(4), 464–466.

Turk, M. A., & Tawaha, A. M., (2001). Influence of rate and method of phosphorus placement to garlic (*Allium sativum* L.) in a Mediterranean environment. *Journal of Applied Horticulture, 3*(2), 115–116.

Turk, M. A., Hameed, K. M., Aqeel, A. M., & Tawaha, A. M., (2003c). Nutritional status of durum wheat grown in soil supplemented with olive mills by-products. *Agrochimica, 5*(6), 209–219.

Turk, M. A., Tawaha, A. M., Samara, N., & Latifa, N., (2003b). The response of six-row Barley (*Hordeum vulgare* L.) to nitrogen fertilizer application and weed control methods in the absence of moisture stress. *Pakistan Journal of Agronomy, 2*(2), 101–108.

Turk, M. A., Tawaha, A. M., Taifor, H., Al-Ghzawi, A., Musallam, I. W., Maghaireh, G. A., & Al-Omari, Y. I., (2003a). Two row barley response to plant density, date of seeding and rate and method of phosphorus application in the absence of moisture stress. *Asian Journal of Plant Science, 2*(2), 180–183.

Van, H. J. W., Nader, K., Hamdy, A., & Mastrorilli, M., (2001). Effect of salinity on yield and nitrogen uptake of four grain legumes and on biological nitrogen contribution from the soil. *Agriculture Water Management, 51*, 87–98.

Vinebrooke, R. D., Cottingham, K. L., Norberg, J., Scheffer, M., Dodson, S. I., Maberly, S. C., & Sommer, U., (2004). Impacts of multiple stressors on biodiversity and ecosystem functioning: The role of species co-tolerance. *Oikos, 104*(3), 451–457.

Wang, G., & Xu, Z., (2013). The effects of biochar on germination and growth of wheat in different saline-alkali soil. *Asian Agricultural Research, 5*, 116–119.

Wang, H., & Jin, J., (2007). Effects of zinc deficiency and drought on plant growth and metabolism of reactive oxygen species in maize (*Zea mays* L.). *Agricultural Sciences in China, 6*, 988–995.

Wang, W., Vinocur, B., & Altman, A., (2003). Plant responses to drought, salinity and extreme temperatures: Towards genetic engineering for stress tolerance. *Planta, 218*(1), 1–14.

Wolfe, A., (2007). *Patterns of Biodiversity*. Ohio State University. doi: 10.1016/j.agee.2006.12.013.

Xu, D., Huang, J., Guo, S., Yang, X., Bao, Y., Tang, H., & Zhang, H., (2008). Over-expression of a TFIII-type zinc finger protein gene ZFP252 enhances drought and salt tolerance in rice (*Oryza sativa* L.). *Federation of European Biochemical Societies Letters, 582*, 1037–1043.

Yang, M. S., Tawaha, A. M., & Lee, Y. D., (2005). Effects of ammonium concentration on the yield, mineral content and active terpene components of *Chrysanthemum coronarium* L. in a hydroponic system. *Research Journal of Agriculture and Biological Science, 1*(2), 170–175.

Yeo, A. R., Yeo, M. E., Flowers, S. A., & Flowers, T. J., (1990). Screening of rice (*Oryza sativa* L.) genotypes for physiological characters contributing to salinity resistance, and their relationship to overall performance. *Theoretical Applied Genetics, 79*, 377–384.

Younis, U., Athar, M., Malik, S. A., RazaShah, M. H., & Mahmood, S., (2015). Biochar impact on physiological and biochemical attributes of spinach *Spinacia oleracea* (L.) in nickel contaminated soil. *Global Journal Environment Science Management, 1*, 245–254.

Zakaria, M., Daniel, S., Murphy, V., & Abbott, L. K., (2012). Biochars influence seed germination and early growth of seedlings. *Plant Soil, 353*, 273–87.

Zargar, M. S., Nagar, P., Deshmukh, R., Nazir, M., Wani, A. A., Masoodi, K. Z., Agrawal, G. K., & Rakwal, R., (2017). Aquaporins as potential drought tolerance inducing proteins: Towards instigating stress tolerance. *Journal of Proteomics, 169*, 233–238.

Zhan, H. X., Chang, Z. J., Wei, A. L., Zhang, X. J., & Li, X., (2011). Impact of drought to wheat physiological index. *Journal of Shanxi Agriculture Sciences, 39*, 1049–1051.

Zheng, W. J., Tawaha, A. M., & Lee, K. D., (2005). *In situ* hybridization analysis of mcMT1 gene expression and physiological mechanisms of Cu-tolerant in *Festuca rubra* cv Merlin. *Bioscience Research, 1*(1), 21–26.

CHAPTER 11

THE ROLE OF ORGANIC MULCHING AND TILLAGE IN ORGANIC FARMING

SHAH KHALID,[1] AMANULLAH,[1] ABDEL RAHMAN M. AL-TAWAHA,[2] NADIA,[1] DEVARAJAN THANGADURAI,[3] JEYABALAN SANGEETHA,[4] SAMIA KHANUM,[5] MUNIR TURK,[6] HIBA ALATRASH,[7] SAMEENA LONE,[8] KHURSHEED HUSSAIN,[8] PALANI SARANRAJ,[9] NIDAL ODAT,[10] and ARUN KARNWAL[11]

[1]*Department of Agronomy, The University of Agriculture, Peshawar, Pakistan*

[2]*Department of Biological Sciences, Al-Hussein Bin Talal University, Maan, Jordan*

[3]*Department of Botany, Karnatak University, Dharwad–580003, Karnataka, India*

[4]*Department of Environmental Science, Central University of Kerala, Kasaragod–671316, Kerala, India*

[5]*Department of Botany, University of the Punjab, Lahore, Pakistan*

[6]*Department of Plant Production, Jordan University of Science and Technology, Irbid, Jordan*

[7]*General Commission for Scientific Agricultural Research, Syria*

[8]*Division of Vegetable Science, SKUAST-Kashmir, Shalimar, Jammu and Kashmir, India*

[9]*Department of Microbiology, Sacred Heart College (Autonomous), Tirupattur–635601, Tamil Nadu, India*

[10]*Department of Medical Laboratories, Al-Balqa Applied University, Al-Salt–19117, Jordan*

[11]*Department of Microbiology, School of Bioengineering and BioSciences, Lovely Professional University, Phagwara, India*

ABSTRACT

The mulches are mainly used for increasing water infiltration, evaporation, soil temperature change, weed control, evaporation prevention, and crop yields increase. Increased biological activities in the soil, changed nutrient levels and contributed to the maintenance and/or increase of soil organic matter (SOM). There would also possibly be a favorable impact in the physical conditions in the soil under conditions of no-tillage or minimal tillage of fine-textured soils. Weeds can be monitored by Mulches in various ways: as physical barrier and related micro-climate transition, pH, C:N soil ratio, nutrient immobilization, allelopathic compounds inhibition, lower visible light reaching the surface of the ground. A large number of works is intended to loosen and homogenize the macroporousness of the soil and the structural uniformity of the tillage region, but some of the work is intended to shape or shape the soil. Some of the tillage effects are deliberate, whereas other effects, such as plow shape, increased vulnerability to compaction and erosion, are unintentional. On the positive hand, tillage has been part of most farm systems in the past since tillage can be used to achieve many agronomic goals.

11.1 INTRODUCTION

This section offers details on the use of organic farming systems (OFSs) of tillage and mulching. It provides a summary of how laboring and paving blends into organic farming standards and structures. It describes forms of tillage, implements tillage equipment, and refers, as used here, to any normal, uninterrupted covering of the soil. It refers to any additional material used as a covering of the ground. Consequently, this term is wide enough to generally include both organic and inorganic materials that are used as soil cover. The key focus in this context is, however, on traditional mulches of organic materials, where plant materials from external sources typically fully cover the soil to different depths. Often referred to are stubble or trash-mulches, where the surface of the soil is normally filled in part only. The results generated are often quite similar but vary in degree, whether the soil is completely or only partially covered. In recent years, much research has shown that a very successful way of using crop residues in many ways is like mulches for growing crops, although sometimes the financial factors are not favorable. The mulch method ensures that the cultivation and soil retain as much time as possible to protect the soil surface. Through the decomposition of crop residues, they eventually move through the activity of organic matter and then through to resistant humus. When soil is kept untouched, much of organic matter, as under virgin or forest conditions, tends to collect on the surface. Where every year's crops are grown, the mulch is disturbed by dirt, and a large part of the mulch is mixed with 2–6 inches of ground every year. In any case, plant residues have the numerous physical, chemical, and biological effects so beneficial for plant development, i.e., the growth rate of good highly valued topsoil is accelerated. The fundamental ways of diversity are the rotation of organic crops. Crop rotation consists of the cultivation of various crops on the same soil in natural succession. Crop rotation offers many benefits to improve the quality of the soil, break disease and cycle of pests and ensure soil and water effects (Altieri et al., 2012). Farmers picked promising plant animals in organic agriculture and saved their offspring year after year to preserve genetic biodiversity. Time and space diversity increases crop resilience in many respects (Zehnder et al., 2007). Careful use of soil nutrients, avoid a chance of plant failure, improve genetic make-up of landraces (Bengtsson et al., 2005; Hole et al., 2005).

11.2 THE ROLE OF TILLAGE AND ORGANIC FARMING

The weakness in organic systems was highly dependent on tillage in order to manage the weeds. However, organic matter and soil structure are also shown to perform as well or better than traditional herbicidal systems with less soil disruption. organic systems generally perform better or better. This is due to imports of organic inputs including manures and compost, organic material recycling in farms and well-designed crop rotations (including decking plants and perennial drilling), which help to compensate for the negative impact of soil structure and organic matter layering. This can include lack of organic matter, increased soil strength, decreased penetration, compaction, and increased erosion. These are negative effects. Experienced organic farmers reduce the negative effects of tillage by taking careful account of the timing of tillage processes, machinery activity, soil conditions and crop rotation. Some crops, such as root vegetables, with intense soil perturbation during harvest and returning low crop residues can be rotated by crops with lower soil perturbations and more crop residues. Starting farmers have a lot to know about the art and science behind productive labor. In terms of the intensity of labor, often greater than traditional systems, policymakers need to understand where and where organic farmers can provide desired ecological services like carbon sequestration.

Increased cooperation among agroecologists, agricultural engineers, and experienced farmers will lead to better tillage systems that enable strategic laying in the context of planned crop systems to more consistently gain and minimize negative side effects.

11.2.1 ORGANIC MANURES AND TILLAGE

Final Regulation of the national organic program (NOP) (United States Department of Agriculture (USDA), 2000) Certified organic farmers are to register their organic system plan tillage activities and procedures. The weed control measures permitted include hand weeding as well as mechanical cultivation. The frequency of application tillage must be recorded in the farm records. This can be accomplished in a variety of ways; the manufacturer should follow, along with other details required by the NOP standard, an organization, easy-to-use system for recording tillage, rotations, and amendment history for each area. The inspector examines the farming plan of a producer to determine if the tillage practices are used, so that the physical, chemical, and biological condition of the soil is maintained or improved and so that soil erosion is minimized.

11.2.2 IMPORTANCE OF TILLAGE

Mechanical soil structure alteration is tillage. Tillage tools change the soil structure through a wide range of interactions between soil and instrument including cutting, milling, crushing, beating, and rebounding. Tillage tools expose the soil structure to compression, shear, and strain. When the stresses exceed the strength of the soil, the soil structure fails to collapse, if the ground is cold, or if the ground is deformed in a plastic condition. The result of interactions between soil and tools differs both with regard to the characteristics of the tillage activity such as depth, distance, speed, and the type of the tilling of the soil, such as texture, structure, humidity, or plasticity. If soil is too damp, it spreads out and produces clods that can last for the duration of the growing season. Clods are usually broken by winter freezing and thawing. A large number of works is intended to loosen and homogenize the macroporousness of the soil and the structural uniformity of the tillage region, but some of the work is intended to shape or shape the soil. Some of the tillage effects are deliberate, whereas other effects, such as plow shape, increased vulnerability to compaction and erosion, are unintentional.

On the positive hand, laying has been part of most agricultural processes over the years, as laying can be used to achieve many agricultural objectives (Amanullah et al., 2016, 2019a). Tillage's value includes:

1. **Residue Management:** The transfer, orientation, or sizing of residues such that harms or crop residues can be minimized and beneficial effects promoted.
2. **Incorporation and Combination:** Placement or distribute, often from a less favorable position to a more favorable geographical distribution, substances such as fertilizers, manures, seeds, and residues.
3. **Soil Conditioning:** Soil-structure changes to promote agricultural processes such as interaction with soils, root proliferation, soil heating, and water penetration.
4. **Crop and Pesticide Control:** Direct end or disturbance of the cycles of weed and pests.
5. **Splitting:** Rock consolidation, root clumps, soil crumb sizes, etc.
6. **Land Formation:** Modifying the soil surface shape; perhaps the simplest variant is leveling; crest, roughing, and furrowing are also examples.
7. **Nutrient Release Stimulation:** Accomplished by aeration and mixture; note that it can be a disadvantage if not targeted with crop uptake.

More basic objectives of tillage include creating seedbeds, stalky seedbeds, minimizing the compression of crops by breaking of soil crusts, cutting, and/or drying of weeds, macerating biofumigant plant cover crops, soil biology stimulation and root crop harvesting. Organic farmers generally use a variety of tools for tillage similar employment, as changes in growth rates of weather, crops, and/or weeds will quickly force them to adjust strategies. Owning or accessing a number of easy to attach and unlock tractors will save time for farmers (Table 11.1).

11.2.3 EFFECTS OF MULCHES

The mulches are mainly used for increasing water infiltration, evaporation, soil temperature change, weed control, evaporation prevention and crop yields increase. There is also a likelihood of a beneficial impact on the physical conditions in the ground under conditions of non-treeing or low tillage of fine-textured soils.

TABLE 11.1 Tillage Implement and Their Purpose

Type of Tillage	Implements	Purpose
Primary tillage	Aggressive equipment including chisel plows, ripers, and subsoilers Aggressive tools like spaders and rotary tillers Moldboard and disk plows	Create a soil state, from which a seedbed with secondary tillage devices can be prepared. The disruption of the soil is usually >6 inches deep. Primary laying is required if current soil conditions hinder the efficiency of secondary instruments.
Secondary tillage	Powered harrows, such as rotavators, rod weeders, and reciprocating harrows Most harrows are draft implements with gangs of tines, disks, rolling baskets, or combinations. Sweeps are used to push residues aside for conservation planting.	The preparation of the seedbed can include the sizing, leveling, and/or burial of residues. Typically, soil preparation is full field but can be concentrated in row zones. Used for seedbed preparation and can do more than draft tools in one pass.
Cultivation	Guided (row) farmers are used to deduct or dislodge weeds between crops that are planted on broad lines (usually >2 feet). Listers/cridge makers and blind farmers come in a wide range of forms and sizes and make beds 6 to 10 inches high, 30 to 40 inches apart, divided by a furrow (interrow). Chain harrows may be used to disperse residues and compost, seeds, and soil surface level.	It is often used pre-emergent or shortly after large crops grow, but is sometimes also used on small or transmitted crops.
Tools to manage surface residue	Mowers and flail choppers are used to manage standing biomass, cutting them into smaller parts to be distributed as a blade or incorporated into primary tillage.	Mechanical herb and residue management. A variety of methods are used to handle residues, mulch, slaughter plants, and distribute soil materials.

11.2.4 MULCH AS A WEED'S CONTROLLER

The synergistic impact of all control methods to the farmer to track the weed populations at the level of a given lot is focused on the integrated weed management. This show will show the benefits of various weed control techniques, including plowing, flooding, nursery transplantation and crop rotation (Al-Tawaha et al., 2010).

Increasingly, mulchery is used as a weed control measure, which is of special importance to organic cropping systems in the cultivation of high quality and safe food production plant raw materials (Bilalis et al., 2003; Petersen and Rover, 2005; Ramakrishna et al., 2006; Jordán et al., 2010; Imranuddin et al., 2017; Khalid et al. 2017; Amanullah et al., 2019a; Al-Tawaha et al., 2020; Amanullah and Khalid, 2020). Agricultural crop processing is used as a plant residue mix and has a multi-sectoral impact on the agroecosystem (Bajorienė et al., 2013). The weed control of mulches (straw, grass, and other) was calculated by Radics and Szné Bognár (2004); Petersen and Rover (2005); and Ramakrishna et al. (2006). Weeds can be monitored by Mulches in various ways. Their suppression of weeds is one of the most significant benefits, preventing cultivation in this way. Now that herbicides provide an alternative way of weed control, this advantage is slightly lower than before. However, in vegetable and flower gardens where herbicides are often dangerous, mulches are more commonly used in order to control weeds. After packing of at least two inches, ideally four inches, most natural mulches must be added to the depth to be successful in the control of weeds. As plant residues on permanent weeds and on many grasses normally do not work, they need to be removed prior to application. Although they can germinate, almost every year weed seedlings cannot develop through a thick mulch (Petersen and Rover, 2005; Ramakrishna et al., 2006). The energy supply of the seeds is too limited to allow significant growth without light. In weed control, some natural mulching materials are not very satisfactory either for tightly packaging and serving as good germination and growth medium or because weed seeds are contaminated by the painting material itself. The majority of artificial materials such as black or plastic paper and aluminum foil for the weed control is satisfactory if the ground is completely protected (Petersen and Rover, 2005; Ramakrishna et al., 2006). Typically, some annual weeds are used in the hollows or cracks as perforated sheets or in strips.

Organic mulch keeps the temperature and humidity in the soil more stable, which leads to a better environment for living organisms in the soil (Ramakrishna et al., 2006). Soil biological properties primarily affect organic farming productivity (Amanullah et al., 2015, 2019a–d, 2020). Mulching is also used to affect the physical properties of the soil. Mulching is necessary for the growth and yield of crops, which helps to lower moisture evaporation, to decrease and sustain a more steady earth temperature (Ji and Unger, 2001). Eventually the natural organic mulch breaks up and returns the organic material to the soil. The slow-releasing of nutrients is more synchronized with the plant needs during the decomposition process (Cherr et al., 2006). The soil appeared to increase the phosphorus and potassium content available in paint and in grass. It was noticed. The rapid decomposition of organic mulch is a significant source of plant nutrients. In grass-mulched plants, significantly greater crop yields were achieved not only because of smothering weeds, but because of increased soil plant nutrients and improved soil physical properties. Better plants are able to kill weeds in greater numbers. Some research indicates that mulching decreases annual weeds, but it does not influence perennial weeds. Plant residues (straws and other) used as a mulch were found in the breakup process to suppress the emergence and growth of weeds. A decrease in the amount of crop residues used for soil mulching in annual weeds has been noticed by many authors (Petersen and Rover, 2005; Ramakrishna et al., 2006; Shah et al., 2019). As the soil cover levels increased, a decrease in weed density was created. Mulching decreases the volume of soil and shears, and increases porosity filled with air. Certain perennial weeds are grown on the ground. Some organic mulches are suitable for vast farms-overgrown wheat, turkeys, sawdust, stroke, and other agricultural residues. We may also use various bio-residues in small farms and gardens for the planting of mulching: grass, sunflower hulls, nuts, coffee beans, and others, regularly cut out from grassplots.

11.2.5 ORGANIC MULCH AND SOIL MICRO FAUNA

The nature of the soil provides the right amount of nutrients for the production of defined plants and the right balance for the development of soil fertility (Turk and Tawaha, 2001, 2002a, b; Tawaha and Turk, 2002; Turk et al., 2003a–c). More research, on the other hand, has shown that low productivity mainly concerns dryland farming activities (Tawaha and Turk, 2001; Turk and Tawaha, 2002b; Abebe et al., 2005, 2005a–d; Assefa et al., 2005; Lee et al., 2005a–d; Nikus et al., 2005a, b; Mesfine et al., 2005; Sulpanjani et al., 2005a–c; Tawaha et al., 2005a, b; Yang et al., 2005; Zheng et al., 2005; Assaf et al., 2006). Bio-mulches boost activity of soil enzymes (Jordán et al., 2010), the biota soil quantity and diversity (Brévault et al., 2007). Earthworms are likely to be provided with soils which are naturally mulched with tree blankets, grass sod, and different forms of plant residue, provided the soil is not too acidic or too dry. The same applies to gardens with residues of plants. One of the key reasons for their satisfactory living conditions is that mulches also provide them with plenty of food. In addition, worms, like most plants find their physical conditions in the soil. The plants in the area have a continuous supply even if the mulch itself is not a satisfactory food source. The lack of soil disturbance also promotes the population of high earthworm, which may stay intact throughout one or more seasons. The channels they produce through the soil. The mulches often keep the soil at a temperature that is more uniform and prevent the soil from freezing. Briefly, the many benefits from earthworms' behaviors in soils which are matched with plant residues appear to reach a near limit. A grass sod is a perfect location for a population of high worms. Many plant materials often encourage the growth of other types of animal living, some of which are beneficial and some of which are harmful. However, harmful insects and some diseases are likely to encounter natural mulches, but this is typically a comparatively small factor in crop production. Multiples of plant materials also have a positive impact on the activities of soil microorganisms. This is partly because of the higher humidity and temperature throughout the whole year, but more particularly because of the ample supply of energy that is always available. In each rain, some energy-generating material is leached into the ground, and the insolvent materials collect on or in the surface layer of the soil as the mulch disintegrates. In addition, earthworms bring uncomposed plant materials continually into the soil. This naturally promotes a significant increase in the amount of soil microorganisms, but also speeds up the loss of the mulch itself.

11.2.6 ORGANIC MULCH AND NUTRIENTS AVAILABILITY

The effect of natural mulches on the amount of nutrients available to plants growing on the ground is mainly determined by the composition of the mulches (Table 11.2) and secondly by the rate of decomposition of the mulches (Amanullah et al., 2019d). This is clear because the floor finally reaches the ground, or the decomposition products that do not escape to the air. The final results are determined by what substance and how much of it is affected and how quickly it enters the ground. The most important factor involved is nitrogen, which is often unpredictable (Imranuddin et al., 2017; Shah, 2017; Khalid et al., 2018). It depends on the carbohydrate's ratio of the mattress material and its ease of decomposition-in particular its lignin content-that the mulch can increase or decrease the available nitrogen under the surface. The vital ratio is typically around 30 for most crop residues (N content 1.4–1.7). The larger the ratio, the closer the rotting residues are to nitrogen. Nitrogen-deficient crop waste used as mulches generally decreases the amount of nitrogen available in the soil, and therefore, unless additional nitrogen is added, the growth of plants growing in the mulched area can degrade. It can be less inhibited than when the mulching materials have been added into the soil.

TABLE 11.2 Various Organic Mulches and Their Chemical Properties

Material	C/N	Ash (%)	Nitrogen (%)	Carbon (%)
Corn cobs	108	1.58	0.45	46.87
Alfalfa hay	18	8.79	2.34	43.15
Pea vines, mature (less pods)	29	8.5	1.5	44.02
Moss peat	58	3.12	0.83	48.29
Ponderosa pine sawdust	1064	0.33	0.05	53.18
Douglas fir sawdust, fresh	996	0.17	0.05	49.8
Wheat straw	373	8.54	0.12	44.7
Rye straw	144	3.51	0.33	47.39
Douglas fir sapwood	548	0.27	0.09	49.36
Cannery waste, beans	10	–	3.17	31.47
Rice hulls	72	19.83	0.55	39.8
Douglas fir needles	58	7.02	0.96	55.75
Red alder sawdust	134	1.21	0.37	49.63
Cannery waste, beets	18	–	2.32	42.85
Douglas fir cones	133	1.51	0.37	49.17
Sewage sludge, digested	10	53.65	2.15	22
Walnut leaves, weathered	26	43.42	1.12	29.48
Pea vines, in bloom	17	10.71	2.69	45.3
Meadow hay (rush and sedge)	43	8.46	1.07	45.6
Douglas fir bark	491	0.45	0.11	53.97
Oak leaves, weathered	26	32.33	1.36	35.11
Bentgrass clippings	13	17.98	3.23	43.22
Western red cedar sawdust	129	0.29	0.07	51.05
Fiber flax, deseeded	373	3.73	0.12	44.7

Vicente-Chandler (1953) notes that the correct type of mat, treated correctly, will provide plants with a cost-effective and permanent supply of nitrogen under heavy rainfalls. This is especially the case when the mulch is annually replaced. In addition to the essential ratio even mulches with carbon-nitrogen ratios will have some existing nitrogen, where the materials would rot very quickly as under humid tropical conditions. The mat of rising roots usually found under such a pail would consume a large part of the nitrogen and other nutrients available before torrential rain can be extracted. Under such conditions, soluble fertilizer nitrogen will not stay in the root area for long. The majority of nitrogen-deficient mulches are also deficient in phosphorus and can need minor uses for both phosphate and nitrogen fertilizers. However, mulches are less likely to deplete phosphates than nitrogen. There is no risk of a natural muzzle becoming defective because of the needs of the microorganisms that decompose mulches for the other major and small components. Once a multitude of plant materials are added, and even before much decomposition occurs, they will be slowed by rain. More and more of these elements are likely to descend into the root zone as the decomposition continues. If the mulch is not discarded, the remaining matter eventually becomes a part of the soil, and the crop assimilates all major and minor elements originally in the mulch.

11.2.7 ORGANIC MULCH AND SOIL ORGANIC MATTER (SOM)

Their organic content is much greater than that of grown soil covered with mulches, which are dept enough to discourage weeds from growing and are not planted for long periods. This is partially due to the absence of cultivation, but if the temperature of the soil is lower in the summer, this is also a factor. Less wetting and drying cycles are also beneficial for the maintenance of organic matter. The major influence of organic mulches on the preservation of humus is the incorporation of organic matter, provided that the mulch is not extracted at intervals. The beneficial impact is very powerful in systems where the soil is maintained mulched and the mulch is filled annually. It is also marked if each spring or fall the old mulch is incorporated in the soil, with fresh mulching materials being changed each year. In the first instance, the upper one or two inches of soil can in organic matter become very high; in the second instance, the whole stuffed or discharged layer benefits from the addition of carbonate substances.

11.2.8 SOIL PHYSICAL PROPERTIES

Soils of medium or fine texture, cut, and stored with plant-residue mulches and scrubbed unusually or hot, typically have excellent physical quality or develop excellent quality. The structure or aggregation of these soils does not suffer from compaction due to tillage and weighting or drying of heavy equipment or aggregate interference. A small proportion of the products of disintegration and the synthetic products of micro-organisms are released into the ground as the organic mulch decomposes. Earthworms are also a bigger factor. These channels help to reduce bulk density and improve ventilation and drainage. The longer organic mat is on the ground, the more it appears to assume the characteristics of a virgin soil, which is the most physically good for many. Such paved soils have a high rate of infiltration and appear to retain the water to the end of the Zhe profile. Good physical characteristics also favor retention of water in the profile before the crop can use it. Factors of a physical nature like rising roots, wetting, and drying, and freezing and thawing appear to be the first in the formation of aggregates; microbe action products are of minor concern, as aggregate forms, and they are helping to keep the clay particles in close proximity.

11.2.9 ORGANIC MULCH AND SOIL EROSION

A large number of plant materials which are not easily transported by running water are extremely successful in erosion prevention. This statement is obvious because poles absorb the energy of the dropping raindrop, increase runoff, allow less water to wash off and improve the soil's quality. This statement is obvious from what is said. Often heavy rain can lead to mulching material and gullies, but this typically does not happen if there is a crop in the soil covering most of the region. Light-weight materials, such as sawdust and peat, can often be floating as gullies, but fibrous materials will not move. Obviously, the straw mulch improved surface sealing and reduced precipitation and runoff. In particular, erosion control is useful for natural mulches as they occur in forests, lands of waste and grassland. Multiples of artificial materials are typically often successful when they are attached enough and allow humidity to easily penetrate the soil.

11.2.10 STUBBLE MULCHES

Stubble-mulch or trash-mulch agriculture was started in semi-arid and wheat-growing areas in the US and Canada between 1930 and 1940. The system was built as a result of the extreme wind erosion at the time and was a major try to reduce as much as possible this harm. The damage caused to the wind in the 1930s was exceptionally bad as a result of the extreme drought shortly after the planting of much land which was long in the native grasses. Stubble-mulch agriculture is generally described as an all-year route to manage residues in croplands in which the residues of the previous crop are selected, grown, seeded-prepared, and planted to leave on or close to the earth's surface. The residues on the surface of the soil decay slower than when buried, and substantial residues can remain from season to season under some conditions. In a stubble-mulch farming method, this of course is quite attractive. The type of residue, temperature, moisture ratios and soil disturbance during laying operations depend on such decline resistance. Working with the stubble-mulch method is usually maintained to the minimum required for the preparation of a proper seedbed and the control of weeds. The sweep or blade is usually used to raise the soil and kill the weeds first. It works at depths up to 6 inches with subsequent tillage at lower depths with different devices. The sweeping cultivator normally has around 70 to 80% more specifications than the moldboard plow at the same depth. The stubble mulch system is particularly suitable for maintaining summer barrier trees and growing small grain crops. In any case, not only should the machinery used leave all crop residues on the soil, it should also be spread consistently. This farming method is also suitable for the cultivation of row crops such as maize or sorghum. The seed is put in the clean furrow of the residues between the ranks. To encourage seeding, a lister operated with a low depth can be used or a planter fitted with a furrow opener. Sweeps or other equipment which undercut residues can be used for cultivation. Weed control is often difficult, particularly of grass, and more crops than would otherwise be considered desirable may be required. When it comes to sodes, the moldboard plow is often used to cause inversion, and then a spring-tooth harrow is applied to get some residues over to the surface. In some ways now, but not for extensive scientific and economic purposes, herbicides are being used.

11.2.11 ORGANIC MULCHING AND SOIL WATER

Mulching affects soil and surface water greatly. Mulching reduces runoff by improving infiltration rate and by improving retention, improves water storage capacity. Low evaporation rates also help increase the time for soil moisture. Mulching dramatically improves the characteristics of the soil water, although various findings were published. Organic mulches on the surface of the soil induce optimum soil conditions for plant growth, retaining, and providing soil water and increasing macroporosity (Martens and Frankenberger, 1907). Many studies have shown that mulch use can increase infiltration and decrease evaporation, leading to more water stowage and less rush (Smika and Unger, 1986). The best way to improve the preservation of water in the soil and to reduce soil evaporation is the wheat straw mulch. High usable water capacities at high mulching rates have been identified, and activities have been reduced or not. The fact that even low mulch concentrations have a major effect on water content available found Mulumba and Lal (2008); and Jordan et al. (2010). Głąb and Kulig (2008), who found no impact on existing water content after applying mulch and various tillage systems, published comparative results. Results can differ from the top to the bottom layers of the soil profile as well.

11.3 CONCLUSION

Mulches are increasing soil biodiversity, increasing penetration by water, reducing evaporation, altering soil temperatures, controlling weeds, preventing evaporation, and increasing crop yield. Mulches can regulate weeds by different means, including a physical barrier and related microclimate changes. On the positive hand, tillage has been part of most farm systems in the past since tillage can be used to achieve many agronomic goals.

KEYWORDS

- **herbicidal systems**
- **mechanical cultivation**
- **organic farming**
- **organic manures**
- **organic mulch**
- **soil erosion**
- **tillage**

REFERENCES

Abebe, G., Assefa, T., Harrun, H., Mesfine, T., & Al-Tawaha, A. M., (2005d). Participatory selection of drought-tolerant maize varieties using mother and baby methodology: A case study in the semiarid zones of the central rift valley of Ethiopia. *World Journal of Agricultural Sciences, 1*(1), 22–27.

Abebe, G., Hattar, B., & Al-Tawaha, A. M., (2005c). Nutrient availability as affected by manure application in cowpea [*Vigna unguiculata* (L.) Walp.] on calcareous soils. *Journal of Agriculture and Social Science, 1*(1), 1–6.

Abebe, G., Sahile, G., & Al-Tawaha, A. M., (2005a). Evaluation of potential trap crops on Orobanche soil seed bank and tomato yield in the central rift valley of Ethiopia. *World Journal of Agricultural Sciences, 1*(2), 148–151.

Abebe, G., Sahile, G., & Al-Tawaha, A. M., (2005b). Effect of soil solarization on Orobanche soil seed bank and tomato yield in the central rift valley of Ethiopia. *World Journal of Agricultural Sciences, 1*(2), 143–147.

Abera, T., Feyisa, D., Yusuf, H., Nikus, O., & Al-Tawaha, A. M., (2005). Grain yield of maize as affected by biogas slurry and N-P fertilizer rate at Bako, Western Oromiya, Ethiopia. *Bioscience Research, 2*(1), 31–38.

Al-Tawaha, A. R. M., & Odat, N., (2010). Use of sorghum and maize allelopathic properties to inhibit germination and growth of wild barley (*Hordeum spontaneum*). *Notulae Botanicae Horti. Agrobotanici Cluj-Napoca, 38*(3), 124–127.

Al-Tawaha, A. R. M., Al-Tawaha, A., Sirajuddin, S. N., McNeil, D., Othman, Y. A., Al-Rawashdeh, I., Amanullah, M., et al., (2020). Ecology and adaptation of legumes crops: A review. *IOP Conference Series: Earth Environ. Sci.,* 012085.

Altieri, M. A., Ponti, L., & Nicholls, C. I., (2012). *Biodiversity and Insect Pests* (pp. 72–84). John Wiley and Sons Ltd.

Amanullah, & Khalid, S., (2020). Agronomy-food security-climate change and the sustainable development goals. *Agronomy-Climate Change and Food Security*. IntechOpen, London.

Amanullah, Iqbal, A., Khan, A., Khalid, S., Shah, A., Parmar, B., Khalid, S., & Muhammad, A., (2019a). Integrated management of phosphorus, organic sources, and beneficial microbes improve dry matter partitioning of maize. *Commun. Soil Sci. Plant Anal., 50*(20), 2544–2569.

Amanullah, Khalid, S., Imran, Khan, H. A., Arif, M., Al-Tawaha, A. R., Adnan, M., Fahad, Sh., & Parmar, B., (2019). Organic matter management in cereals-based system: Symbiosis for improving crop productivity and soil health. In: Lal, R., & Francaviglia, R., (eds.), *Sustainable Agriculture Reviews 29: Sustainable Soil Management: Preventive and Ameliorative Strategies* (pp. 67–92). Cham: Springer International Publishing.

Amanullah, Khalid, S., Khalil, F., & Imranuddin, (2020). Influence of irrigation regimes on competition indexes of winter and summer intercropping system under semi-arid regions of Pakistan. *Sci. Rep., 10*(1), 1–21.

Amanullah, Khan, I., Jan, A., Jan, M. T., Khalil, S. K., Shah, Z., & Afzal, M., (2015). Compost and nitrogen management influence productivity of spring maize (*Zea mays* L.) under deep and conventional tillage systems in semi-arid regions. *Commun. Soil Sci. Plant Anal., 46*(12), 1566–1578.

Amanullah, Khan, N., Khan, M. I., Khalid, Sh., Iqbal, A., & Al-Tawaha, A. R., (2019). Wheat biomass and harvest index increases with integrated use of phosphorus, zinc, and beneficial microbes under semiarid climates. *J. Microbiol. Biotech. Food Sci., 242*–247.

Amanullah, Zahid, A., Iqbal, A., & Ikramullah, (2016). Phosphorus and tillage management for maize under irrigated and dryland conditions. *Ann. Plant Sci., 5*(3), 1304–1311.

Assaf, T. A., Hameed, K. M., Turk, M. A., & Tawaha, A. M., (2006). Effect of soil amendment with olive mill by-products under soil solarization on growth and productivity of faba bean and their symbiosis with mycorrhizal fungi. *World Journal of Agricultural Sciences, 2*(1), 21–28.

Assefa, T., Abebe, G., Fininsa, C., Tesso, B., & Al-Tawaha, A. M., (2005). Participatory bean breeding with women and smallholder farmers in eastern Ethiopia. *World Journal of Agricultural Sciences, 1*(1), 28–35.

Bajorienė, K., Jodaugienė, D., Pupalienė, R., & Sinkevičienė, A., (2013). Effect of organic mulches on the content of organic carbon in the soil. *Est. J. Ecol., 62*(2), 100.

Bengtsson, J., Ahnström, J., & Weibull, A. C., (2005). The effects of organic agriculture on biodiversity and abundance: A meta-analysis. *J Appl Ecol., 42*(2), 261–269.

Bilalis, D., Sidiras, N., Economou, G., & Vakali, C., (2003). Effect of different levels of wheat straw soil surface coverage on weed flora in *Vicia faba* crops. *J. Agron. Crop Sci., 189*(4), 233–241.

Brévault, T., Bikay, S., Maldès, J. M., & Naudin, K., (2007). Impact of a no-till with mulch soil management strategy on soil macrofauna communities in a cotton cropping system. *Soil Till. Res., 97*(2), 140–149.

Cherr, C. M., Scholberg, J. M. S., & McSorley, R., (2006). Green manure approaches to crop production: A synthesis. *Agron. J., 98*(2), 302–319.

Głąb, T., & Kulig, B., (2008). Effect of mulch and tillage system on soil porosity under wheat (*Triticum aestivum*). *Soil Til. Res., 99*(2), 169–178.

Hole, D. G., Perkins, A. J., Wilson, J. D., Alexander, I. H., Grice, P. V., & Evans, A. D., (2005). Does organic farming benefit biodiversity? *Biol. Conserv., 122*(1), 113–130.

Imranuddin, Arif, M., Khalid, S., Nadia, Saddamullah, Idrees, M., & Amir, M., (2017). Effect of seed priming, nitrogen levels, and moisture regimes on yield and yield components of wheat. *Pure Appl. Biol., 6*(1), 369–377.

Ji, S., & Unger, P. W., (2001). Soil water accumulation under different precipitation, potential evaporation, and straw mulch conditions. *Soil Sci Soc Am J., 65*(2), 442–448.

Jordán, A., Zavala, L. M., & Gil, J., (2010). Effects of mulching on soil physical properties and runoff under semi-arid conditions in southern Spain. *Catena, 81*(1), 77–85.

Khalid, S., Imranuddin, N., Nadeem, F., Saddamullah, A. M., & Ghani, F., (2017). Allelopathic effect of parthenium liquid extract on mung bean germination ability and early growth. *Inter. J. Agron. Agri. Res., 11*(4), 31–36.

Khalid, S., Imranuddin, Nadeem, F., Azam, S., Ali, S., Nadia, Ikramullah, M., et al., (2019). Influence of source limitation on yield and yield components of wheat. *Inter. J. Biol., 14*(2), 1–12.

Khalid, S., Mhammad, Z. A., Imranuddin, Nadia, Faisal N., Muhammad, A., & Saddamullah, (2017). Effect of sulfur foliar fertilization on reproductive growth and development of canola. *Inter. J. Agron. Agri. Res., 11*(3), 61–67.

Khalid, S., Munsif, F., Imranuddin, Nadia, Nadeem, F., Ali, S., Ghani, F., & Idrees, M., (2018). Influence of source limitation on physiological traits of wheat. *Pure Appl. Biol., 7*(1), 85–92.

Lee, K. D., & Tawaha, A. R. M., & Supanjani, (2005c). Antioxidant status, stomatal resistance and mineral composition of hot pepper under salinity and boron stress. *Bioscience Research, 2*(3), 148–154.

Lee, K. D., Sulpanjani, Tawaha, A. M., & Yang, S. M., (2005a). Effect of Phosphorus application on yield, mineral contents and active components of *Chrysanthemum coronarium* L. *Bioscience Research, 2*(3), 118–124.

Lee, K. D., Turk, M. A., & Tawaha, A. M., (2005b). Nitrogen fixation in rice-based farming system. *Bioscience Research, 2*(3), 130–138.

Martens, D. A., & Frankenberger, W. T., (1907). Modification of infiltration rates in an organic-amended irrigated. *Agron. J., 84*(4), 707–717.

Mesfine, T., Abebe, G., & Al-Tawaha, A. M., (2005). Effect of reduced tillage and crop residue ground cover on yield and water use efficiency of sorghum (*Sorghum bicolor* (L.) Moench) under semi-arid conditions of Ethiopia. *World Journal of Agricultural Sciences, 1*(2), 152–160.

Mulumba, L. N., & Lal, R., (2008). Mulching effects on selected soil physical properties. *Soil and Tillage Research, 98*(1), 106–111.

Nikus, O., Abebe, G., Takele, A., Harrun, H., Chanyalew, S., Al Tawaha, A. M., & Mesfin, T., (2005b). Yield response of tef (*Eragrostis tef* (Zucc.) Trotter) to NP fertilization in the semi arid zones of the central rift valley in Ethiopia. *European Journal of Scientific Research, 4*(4), 49–60.

Nikus, O., Nigussie, M., & Al Tawaha, A. M., (2005a). Agronomic performance of maize varieties under irrigation in Awash valley, Ethiopia. *Bioscience Research, 2*(1), 26–30.

Petersen, J., & Rover, A., (2005). Comparison of sugar beet cropping systems with dead and living mulch using a glyphosate-resistant hybrid. *J. Agron. Crop Sci., 191*(1), 55–63.

Radics, L., & Szné, B. E., (2004). Comparison of different mulching methods for weed control in organic green bean and tomato. *Acta Hortic., 638*, 189–196.

Ramakrishna, A., Tam, H. M., Wani, S. P., & Long, T. D., (2006). Effect of mulch on soil temperature, moisture, weed infestation and yield of groundnut in northern Vietnam. *Field Crops Res., 95*(2–3), 115–125.

Smika, D. E., & Unger, P. W., (1986). Effect of surface residues on soil water storage. In: *Advances in Soil Science* (pp. 111–138). Springer, New York.

Sulpanjani, A., Yang, M. S., Tawaha, A. R. M., & Lee, K. D., (2005a). Effect of magnesium application on yield, mineral contents and active components of *Chrysanthemum coronarium* L. under hydroponics conditions. *Bioscience Research, 2*(2), 73–79.

Supanjani, Tawaha, A. M., Yang, M. S., & Lee, K. D., (2005b). Role of calcium in yield and medicinal quality of *Chrysanthemum coronarium* L. *Journal of Agronomy, 4*(3), 188–192.

Supanjani, Tawaha, A. M., Yang, M. S., & Lee, Y. D., (2005c). Calcium effects on yield, mineral uptake and terpene components of hydroponic *Chrysanthemum coronarium* L. *Research Journal of Agriculture and Biological Science, 1*(1), 146–151.

Tawaha, A. M., & Turk, M. A., (2001). Effects of dates and rates of sowing on yield and yield components of carbon vetch under semi-arid conditions. *Acta Agronomica Hungarica, 49*(1), 103–105.

Tawaha, A. M., & Turk, M. A., (2002). Lentil (*Lens culinaris* Medic.) productivity as influenced by rate and method of phosphate placement in a Mediterranean environment. *Acta Agronomica Hungarica, 50*(2), 197–201.

Tawaha, A. M., Turk, M. A., Lee, K. D., Supanjani, Nikus, O., Al-Rifaee, M., & Sen, R., (2005a). Awnless barley response to crop management under Jordanian environment. *Bioscience Research, 2*(3), 125–129.

Tawaha, A. R. M., Turk, M. A., & Lee, K. D., (2005b). Adaptation of chickpea to cultural practices in a Mediterranean type environment. *Research Journal of Agriculture and Biological Science, 1*(2), 152–157.

Turk, A. M., & Tawaha, A. M., (2002a). Impact of seeding rate, seeding date, rate and method of phosphorus application in faba (*Vicia faba* L. Minor) in the absence of moisture stress. *Biotechnology, Agronomy, Society and Environment, 6*(3), 171–178.

Turk, A. M., & Tawaha, A. M., (2002b). Response of winter wheat to applied N with or without ethrel spray under irrigation planted in semi-arid environments. *Asian Journal of Plant Sciences, 1*(4), 464–466.

Turk, M. A., & Tawaha, A. M., (2001). Influence of rate and method of phosphorus placement to garlic (*Allium sativum* L.) in a Mediterranean environment. *Journal of Applied Horticulture, 3*(2), 115–116.

Turk, M. A., Hameed, K. M., Aqeel, A. M., & Tawaha, A. M., (2003c). Nutritional status of durum wheat grown in soil supplemented with olive mills by-products. *Agrochimica, 5*, 209–219.

Turk, M. A., Tawaha, A. M., Samara, N., & Latifa, N., (2003b). The response of six-row barley (*Hordeum vulgare* L.) to nitrogen fertilizer application and weed control methods in the absence of moisture stress. *Pakistan Journal of Agronomy, 2*(2), 101–108.

Turk, M. A., Tawaha, A. M., Taifor, H., Al-Ghzawi, A., Musallam, I. W., Maghaireh, G. A., & Al-Omari, Y. I., (2003a). Two-row barley response to plant density, date of seeding and rate and method of phosphorus application in the absence of moisture stress. *Asian Journal of Plant Science, 2*(2), 180–183.

Yang, M. S., Tawaha, A. M., & Lee, Y. D., (2005). Effects of ammonium concentration on the yield, mineral content and active terpene components of *Chrysanthemum coronarium* L. in a hydroponic system. *Research Journal of Agriculture and Biological Science, 1*(2), 170–175.

Zehnder, G., Gurr, G. M., Kühne, S., Wade, M. R., Wratten, S. D., & Wyss, E., (2007). Arthropod pest management in organic crops. *Annu. Rev. Entomol., 52*(1), 57–80.

Zheng, W. J., Tawaha, A. M., & Lee, K. D., (2005). In situ hybridization analysis of mcMT1 gene expression and physiological mechanisms of Cu-tolerant in *Festuca rubra* cv Merlin. *Bioscience Research, 1*(1), 21–26.

CHAPTER 12

WEED MANAGEMENT IN ORGANIC CROPPING SYSTEMS

ABDEL RAHMAN M. AL-TAWAHA,[1] ZAHRA FARROKHI,[2] NANDHINI YOGA,[3] POONAM ROSHAN,[4] IMRAN,[5] AMANULLAH,[5] ABDEL RAZZAQ M. AL-TAWAHA,[6] ALLA ALEKSANYAN,[7] SAMIA KHANUM,[8] DEVARAJAN THANGADURAI,[9] JEYABALAN SANGEETHA,[10] ABDUR RAUF,[11] SHAH KHALID,[5] PALANI SARANRAJ,[12] ABDUL BASIT,[13] AYŞE YEŞILAYER,[14] HIBA ALATRASH,[15] MAZEN A. ATEYYA,[16] MUNIR TURK,[17] ARUN KARNWAL,[18] SAMEENA LONE,[19] and KHURSHEED HUSSAIN[19]

[1]*Department of Biological Sciences, Al-Hussein Bin Talal University, P.O. Box 20, Maan, Jordan*

[2]*College of Agriculture and Natural Resources, University of Tehran, Iran*

[3]*Department of Agronomy, Tamil Nadu Agricultural University, Coimbatore–641003, Tamil Nadu, India*

[4]*Department of Biotechnology, Guru Nanak Dev University, Amritsar, Punjab–143005, India*

[5]*Department of Agronomy, The University of Agriculture, Peshawar, Pakistan*

[6]*Department of Crop Science, Faculty of Agriculture, University Putra Malaysia, Serdang–43400, Selangor, Malaysia*

[7]*Institute of Botany aft. A.L. Takhtajyan NAS RA/Department of Geobotany and Plant Eco-Physiology, Yerevan, Armenia*

[8]*Department of Botany, University of the Punjab, Lahore, Pakistan*

[9]*Department of Botany, Karnatak University, Dharwad–580003, Karnataka, India*

[10]*Department of Environmental Science, Central University of Kerala, Kasaragod–671316, Kerala, India*

[11]*Department of Chemistry, University of Swabi, Anbar, Khyber Pakhtunkhwa, Pakistan*

[12]*Department of Microbiology, Sacred Heart College (Autonomous), Tirupattur–635601, Tamil Nadu, India*

[13]*Department of Plant Pathology, Agriculture College, Guizhou University, Guiyang–550025, P.R. China*

[14]*Faculty of Agriculture, Tokat Gaziosmanpasa University, Tokat, Turkey*

[15]*General Commission for Scientific Agricultural Research, Syria*

[16]*Faculty of Agricultural Technology, Al Balqa Applied University, Al-Salt–19117, Jordan*

[17]*Department of Plant Production, Jordan University of Science and Technology, Irbid, Jordan*

[18]*Department of Microbiology, School of Bioengineering and BioSciences, Lovely Professional University, Phagwara, Punjab, India*

[19]*Division of Vegetable Science, SKUAST-Kashmir, Jammu and Kashmir, India*

ABSTRACT

Global climate changes such as increasing CO_2 levels, temperature increases, drought, etc., have been shown to adversely affect agricultural products. Weeds can be seen as a major problem because they seem to fight at the expense of the crop for resources such as light, water, and nutrients. Weed species are commonly favored in crops that are more closely associated with them. The association between weeds and crops is due to their similar growth habits. The basic strategy should be considered as the core of the organic weed control strategy: crop stand, crop rotation, crop cover, variety selection, clean seeds, soil health, soil structure, spring tillage, delayed planting, post-emergence tillage, hand weeding, mulches, and organic herbicides. This chapter focuses on: (i) classification of weed species; and (ii) weed management strategies under climate change.

12.1 INTRODUCTION

Organic farming is a natural way of producing food while respecting the land and animals, and avoiding methods that are potentially harmful to the environment and human health. It is an integrated agricultural production system, based on ecological principles, which seeks to respect life and natural cycles. The use of synthetic pesticides, genetically modified organisms, synthetic fertilizers, and animal growth hormones is prohibited in organic farming. Weeds can be considered a significant problem because they tend to compete for resources such as light, water, and nutrients, at the expense of the crop (Tawaha et al., 2001; Tawaha and Turk, 2001, 2001c, 2002; Turk and Tawaha, 2001a, b, 2002a, b). Weeds are often documented as the most severe threat to organic crop production (Penfold et al., 1995; Stonehouse et al., 1996; Clark et al., 1998). The foundation of an organic weed control strategy should take into consideration the following basic strategy: crop stand, crop rotation, cover crops, variety selection, clean seed, soil health, soil structure, spring tillage, delayed planting, pre-emerge tillage, post-emerge tillage, hand weeding, mulches, organic-based herbicides. This chapter discusses: (i) classification of weed species; (ii) strategies for weed management in organic farming system (OFS); and (iii) weed management strategies under climate change.

12.2 CLASSIFICATION OF WEED SPECIES

There are about 250 true weed species around the world that are classified by using several classification methods. These methods are based on habitat, physiology, morphology, origin, soil pH, and life cycle (Zimdahl, 2012). The lifecycle-based classification method has been widely exploited by many weed management strategies. This method classifies weeds into three classes of annual, biennial, and perennial, taking into account weed's life span, growing season, the timing of growth and reproduction, and the reproductive form (Zimdahl, 2012). Such information may determine which weed species are favored in which crops (Ziska and Dukes, 2011), being a prerequisite for making the appropriate selection for the most effective control method (Monaco et al., 2002).

12.2.1 ANNUAL SPECIES

Weeds can be considered a major problem because they appear to fight for resources like light, water, and nutrients at the cost of the crop (Turk and

Tawaha, 2002c–e, 2003a, b; Turk et al., 2003). Annuals are weeds that all their growth stages, including both vegetative and reproductive growth, occur within one growing season (less than one year) (Ziska and Dukes, 2011). They do not have any vegetative structures for reproduction and spread, and reproduce only by their seeds. High persistence of these species is due to their quick growth, short life span, and production of a significant number of dormant seeds (Ziska and Dukes, 2011). Annual weeds may find in different crop fields; however, they tend to be more abundant in annually tilled fields. Usually, eradication, prevention, and controlling of annual weeds are easier than perennial ones. The most appropriate time to control these weeds is during their seedling stage by clipping or pulling up (Zimdahl, 2012).

Annual weeds are usually categorized into summer and winter annuals. Summer annuals tend to start growing in the spring (April to May), go to flower and produce seeds through the summer, then they die in the autumn. The weeds foxtails, goosegrass, common cocklebur, purse lane, water hemp, morning glories, pigweeds, common lambsquarters, crabgrass, common ragweed, and wild buckwheat are summer annual weeds. The seeds remain dormant in the soil during the winter months. Summer annual weeds are often problematic in summer-growing crops such as soybeans, corn, cotton, peanuts, sorghum, and many vegetables (Monaco et al., 2002). In contrast, winter annual weeds begin to grow in the autumn, form rosette (a form of plant with no central stem and of leaf-like structures) in the beginning of the cold period, then bloom and produce seeds in the spring, and die in the middle of the summer. Soil temperatures of 125°F or higher cause dormancy in the seeds of these weed species, preventing seed germination (Monaco et al., 2002). Some examples of winter annual weeds are downy brome, chickweed, pinnate tansy mustard, shepherd's-purse, flixweed, hairy cress, cheat, field pennycress, corn cockle, cornflower, and henbit. They are troublesome mostly in winter crops such as winter cereal, early spring grains, fall-seeded crops, pastures, no-till fields, and in alfalfa, a perennial (Zimdahl, 2012).

12.2.2 BIENNIAL SPECIES

Biennials complete all their growth stages (vegetative and reproductive growth) in around 2 years. Their biological life cycle has two phases. During the first phase of growth, seeds germinate, and seedlings grow and form rosette at the beginning of the cold period. In the second phase, it elongates a flowering stalk, sets seed, and then it dies (Monaco et al., 2002). Because of having a lifespan of about two calendar years, they are sometimes confused by winter annual weeds. However, biennial weeds live for more than 1 year (12 months) and less than 2 years. In addition, they are usually longer and larger than annuals at maturity and have fleshy and thick roots (Radosevich et al., 2007). Like annuals, biennial weeds reproduce only sexually, and hence preventing seed production is the most effective way to control this class of weed species (Zimdahl, 2012). Biennial weed species are relatively few compared to the other weed species. Some examples include musk thistle, wild lettuce, bull thistle, common mullein, common burdock, and wild carrot. These weeds are typically inhabiting undisturbed fields for at least 2 years (min- or no-till fields or in perennial crops), and in pastures and non-crop areas like along roadsides, and fencerows (Monaco et al., 2002).

12.2.3 PERENNIAL SPECIES

Perennials are weeds that live for several years through regrown from the underground perennating structures (Radosevich et al., 2007). Unlike annuals and biennials, the life cycle of these species does not finish after flowering and can reproduce by vegetative organs, as well as by seeds. The perennials mainly find in no-till fields, pastures, roadsides, and sometimes in tilled fields. Based on the method of reproduction, these species are classified into two classes: stationary and creeping perennials. In the following, characteristics of the two mentioned classes are explained.

12.2.3.1 STATIONARY PERENNIALS

This group of perennials normally reproduces by seeds, and they have no normal means for vegetative reproduction. However, if their roots are cut or damaged, new plants can be regenerated from these damaged pieces (Zimdahl, 2012). Since the first infestation of stationary perennials relies on seed (as a major reproduction strategy), and therefore, many of these species can be removed by preventing seed production (Monaco et al., 2002). They have usually fleshy and deep taproots that grow vertically downward into soil. The root systems are separate and there is no underground joining among these plants (https://ag.umass.edu/sites/ag.umass.edu/files/fact-sheets/pdf/weed_life_cycles). Some examples of simple perennials include broadleaf plantain, common pokeweed, broadleaf/curly dock, and dandelion.

12.2.3.2 CREEPING PERENNIALS

These perennial weeds establish by seeds and by vegetative reproduction. These weeds store carbohydrates in their underground overwintering structures allowing them to regrow without seed in the spring. Some examples of these vegetative organs include: above-ground (stolon) and below-ground (rhizomes) creeping stems, bulbs, corms, budding roots, and tubers. Regardless of management strategies, controlling creeping perennials is most difficult because they reproduce by vegetative structures as well as seeds. These structures are spread by tillage equipment from one field to another. In order to adequate control of these perennial weeds, management practices such as cultivation, mowing, and herbicides must be repeated or combined (Monaco et al., 2002). Most aquatic weeds (except algae), field bindweed, Johnson grass, leafy spurge, quack grass, purple, and yellow nutsedge, Canada thistle, Russian knapweed are some examples of creeping perennial weeds (Radosevich et al., 2007).

It is to be noted that the lifecycle-based classification sometimes may be affected by local climate conditions. For example, in temperate climates, some biennial weeds may behave more like simple perennials, and under particular conditions, some annual weeds may act like more biennial weed (Radosevich et al., 2007).

12.2.4 WHICH WEED SPECIES ARE FAVORED IN WHICH CROPS?

Weed species are commonly favored in crops that have more association with them. The association between weed and crop are because of their similarity in growth habit (such as growth form, height), life cycle, etc., (Zimdahl, 2012). For example, summer annual weeds prefer summer annual crops, and/or perennial weeds are favored in perennial crops. This similarity in growth habits makes weed management very difficult, provides an opportunity for weeds to disperse their seeds, and makes difficult separation of weed and crop seeds due to the similar seed morphology (for example, nightshades in potatoes, tomatoes, and beans, Kochia, and lambs quarters in sugar beets, barnyard grass in rice, wild oat in wheat, or little seed canary grass in wheat) (Gbèhounou, 2013).

Sometimes the tendency of a weed to a certain crop can be induced by different selection pressures including cultural practices (e.g., tillage, mowing, time of planting, irrigation, rotation, and soil preparation practices, and even post-harvest selection pressures such as winnowing and threshing), cropping system (mixed or monoculture), resistance or adaptation to imposed weed control tactics, and altering weed control methods over time (Tominaga and Yamasue, 2004). Weeds are evolving in order to adapt to the control measures or environment conditions (vegetative and seed mimicry or evade) under different selection pressures. Implementation of any cultural practice as selection pressure not only may favor the growth conditions of crop, but also can favor the growth conditions of specific weeds (Harlan, 1982). Accordingly, some types of the relevant weeds are able to persist on a particular field, whereas they are rarely able to grow under another field conditions (crop-specific weeds) (Tominaga and Yamasue, 2004).

A common example of selection pressure is applying herbicides. Because of more application of graminicides by farmers, broad-leaved weeds succession occurs in broad-leaved crops. Thus, these weeds become dominant over time and maybe more associated with that certain crop (https://www.researchgate.net/post/Why_weeds_are_crop_specific). Another example of exerted changes in weed communities related to selection pressure is tillage. A periodic soil disturbance in conventional tillage systems is prevailed in the annuals because they can reproduce earlier and produce a high number of seeds. In contrast, no-tillage farming dominates the perennial weeds (Carr et al., 2013a; Melander et al., 2013) because of reduction in physically damaging of underground organs and in exposing them to unfavorable weather conditions. The more example is a post-harvest selection pressure such ash and weeding that is evolved seed and vegetative mimicry of *Lolium temulentum* to barley and wheat, and the vegetative mimicry and physiological traits of *Echinochloao ryzicola* in rice (Tominaga and Yamasue, 2004).

12.3 STRATEGIES FOR WEED MANAGEMENT

12.3.1 CULTURAL METHODS

In cultural methods of weed control, the cultural practices generally followed in a crop is employed, such as tillage, fertilizer application, crop rotation, allelopathy, etc. Some of them are discussed in subsections (Table 12.1).

12.3.1.1 WEED-CROP COMPETITION

Weeds compete with crops mainly for the water, light, space, and nutrients which affect the growth of the plants (Tables 12.2 and 12.3).

TABLE 12.1 Potential Allelopathy Crops with Their Allelochemicals and Their Usage in the Cropping System

Name of the Crop	Allelochemicals Present	Ways to Use in Cropping Systems
Rye (Tabaglio et al., 2013)	ß-phenyllactic acid, protocatechuic acid, DIBOA (glucoside), vanillic acid, apigenin-glycosides, syringic acid, luteolinglucuronides, ρ-hydroxybenzoic acid, ρ-coumaric acid, benzoxazolinones BOA, cyanidin glycosides, ß-hydroxybutyric acid, isovitexinglucosides, DIMBOA (glucoside), gallic acid, and ferulic acid/conjugates	Used in a cropping system as a rotational crop, cover crop, or mulch
Sorghum (Weston et al., 2013)	Hydrophobic ρ-benzoquinone (sorgoleone), phenolics, and acyanogenic glycoside (dhurrin)	Used by planting allelopathic cultivars, applying sorghum residues as mulch, using sorghum as cover crop and intercrop, or including sorghum cultivars in a crop rotation
Brassicas (Fahey et al., 2001)	Glucosinate	As cover crops, intercropping brassica crops with the main crop, crop rotation, or the use of brassica litter as mulch
Sunflower (Alsaadawi et al., 2012)	16 allelochemicals (phenolic acids)	Used either by growing allelopathic sunflower cultivars in a mixture with weeds, or applying the residues of sunflower cultivars to the wheat crop and its weeds

TABLE 12.2 Insects Selected for the Control of Specific Weeds

Weeds	Insects Selected for the Control
Prickly pear cactus (*Opuntia inermis*), Spiny prickly pear (*Opuntia stricta*)	*Cactoblastis cactorum*
Hypericum perforatum	*Chrysolina quadrigemina*
Parthenium hysterophorous	*Zygogramma bicolorata*
Eichhornia crassipes	Weevils (*Neochetina eichhorniae, N. bruchi*)
Salvinia molesta	*Paulina acuminata*

TABLE 12.3 Using Fungi on Specific Weed Control

Pathogen	Weed Controlled
Alternaria sp.	*Crissum avenae*
A. helianthi	*Xanthium strumarium*
A. crassa	*Datura stramonium*
Bipolaris halopense	*Sorghum halepense*
Colletotrichum furariodes	*Asclepias sericea*
Phomopsis convolvulus	*Convolvulus arvensis*

12.3.1.1.1 Competition for Light

Light is an important factor for the rapid growth of crop plants as well as weeds. Photosynthesis of the plants is dependent upon the light. Broadleaved weeds establish prior to the crop plants and restrict light to the crop plants through shading effect, thereby hindering the crop growth. It is estimated that weed competition reduces light intensity up to 85% in onions and beets, thus reducing yield by 60%. If there is adequate moisture and fertility levels, the competition will be more for light. Plants with higher leaf density are benefitted as it has higher leaf area index. The presence of higher leaf density affects the quality and quantity of light obtained to weeds.

12.3.1.1.2 Competition for Water

Weeds generally absorb and transpire more water than the crop plants (four times higher transpiration rate). Hence many weeds are considered as "water wasters." During water stress conditions, weeds cause severe moisture depletion and transpire the moisture rapidly, up to 50% yield loss occurs through moisture competition alone. Rather than the above ground biomass, the below-ground biomass contributes more for the water extraction from soil through the root zone. Most of the perennial weeds have deeper rooting system, which is less affected by drought as it can acquire water from the deeper zones of soil. On the other hand, annual weeds are prone to drought, as it has a shallow root system. Some of the weeds like ragweed (*Parthenium hysterophorus*), and common ragweed (*Ambrosia* sp.) require water to the extent of about three times than that of millets. Among the problematic weeds around the world, 14 weeds are C_4 plants and 76% of C_3 crops are the cultivated crop area. So under elevated CO_2 on drought conditions, C_4 crops has an advantage over the C_3 crops.

12.3.1.1.3 Competition for Nutrients

Weeds remove mineral nutrients from soil like nitrogen, phosphorous, and potassium (NPK) more rapidly and in large quantities than the crop plants and reduces the availability of these nutrients, especially nitrogen and potassium to the crops. Certain weeds have very deep and prolific root system than the crops and absorb nutrients causing a reduction in the crop yield. Some parasitic weeds like dodder (*Cuscuta* sp.), *Loranthus* sp., etc., absorb mineral nutrients directly from the host crops and destroy them. *Amaranthus virdis*, *Chenopodium album*, *Portulaca quadrifida* and *Celosia argentea* can remove 3.5 to 4.5% K_2O and 2.4 to 3.1% N from the soil during the crop period and become a potential competitor for nutrients.

12.3.1.2 CROP ROTATIONS

Crop rotation is the yearly sequence and arrangement of crops. The crop sequences will suppress and remove the weeds from the field. Also, the problematic weeds associated with crops have been reduced to some extent, but the overall weed diversity was increased. By growing the same crop year after year will increase the weed intensity, changing the crops grown year after year will reduce the weed intensity. Adeux et al. (2019) reported that changing the crop sequence reduced the usage of herbicides and increased the productivity. Different crops have different requirements, so it will disrupt the weed phase in a crop rotation.

12.3.1.3 ALLELOPATHY

Allelopathy is a biological phenomenon by which an organism produces one or more biochemicals that influence the germination, growth, survival, and reproduction of other organisms from the same community. Allelochemicals are derived either from root exudates or leaching effect or through volatilization. The effect of the biochemicals may be direct or indirect on the plant species, affecting the germination, plant growth, survival, and reproduction of other organisms. The allelopathic plants can be used to control the weed by raising potential allelopathic cultivars, by intercropping them in the cropping system, as cover crops, and as well as plant residues.

12.3.1.4 WEED ON WEED

Cogon grass inhibits the emergence and growth of buttonweed by exudation of allelochemicals through rhizomes; Johnson grass showed an effect on giant foxtail and large crabgrass by inhibiting the growth through release of chemicals from living and decaying rhizomes and leaves.

12.3.1.5 BIOLOGICAL CONTROL

This method is one of the important and practically feasible ones on controlling weeds. Biological control means, "the utilization of any living organism for the control of insect pests, diseases, and weeds." It means controlling the pest or weed population using the biotic agents either directly or indirectly. In a direct control, the biocontrol agent bores into the plant, weakens the plant structure causing it to collapse. The collapsed plants are destroyed by consuming the vital parts of the plant. On the other hand, indirect control reserves the competitive weed ability of weed over other plants and paves the way for favorable conditions for plant growth. There are natural enemies in the crop environment which will work against the crop pests that suppress the growth of the plant, while aiming on increasing these natural enemies both quantitatively and qualitatively is termed as biological control.

12.3.1.5.1 Insects

On selecting insects for biocontrol of weeds, the selected insect should target specifically the weed and should not feed on the crop plants for its survival.

12.3.1.5.2 Plant Pathogens

Using of plant pathogens to control weeds, especially fungi, is now becoming an important aspect of biological weed control. The pathogens are formulated as mycoherbicides on controlling the weed population. From these mycoherbicides, toxic compounds are released on targeted plants and it will degrade the cell wall. Also, it has the ability to reproduce on its own and can thrive in soil for years (Table 12.4).

TABLE 12.4 Mycoherbicides, Their Formulations to Control Weeds

Mycoherbicide	Formulation of Fungus	Weed Controlled
Devine	*Phytophthora palmivora*	*Morrania odorata*
Collego	*Colletotrichum gleosporioides* f. sp. *aeschynomene*	*Aeschynomene verginicia*
Biomal	*C. gleosporioides* f. sp. *malvae*	*Malvae pusilla*
Luboa 2	*C. gleosporioides* f. sp. *cuscutae*	*Cuscuta* sp.
Velgo	*C. coccodes*	Velvetleaf
Caset	*Alternaria cassia*	*Cassia obtusifolia*
ABG-5003	*Cercospora rodmanii*	*Eichornia crassipes* (Water hyacinth)

12.3.2 PHYSICAL WEED CONTROL METHODS

Physical or mechanical control of weeds is the removal of weeds by implements. Some of the physical weed control methods are discussed in subsections.

12.3.2.1 STUBBLE CULTIVATION

After harvest of the grain crops, the stubbles are left in the field for a few days. Then the tillage operation is carried out to reduce the soil weed seed bank and perennial weeds by causing mechanical damage (Pekrun and Claupein, 2006). Stubble tillage additionally helps in controlling unnecessary evaporation and incorporation of crop residues for a quicker decomposition. Chikowo et al. (2009) reported that stubble cultivation along with crop diversification and delayed sowing reduced the use of chemical herbicides up to 77%. Melander et al. (2013) suggested that cover crops and stubble management practices strengthen the crop growth and reduced the weed growth.

12.3.2.2 PLOUGHING

Ploughing is generally practiced to control weeds in the summer season. It will uproot the weeds that are in the sub soil layer, exposed roots and stems are desiccated, and the plant will eventually die. The commonly used plow for controlling the weeds is moldboard plow. While seeds in the surface layer are buried deep down in the soil by moldboard plow. Usually plowing is used to break seed dormancy and initiate seed germination due to exposure of sunlight. This mechanism is taken into favor to control weeds by stale seedbed method. The weed seeds are made to germinate earlier before cultivating the crop and the weeds are eliminated by subsequent plowing operations. The number of weeds in a cropped area and the reduction of weed seeds in the soil seed bank is reduced through the stale seedbed technique (Lampkin, 2002).

12.3.2.3 HARROWING AND SEED BED PREPARATION

Harrowing is a tillage operation, where the implement consists of tines drill into the soil creating favorable conditions for crop growth. Generally, a shallow harrowing will reduce weed intensity. In seedbed preparation, stale seedbed or false seedbed reduces the weed intensity during the cropping period. The seedbed is prepared few days before the original planting, due to optimum moisture condition in the field, weeds will germinate and emerge. Then the emerged weeds are removed either by harrowing, flaming, or drilling.

12.3.2.4 WEED HARROWING

Weed harrowing is mechanical control of weeds applied to weed plants. Here harrows are being used as a weeder which will work both in intra-row and inter-row. But it cannot be used in the earlier stages of crop growth. The non-hydraulic drag harrow is used in vegetable crops, whereas the modern spring tine harrows are used in dicot and monocot crops. However, careful harrowing should be done in dicot crops in all the stages of the crop growth. Two types of harrowing such as pre-emergence and post-emergence harrowing are practiced. The pre-emergence harrowing is used for deep rooted crops, and the post-emergence harrowing is used after raising of crops to control small weeds in the early stages of crop growth. In the study of reducing quinoa weed density, harrowing, and hoeing showed lesser quinoa weed density. Also, it is suggested that harrowing can be used as a supplement in inter-row hoeing (Jacobsen et al., 2010).

12.3.2.5 INTER-ROW CULTIVATORS

Inter-row cultivators are selective mechanical weed control methods. In harrowing, ineffective weed control and injuries to crop paved the way for specific technology like inter-row cultivators. In this, only the weeds are damaged without any impact on the crop. This is achieved by increasing the inter-row spacing of the crop to allow the hoe blade, which will effectively control the weeds by addition of cutting action with uprooting. The inter-row weeds are completely killed while the intra-row weeds are partially affected. For minimizing the intra-row weeds, increasing the population density will cause crop competition against weeds and suppress the intra-row weeds (Melander and Rasmussen, 2001).

12.3.2.6 MOWING

Mowing is cutting of grass and crops with a hand implement or a machine. It is mainly used in pastures, along roadside and also in waste places. It is primarily used to reduce weed seed production and to restrict weed growth. Mowing alone will not control the weeds effectively, but along the combination with other management methods it shows effectiveness. An even soil surface without any rocks is required for mowing the field. Mowers can be connected to a tractor with three-point hitches and used for cutting the weeds. With a high-speed velocity, rotary or disc mower will cut the plant tissue. The importance of mowing in the cropping system is maintaining the pastures, rangeland, cover crops, grassed waterways, field margins, orchards, tree plantations, vineyards, horticultural ornamentals, woody perennial, and lawns. Mowing is highly practiced in areas where there are no further soil disturbances.

12.3.2.7 BRUSH WEEDING

A mechanical instrument called a brush weeder is used for controlling the weeds through cutting action. It consists of rotating brushes placed in a vertical axis that revolves in higher speed velocity (Melander, 1998). Mostly, brush weeders are used to control weeds in the inter-rows between the plants. Horticultural crops use vertical brush weeders use rotating nylon bristles to pull weeds and aerate soil about 1 inch deep.

12.3.2.8 RIDGING (POTATOES AND OTHER ROW CROPS)

Ridging or earthing up is mounding the soil around the crop taken from between the rows of crop. By ridging, the plant will be in an upright position, and it also helps to control the weeds. It can be followed in all row crops to prevent the growth of weeds, even though the main aim of ridging is to loosen the soil for aerating the root. Maize, pearl millet, groundnut, sugarcane, potato, ginger, and turmeric are some examples of crops which are following earthing up. In sugarcane, ridging is practiced three times at 45, 120, and 180 days after planting. In potato crop, an early ridging is done and also further once or twice it can be done.

12.3.2.9 FLAMING

Using fire to control weeds is one of the physical measures to reduce the weed intensity. A directed flame from liquid propane increases the temperature of the weeds and causes to rupture the cell walls. In organic weed control practice, flaming, and rotary hoeing offered the same yield in corn crop. Also flaming is less invasive and preserves the soil structure (Mutch, 2008).

12.4 WEED MANAGEMENT STRATEGIES UNDER CLIMATE CHANGE

Climate change is represented as shifting of weather events (IPCC, 2014) due to fluctuations in essential climate variables (ECV). Bojinski et al.

(2014) described ECVs as physical, chemical, or biological parameters that are critical for the characterization of climate or climate related changes. Some key parameters such as temperature, greenhouse gases (GHGs) (carbon dioxide (CO_2) and methane), rainfall, wind, precipitation, drought, water availability and light impart a pronounced role in the growth of plant species. Elevated levels of CO_2 and extreme weather events have emerged as serious threats to agricultural production, thereby posing a negative impact on the food security (IPCC, 2019). The changing climatic conditions also found to affect the growth and distribution pattern of weed species. Weeds are recognized as one of the key biological constraints that scuffle with crops for physical resources like water, sunlight, nutrients, aeration, etc., hence, reducing global crop productivity (Varanasi et al., 2016; Ray et al., 2019). Moreover, they act as reservoir hosts for various destructive pathogens such as bacteria, viruses, fungi, etc., and their insect vectors (Agrios, 2005; Roshan et al., 2019). As a consequence, researchers, and scientists are concerned to develop ingenious, cost-effective, and sustainable weed management system to tackle the weed related challenges in agriculture.

12.4.1 CROSSTALK BETWEEN CLIMATE CHANGE AND CROP-WEED COMPETITION

Global changes in the climatic factors such as increased CO_2 concentration, elevated temperature, drought, solar radiation pattern, etc., found to negatively influence the agriculture productivity. Furthermore, these factors alter the competitiveness of weeds, leading to a reduction in the crop yields (Ziska et al., 2011).

It is evident that agriculture based activities contribute (10–12%) for the emission of GHGs such as CO_2, nitrous oxide (NO_2), and methane (CH_4) (Smith et al., 2014). In context to plant species, elevated CO_2 (e[CO_2]) can have adverse consequences in determining the invasiveness of C3 weeds species over C4 crops (Ziska, 2004). The most probable reason for this change is different modes of CO_2 enrichment by ribulose-1,5-biphosphate (RuBisCO) enzyme, and C3 plants have been reported to efficiently fix the higher concentrations of CO_2 (600–800 ppm) during photosynthesis. However, C4 plants have a lesser tendency to tolerate an increased amount of atmospheric CO_2 because their innate machinery itself can accumulate 2000 ppm of CO_2, which is sufficient to saturate RuBisCO enzyme (Ziska et al., 2011; Korres et al., 2016). Additionally, e[CO_2] levels lead to enhancement of leaf area and biomass that are unequivocally correlated with the reproductive capacity of C3 weed species (Chandrasena, 2009). For instance, a C3 weed common cocklebur outcompetes C4 crop, sorghum in terms of biomass and leaf area under e[CO_2] concentrations (Ziska, 2001). Ziska (2003) has also reported that e[CO_2] concentrations enhanced 70% growth of a North American weed in comparison to native species. The effect of e[CO_2] on weed competitiveness is also strengthened by some reports in the Asian subcontinent, where a C3 weed species, *Phalaris minor* Retz. affect the growth of wheat crop (Ramesh et al., 2017). Contrarily, in a field of C3 crop and C4 weed, the former is being benefitted by the increased CO_2 rates (Ziska, 2013).

Together with e[CO_2], atmospheric temperature is considered as an important attribute in determining competitiveness and distribution of weed species in a geographical area. Under temperature stress, both C3 and C4 species experience variations in the photosynthetic stimulation and growth. Beyond 25°C, C3 species undergo increased photorespiration accompanied with inhibition of CO_2 assimilation, whereas C4 species sustain low photorespiration rates in response to temperature change. This mechanism favors the growth of C4 species at different atmospheric temperatures in comparison to C3 species (Sage and Kubein, 2007). Moreover, C4 weeds tend to expand their geographical territories (range shift) because higher latitudes are convenient for plant growth at elevated temperatures (Clements and Ditommaso, 2011). In a maize field, germination of a C4 weed, *Setaria viridis* occurs in the latter stage, under warmer temperatures. Recently, due to climate change events and elevated temperature, the germination of this weed species is synchronized with maize, thus becoming as a competitor to maize (Peters and Gerowitt, 2014).

Elevated CO_2 and temperature events are correlated with the increased aridity and global alteration in the rainfall pattern. As a consequence, the monsoon fed geographical area would dried up, leading to increased drought spells and favor growth of C4 over C3 weed species (Valerio et al., 2011). Weed species, such as *Striga hermonthica* and *Lantana* sp. adapt to thrive under drought conditions (trait shift) by shortening their life cycle. On the other hand, weeds like *Ramphicarpa fistulosa* are benefitted by moisture abundance (Rodenburg et al., 2010; Taylor et al., 2012). In south Asia, direct-seeded rice (DSR) technology was implemented to achieve higher water productivity, but the major limitation of this approach was serious weed infestation and competition that resulted in 30–80% yield loss (Matloob et al., 2015). Karkanis et al. (2018) postulated that moisture fed weeds species could shift to northern Europe due to decreased rainfall content in the southern part, and under such circumstances; drought can increase weed derived crop losses throughout Europe.

12.4.2 EFFECTIVE STRATEGIES OF WEED MANAGEMENT UNDER CLIMATE CHANGE

It is quite evident from our previous discussion that changing ECVs can potentially affect weed invasiveness and distribution of a geographical region. Hence, at a realistic level, systemic, and comprehensive research of weed biology and ecology accompanied with higher crop yield goals is a prerequisite for the development of sustainable weed management programs.

There is a pressing need to identify geographical ranges vulnerable to weed invasion and expansion under changing climatic conditions. The development of high throughput bioclimatic prediction models integrated with understanding of weed biology and ecology would be helpful in making weed management decisions and precise application of control measures (Chauhan et al., 2017). These simulation models are based on combinatorial approaches such as niche-based species distribution, spatial distribution, geographic information system (GIS) and successfully utilized to study distribution of weed species, (*Avena sterilis* L. and *Ambrosia artemisiifolia* L.) and assessment of crop/weed competition (Richter et al., 2013; Castellanos-Frías et al., 2014; Andrew and Storkey, 2017).

Weed species retain their seeds in soil and serve as inoculums for next season weed emergence. These seeds can remain dormant and viable for longer periods, thus escaping the effect of changing climatic conditions and germinate on arrival of favorable climate. The most effective way to prevent the addition of weed seeds into soil is by implementation of harvest weed seed control methods (HWSC). As a spinoff of this concept, several weed control methods like chaff carts, narrow window burning, bale direct system and recently, a weed management device called 'Harrington seed destructor' is coupled with grain harvesters to destroy weed seeds (Walsh et al., 2012, 2017). Seeds of some weeds display greater extent of dormancy while remaining in the soil, causing hindrance in weed control practices. In such a case, an application of smoke derived stimulant, namely Karrikin (KAR1) is helpful in the synchronized germination of *Avena fatua* seeds (Kępczyński, 2018).

The application of herbicides has been considered as most economical way for weed management in agriculture. However, continuous use of concentrated herbicide formulations on crops under elevated temperature, rendered the evolution of herbicide resistant weeds species (Heap, 2014). Taking into account of abovementioned limitations of chemical herbicides, the need of the hour is to implement the usage of natural phytotoxins as non-chemical herbicides for controlling the weed population in a sustainable manner (Dayan and Duke, 2014).

Some crop cultivars possess complex traits which are responsible for suppression/allelopathy of weed growth, therefore, identification of such traits is a crucial step in organic plant breeding (OPB) program. Location and mapping of genes controlling high yield, early vigor, height, and allelopathy in crop cultivars would be important in weed management. So far, several breeding studies have been conducted between high yielding and weed suppressive cultivars to develop superior and competitive cultivars in rice, wheat, and canola (Worthington et al., 2015; Dimaano et al., 2017; Mwendwa et al., 2018).

In the 21st century, initiatives must be taken to educate farmers about the weed risk assessment amid climate change, and the association of farmers with weed scientists will encourage them to adopt the latest technology to control weeds.

12.5 CONCLUSION

Weeds can be considered a major problem because at the cost of crops they appear to fight for resources including light, water, and nutrients. The cornerstone of an organic weed control strategy should consider the following basic strategy: crop stand, crop rotation, crop cover, variety selection, clean seed, soil health, soil structure, spring tillage, delayed planting, post-emergence tillage, hand weeding, mulches, organic herbicides.

KEYWORDS

- **delayed planting**
- **hand weeding**
- **mulches**
- **organic herbicides**
- **post-emergence tillage**
- **spring tillage**
- **weed**

REFERENCES

Adeux, G., Munier-Jolain, N., Meunier, D., Farcy, P., Carlesi, S., Barberi, P., & Cordeau, S., (2019). Diversified grain-based cropping systems provide long-term weed control while limiting herbicide use and yield losses. *Agronomy for Sustainable Development, 39*(4), 42.

Alsaadawi, I. S., Sarbout, A. K., & Al-Shamma, L. M., (2012). Differential allelopathic potential of sunflower (*Helianthus annuus* L.) genotypes on weeds and wheat (*Triticum aestivum* L.) crop. *Archives of Agronomy and Soil Science, 58*(10), 1139–1148.

Al-Tawaha, A. R. M., & Nidal, O., (2010). Use of sorghum and maize allelopathic properties to inhibit germination and growth of wild barley (*Hordeum spontaneum*). *Notulae Botanicae Horti. Agrobotanici Cluj-Napoca, 38*(3), 124–127.

Andrew, I. K., & Storkey, J., (2017). Using simulation models to investigate the cumulative effects of sowing rate, sowing date and cultivar choice on weed competition. *Crop Protection, 95*, 109–15.

Bojinski, S., Verstraete, M., Peterson, T. C., Richter, C., Simmons, A., & Zemp, M., (2014). The concept of essential climate variables in support of climate research, applications, and policy. *Bulletin of the American Meteorological Society, 95*(9), 1431–43.

Carr, P. M., Gramig, G. G., & Liebig, M. A., (2013). Impacts of organic zero tillage systems on crops, weeds, and soil quality. *Sustainability, 5*(7), 3172–3201.

Castellanos-Frías, E., García De, L. D., Pujadas-Salva, A., Dorado, J., & Gonzalez-Andujar, J. L., (2014). Potential distribution of *Avena sterilis* L. in Europe under climate change. *Annals of Applied Biology, 165*(1), 53–61.

Chandrasena, N., (2009). How will weed management change under climate change? Some perspectives. *Journal of Crop and Weed, 5*(2), 95–105.

Chauhan, B. S., Matloob, A., Mahajan, G., Aslam, F., Florentine, S. K., & Jha, P., (2017). Emerging challenges and opportunities for education and research in weed science. *Frontiers in Plant Science, 8*, 1537.

Clements, D. R., & Ditommaso, A., (2011). Climate change and weed adaptation: Can evolution of invasive plants lead to greater range expansion than forecasted? *Weed Research, 51*(3), 227–240.

Climate Change, (2014). *Impacts, Adaptation, and Vulnerability-Contribution of Working Group II to the Fifth Assessment Report of the Intergovernmental Panel on Climate Change.* Environmental Migration Portal. https://environmentalmigration.iom.int/climate-change-2014-impacts-adaptation-and-vulnerability-contribution-working-group-ii-fifth (accessed on 12 July 2021).

Dayan, F. E., & Duke, S. O., (2014). Natural compounds as next-generation herbicides. *Plant Physiology, 166*(3), 1090–1105.

Derksen, D. A., Anderson, R. L., Blackshaw, R. E., & Maxwell, B., (2002). Weed dynamics and management strategies for cropping systems in the northern great plains. *Agronomy Journal, 94*(2), 174–185.

Dimaano, N. G. B., Ali, J., Cruz, P. C. S., Baltazar, A. M., Diaz, M. G. Q., Acero, B. L., & Li, Z., (2017). Performance of newly developed weed-competitive rice cultivars under lowland and upland weedy conditions. *Weed Science, 65*(6), 798–817.

Fahey, J. W., Zalcmann, A. T., & Talalay, P., (2001). The chemical diversity and distribution of glucosinolates and isothiocyanates among plants. *Phytochemistry, 56*(1), 5–51.

Gbèhounou, G., (2013). *Guidance on Weed Issues and Assessment of Noxious Weeds in a Context of Harmonized Legislation for Production of Certified Seeds*. Plant Production and Protection Division, Food and Agriculture Organization of the United Nations.

Harlan, J. R., (1982). Relationships between weeds and crops. In: *Biology and ecology of weeds* (pp. 91–96). Springer, Dordrecht.

Heap, I., (2014). Global perspective of herbicide-resistant weeds. *Pest Management Science, 70*(9), 1306–1315.

Jabran, K., Mahajan, G., Sardana, V., & Chauhan, B. S., (2015). Allelopathy for weed control in agricultural systems. *Crop Protection, 72*, 57–65.

Jacobsen, S. E., Christiansen, J. L., & Rasmussen, J., (2010). Weed harrowing and inter-row hoeing in organic grown quinoa (*Chenopodium quinoa* Willd.). *Outlook on Agriculture, 39*(3), 223–227.

Karkanis, A., Ntatsi, G., Alemardan, A., Petropoulos, S., & Bilalis, D., (2018). Interference of weeds in vegetable crop cultivation, in the changing climate of Southern Europe with emphasis on drought and elevated temperatures: A review. *The Journal of Agricultural Science, 156*(10), 1175–1185.

Kępczyński, J., (2018). Induction of agricultural weed seed germination by smoke and smoke-derived karrikin (KAR 1), with a particular reference to *Avena fatua* L. *Acta Physiologiae Plantarum, 40*(5), 87.

Korres, N. E., Norsworthy, J. K., Tehranchian, P., Gitsopoulos, T. K., Loka, D. A., Oosterhuis, D. M., Gealy, D. R., et al., (2016). Cultivars to face climate change effects on crops and weeds: A review. *Agronomy for Sustainable Development, 36*(1), 1–22.

Lampkin, P., (2002). *Organic Farming* (p. 701). Old Pond Publishing, Ipswich, UK.

Matloob, A., Khaliq, A., & Chauhan, B. S., (2015). Weeds of direct-seeded rice in Asia: Problems and opportunities. *Advances in Agronomy, 130*, 291–336.

Melander, B., & Rasmussen, G., (2001). Effects of cultural methods and physical weed control on intrarow weed numbers, manual weeding and marketable yield in direct-sown leek and bulb onion. *Weed Research, 41*(6), 491–508.

Melander, B., (1998). Interactions between soil cultivation in darkness, flaming and brush weeding when used for in-row weed control in vegetables. *Biological Agriculture and Horticulture, 16*(1), 1–14.

Melander, B., Munier-Jolain, N., Charles, R., Wirth, J., Schwarz, J., Van, D. W. R., Bonin, L., et al., (2013). European perspectives on the adoption of nonchemical weed management in reduced-tillage systems for arable crops. *Weed Technology, 27*, 231–240.

Monaco, T. J., Weller, S. C., & Ashton, F. M., (2002). *Weed Science: Principles and Practices* (p. 671). John Wiley and Sons Inc., New York, USA.

Mutch, D. R., (2008). *Flaming as a Method of Weed Control in Organic Farming Systems* (pp. 1–4). Michigan State University Extension Bulletin E-3038.

Mwendwa, J. M., Brown, W. B., Wu, H., Weston, P. A., Weidenhamer, J. D., Quinn, J. C., & Weston, L., (2018). A. The weed suppressive ability of selected Australian grain crops; case studies from the Riverina region in New South Wales. *Crop Protection, 103*, 9–19.

Pekrun, C., & Claupein, W., (2006). The implication of stubble tillage for weed population dynamics in organic farming. *Weed Research, 46*(5), 414–423.

Peters, K., & Gerowitt, B., (2014). Important maize weeds profit in growth and reproduction from climate change conditions represented by higher temperatures and reduced humidity. *Journal of Applied Botany and Food Quality, 87*, 234–42.

Radosevich, S. R., Holt, J. S., & Ghersa, C. M., (2007). *Ecology of Weeds and Invasive Plants: Relationship to Agriculture and Natural Resource Management* (p. 471). John Wiley and Sons, New York.

Ramesh, K., Matloob, A., Aslam, F., Florentine, S. K., & Chauhan, B. S., (2017). Weeds in a changing climate: Vulnerabilities, consequences, and implications for future weed management. *Frontiers in Plant Science, 8*, 1–12.

Ray, D. K., West, P. C., Clark, M., Gerber, J. S., Prishchepov, A. V., & Chatterjee, S., (2019). Climate change has likely already affected global food production. *PLoS One, 14*(5), e0217148.

Richter, R., Berger, U. E., Dullinger, S., Essl, F., Leitner, M., Smith, M., & Vogl, G., (2013). Spread of invasive ragweed: Climate change, management and how to reduce allergy costs. *Journal of Applied Ecology, 50*(6), 1422–1430.

Roshan, P., Kulshreshtha, A., & Hallan, V., (2019). Global weed-infecting Gemini viruses. In: *Gemini Viruses* (103–121). Springer, Cham.

Sage, R. F., & Kubien, D. S., (2007). The temperature response of C3 and C4 photosynthesis. *Plant, Cell and Environment, 30*(9), 1086–1106.

Smith, P., Clark, H., Dong, H., Elsiddig, E. A., Haberl, H., Harper, R., House, J., et al. (2014). *Agriculture, Forestry and Other Land Use*. Cambridge University Press, pp. 811–922.

Tabaglio, V., Marocco, A., & Schulz, M., (2013). Allelopathic cover crop of rye for integrated weed control in sustainable agroecosystems. *Italian Journal of Agronomy, 8*(5), 10.

Tawaha, A. M., & Turk, M. A., (2001). Crop-weed competition studies in faba bean (*Vicia faba* L.) under rainfed conditions. *Acta Agronomica Hungarica, 49*(3), 299–303.

Tawaha, A. M., & Turk, M. A., (2002). Response of *Tetragonolobus palaestinus* Boiss to several frequencies of hand weeding. *Acta Agronomica Hungarica, 50*(1), 91–93.

Tawaha, A. M., & Turk, M. A., (2003). Allelopathic effects of black mustard (*Brassica nigra*) on germination and growth of wild barley (*Hordeum spontaneum*). *Journal of Agronomy and Crop Science, 189*(5), 298–303.

Tawaha, A. R., Turk, M. A., & Maghaereh, G. A., (2001). Field pea response to several frequencies of hand weeding under Mediterranean environment. *Crop Research, 22*(2), 161–163.

Taylor, S., Kumar, L., Reid, N., & Kriticos, D. J., (2012). Climate change and the potential distribution of an invasive shrub, *Lantana camara* L. *PLoS One, 7*(4), e35565.

Tominaga, T., & Yamasue, Y., (2004). Crop-associated weeds. In: *Weed Biology and Management* (pp. 47–63). Springer, Dordrecht.

Turk, A. M., & Tawaha, A. M., (2001a). Wheat response to 2,4-D application at two growth stages under semi-arid conditions. *Acta Agronomica Hungarica, 49*(4), 387–391.

Turk, A. M., & Tawaha, A. M., (2001b). Effect of time and frequency of weeding on growth, yield and economics of chickpea and lentil. *Research on Crop, 2*(2), 103–107.

Turk, A. M., & Tawaha, A. M., (2002). Inhibitory effects of aqueous extract Black mustard on germination and growth of lentil. *Pakistan Journal of Agronomy, 1*, 28–30.

Turk, A. M., & Tawaha, A. M., (2002a). Response of sorghum genotypes to weed management under Mediterranean conditions. *Pakistan Journal of Agronomy, 1*(1), 31–33.

Turk, A. M., & Tawaha, A. M., (2002b). Awnless barley (*Hordeum vulagre* L.) response to hand weeding and 2,4-D application at two growth stages under Mediterranean environment. *Weed Biology and Management, 2*, 163–168.

Turk, A. M., & Tawaha, A. M., (2002c). Crop-weed competition studies in garlic (*Allium sativum* L.) under irrigated condition. *Crop Research, 23*(2), 321–323.

Turk, A. M., & Tawaha, A. M., (2002d). Lentil response to several frequency of hand weeding. *Indian Journal of Agricultural Research, 36*(2), 137–140.

Turk, M. A., & Tawaha, A. M., (2003). The response of wild oats (*Avena fatua* L.) to sowing rate and herbicide application. *African Journal of Range and Forage Science, 20*(3), 239–242.

Turk, M. A., & Tawaha, A. R. M., (2002e). Inhibitory effects of aqueous extracts of black mustard on germination and growth of lentil. *Pakistan Journal of Agronomy, 1*(1), 28–30.

Turk, M. A., & Tawaha, A., (2003). Allelopathic effect of black mustard (*Brassica nigra* L.) on germination and growth of wild oat (*Avena fatua* L.). *Crop Protection, 22*(4), 673–677.

Turk, M. A., Shatnawi, M. K., & Tawaha, A. M., (2003). Inhibitory effects of aqueous extracts of black mustard on germination and growth of alfalfa. *Weed Biology and Management, 3*(1), 37–40.

Valerio, M., Tomecek, M. B., Lovelli, S., & Ziska, L. H., (2011). Quantifying the effect of drought on carbon dioxide-induced changes in competition between a C3 crop (tomato) and a C4 weed (*Amaranthus retroflexus*). *Weed Research, 51*(6), 591–600.

Varanasi, A., Prasad, P. V., & Jugulam, M., (2016). Impact of climate change factors on weeds and herbicide efficacy. In: *Advances in Agronomy* (Vol. 135, pp. 107–146).

Walsh, M. J., Harrington, R. B., & Powles, S. B., (2012). Harrington seed destructor: A new nonchemical weed control tool for global grain crops. *Crop Science, 52*(3), 1343–47.

Walsh, M., Ouzman, J., Newman, P., Powles, S., & Llewellyn, R., (2017). High levels of adoption indicate that harvest weed seed control is now an established weed control practice in Australian cropping. *Weed Technology, 31*(3), 341–47.

Weston, L. A., Alsaadawi, I. S., & Baerson, S. R., (2013). Sorghum allelopathy-from ecosystem to molecule. *Journal of Chemical Ecology, 39*(2), 142–153.

Worthington, M., Reberg-Horton, S. C., Brown-Guedira, G., Jordan, D., Weisz, R., & Murphy, J. P., (2015). Relative contributions of allelopathy and competitive traits to the weed suppressive ability of winter wheat lines against Italian ryegrass. *Crop Science, 55*(1), 57.

Zimdahl, R. L., (2012). *Fundamentals of Weed Science* (p. 661). Academic Press.

Ziska, L. H., & Dukes, J. S., (2011). *Weed Biology and Climate Change* (p. 235). Wiley.

Ziska, L. H., (2001). Changes in competitive ability between a C4 crop and a C3 weed with elevated carbon dioxide. *Weed Science, 49*(5), 622–27.

Ziska, L. H., (2003). Evaluation of the growth response of six invasive species to past, present and future atmospheric carbon dioxide. *Journal of Experimental Botany, 54*(381), 395–404.

Ziska, L. H., (2004). Rising carbon dioxide and weed ecology. In: *Weed biology and Management* (pp. 159–176). Springer, Dordrecht.

Ziska, L. H., Blumenthal, D. M., Runion, G. B., Hunt, E. R., & Diaz-Soltero, H., (2011). Invasive species and climate change: An agronomic perspective. *Climatic Change, 105*(1), 13–42.

Ziska, L., (2013). Observed changes in soybean growth and seed yield from *Abutilon theophrasti* competition as a function of carbon dioxide concentration. *Weed Research, 53*(2), 140–145.

PART III
Organic Agriculture for Food Safety

CHAPTER 13

ORGANIC PRODUCTION TECHNOLOGY OF RICE

SHAH KHALID,[1] AMANULLAH,[1] NADIA,[1] IMRANUDDIN,[2] MUJEEB UR RAHMAN,[2] ABDEL RAHMAN M. AL-TAWAHA,[3] DEVARAJAN THANGADURAI,[4] JEYABALAN SANGEETHA,[5] SAMIA KHANUM,[6] MUNIR TURK,[7] HIBA ALATRASH,[8] SAMEENA LONE,[9] KHURSHEED HUSSAIN,[9] PALANI SARANRAJ,[10] and ARUN KARNWAL[11]

[1]*Department of Agronomy, The University of Agriculture, Peshawar, Pakistan*

[2]*Department of Horticulture, The University of Agriculture, Peshawar, Pakistan*

[3]*Department of Biological Sciences, Al-Hussein Bin Talal University, Maan, Jordan*

[4]*Department of Botany, Karnatak University, Dharwad–580003, Karnataka, India*

[5]*Department of Environmental Science, Central University of Kerala, Kasaragod–671316, Kerala, India*

[6]*Department of Botany, University of the Punjab, Lahore, Pakistan*

[7]*Department of Plant Production, Jordan University of Science and Technology, Irbid, Jordan*

[8]*General Commission for Scientific Agricultural Research, Syria*

[9]*Division of Vegetable Science, SKUAST-Kashmir, Shalimar, Jammu and Kashmir, India*

[10]*Department of Microbiology, Sacred Heart College (Autonomous), Tirupattur–635601, Tamil Nadu, India*

[11]*Department of Microbiology, School of Bioengineering and BioSciences, Lovely Professional University, Phagwara, Punjab, India*

ABSTRACT

Bio-based modifications such as compost, farms manure, slurry biogas, vermicompost, green manures (GMs), biofertilizer, crop residues and crop cover are a valuable nutrient source to boost the attributes of growing and yielding, yield, nutrient absorption, rice quality, and soil fertility. Organic farming improves soil organic carbon, phosphorus content available, and enzymatic/microbial population behavior of the soil to make organic farming sustainable. The organic system becomes sustainable in the long term by preserving the soil and improving its fertility by guaranteeing future generations production potential. The use of farmyard manure (FYM) will increase the soil's microbial activity. This microbial soil behavior is dependent on the soil's ability to release, store, and supply plant nutrients. Product consistency is also deteriorated as a result of chemical contaminants entering the plant body and the food chain. The emerging scenario requires the adoption of practices that preserve soil health, maintain a sustainable production system, and provide quality food to meet people's nutritional needs. Organic agriculture is a method to improve the sustainability of the rice-white cultivation system without detrimental impacts on natural resources and the environment.

13.1 INTRODUCTION

The cultivation method for rice (*Oryza sativa* L.)-wheat (*Triticum aestivum* L.) covers some 28.8 million hectares, spread primarily over five Asian countries, namely India, Pakistan, Nepal, Bangladesh, and China (Prasad, 2005; Amanullah and Inamullah, 2016). The rice-wheat production mechanism accounts for around one-fourth of Southeast Asia's total food grain production. This means that the rice-wheat cultivation system contributes to meeting the food requirements of the country. The rice-wheat crop system, which is considered the cornerstone of food autosufficiency, faces a sustainability issue as a result of modern production methods using chemical fertilizers and pesticides indiscriminately (Amanullah and Inamullah, 2016; Amanullah et al., 2019a, c; Amanullah and Khalid, 2020). Many studies have concluded that low productivity mainly concerns dryland farming management practices (Tawaha and Turk, 2001; Turk and Tawaha, 2002b; Abebe et al., 2005a–d; Abera et al., 2005; Assefa et al., 2005; Lee et al., 2005a–d; Nikus et al., 2005a, b; Mesfine et al., 2005; Sulpanjani et al., 2005a–c; Tawaha et al., 2005a, b; Yang et al., 2005; Zheng et al., 2005; Assaf et al., 2006). The effects of modern rice-wheat manufacturing systems involving unequaled and harmful use of chemical fertilizers and pesticides are the decrease in productivity, depletion of soil organic carbon and mineral-nutrient contents (Shah Khalid, 2017: Imranuddin et al., 2017; Amanullah et al., 2019d). The adverse effects of these chemicals on the soil structure, microflora, water quality, food, and fodder are clearly evident. Organic farming is one of the practices for making the rice-wheat cultivation method viable without adverse environmental and natural resources consequences (Hidayatullah et al., 2013; Hidaytullah, 2015; Khalid et al., 2018a). According to a global survey conducted by Ockologie and Landbau (SOUL), organic food was only grown for 5% of the world's cultivated area in 2003. Austria's organic farming sector was the highest percentage, followed by Switzerland, Italy, Finland, Denmark, Sweden, and the Czech Republic. In recent years, awareness of improved quality of food goods, health risks and environmental concerns both globally and nationally has increased. There is a strong demand on the international market for high quality products and organically grown products and can capitalize on its ability to be widely used in organic farming. Farmers need to be educated on the scientific methods of organic farming to eventually increase their profits. Bio-farming is also favored due to increased demand from consumers for natural, high-quality, ethical organic foods. Organically, it also produces strong yields.

13.2 COMPONENTS OF ORGANIC FARMING

13.2.1 NUTRIENT SOURCES

Soil fertility is defined as the soil quality which provides the right amount of nutrients for the production and balance of specified plants or plants (Turk and Tawaha, 2001, 2002a, b; Tawaha and Turk, 2002; Turk et al., 2003a–c). The objective of organic nutrient management is to maximize the use of on-farm capital and to minimize losses. Organic materials such as field manure, compost, vermicompost, biogas fog, green manures (GMs), crop residues, biofertilizers, and cover crops constitute a valuable source of nutrients to enhance qualities of growth and yield, yield, absorption of nutrients and grain and soil fertility. The values of these nutrient sources are examined for rice and wheat separately. Farmyard manure (FYM) is voluminous organic manure arising from the decomposed dung and urine mixture and litter (bedding material). Average, well-rotted FYM contains 0.5–1.0% N, 0.15–2.0% P_2O_5 and 0.5–0.6% K20. The desired FYM C:N ratio should not be higher than 15 to 20 (Bhattacharyya et al., 1986; Mäder et al., 2002; Ghimire et al., 2017). The use of FYM is more important for sustainability because it has great potential not only to provide nutrients, but to improve the physical properties of soil. Each of these physical properties has a great practical impact on the maintenance of soils as a production medium and a major role in stopping environmental degradation. It is important to maintain them and improve them in the long term for the sustainability of the environment.

Farmyard microbial manure includes large numbers of microbial populations. Microbial activity on the soil can be increased by using agricultural manure both to improve microbial activity and to enable microbials to multiply. In compliance with the biological request of soil, the application of agricultural manure contributes to the sustainable production of the soil. The soil has the capacity to release, store, and provide nutrients of plants for this microbial activity.

Many studies show the improved microbial behavior by the use of farms manure are available in the literature. Application of FYM has shown that the increased number of bacteria, actinomycetes, and fungi is increased in organic carbon. Like many other organic dung, agricultural dung leaves some positive residual impacts on the next crop behind. The total nutrients which become labile in the beneficial residual effects of the following year are derived by: A) part of the total nutrient intake of labile type in the following years is obtained by: b) Improved physical properties aid in the enrichment of the soil and enhanced physical properties of the soil that eventually contribute to crop production in various ways. In experiments involving crop rotation, improved yield of the successive crop by FYM for preceding crop was recorded at Pusa in 1932. Many other jobs have recently identified beneficial FYM application residual effects.

13.2.2 EFFECT OF FARMYARD MANURE (FYM) ON RICE

Shanmugam and Veeraputhran (2000) revealed that application of FYM at 12.5 t ha^{-1} significantly increased the growth and yield attributes and yield of rice. Bridgit and Potty (2006) found that increasing the FYM level increased the number of roots per plant and average root length. Maximum correlation was observed between root number per plant and grain yield, straw yield, and total biomass, followed by the difference in total dry weight between flowering and panicle initiation stages. Bhattacharya (2003) recorded the highest plant height at 45 and 90 days after transplanting with 9.0 t FYM ha^{-1}. The application of 7.0 t FYM ha^{-1} resulted in the highest dry matter accumulation at 45 (327.1 and 319.8 g m^{-2}) and 90 days after transplanting (648.4 and 651.1 g m^{-2}) and the dry weight at tillering and flowering growth stages. The beneficial effects of organic manure on grain and straw yield have been reported by many workers (Badgley et al. 2007; Leifeld et al. 2013; Lori et al. 2017; Amanullah et al., 2019a–d). Summarizing the work done in China, where FYM is widely used, FAO (1978) reported that application of 30–40 tons FYM ha^{-1} increased the yield ranging from 24 to 89% as compared to control. It was shown by Meelu and Morris (1984) that applying 12 tons FYM ha^{-1} on the rice method leads to a 40 kg/ha saving on rice and 20 kg P_2O_5 and 30 kg K_2O ha^{-1} in the successor wheat. 10 t FYM ha^{-1} was significantly increased in N, P, and K absorption in the rice-wheat system by 4.0, 7.8, and 7.6% compared to controls (Singh et al., 2018). The impact on sustainability of rice (*Oryza sativa*)-wheat (*Triticum aestivum*) cropping system from the nutrient management studied by Hidayatullah et al. (2013) showed that the grain yield of rice-wheat system was increased by the FYM @ 10 t ha^{-1}. FYM's positive role in improving the physical, chemical, and biological characteristics of soil (Amanullah et al., 2019a, b; Amanullah and Khalid, 2020). Application of FYM significantly increased organic carbon available N and available P and K (Hobson, 2005). Application of FYM and fertilizer-N increased alkaline permanganate oxidizable N. FYM application has been documented to increase plant growth through the provision of plant nutrients including micronutrients and soil enhancement of physical, chemical, and biological properties. The rise was between 19.3% and 27.4% compared to NPK alone. They also saw FYM's application as the highest in the WHC, i.e., 46.4%, followed by mushroom expended, rice straw compost, and coir pith. By improving the soil structure, FYM provides a better environment for root growth. Ibrahim et al. (2010) recorded significant increases in the length and volume of the rice root with the FYM application that would enable the plant to exploit more water under water stress conditions by improving root growth. Other advantages of FYM soil modifications are the faster rate of water penetration due to increased soil aggregation. Thus, water is available for rice plants for a longer period of time. Plants supplied with FYM would take a longer time to wilt in conditions of drought than plants not supplied with organic fertilizer. Prasad and Misra (2001) reported that application of FYM increased the available NPK of organic carbon in comparison to regulation. This may be because organic matter is decomposed and mineralized.

13.2.3 GREEN MANURING

Organic farming is based on soil health and the use of natural processes through soil cycling of nutrients. GMs fulfil, in addition to the addition of animal manures, the essential role of fertilization. The use of GM has been found to be very promising for improving crop yield and saving fertilizer. GM means that any GM plants are plowed under or soil when they are green, or soon after they have flowered. GMs are forages or legumes grown for their leafy substances necessary to preserve the soil. It has been found that about 18 species of grain legumes are essential to GM in various rice farms in Asia. In organic farming systems (OFSs) the following two types of GMs can be used:

1. **Green Manuring *In Situ*:** GM plants in this method are cultivated and buried in the same GM area. Sesbania (*Sesbania aculeata*) is a common leguminous green-manure crop called dhaincha or sawri. The branches and leaf stalks are armed with thin, weak prickles, and grows to a height of several feet per year. The pinnate leaves have 40 to 81 narrow leaflets, and 3 to 6 loose racemes have a light-yellow dotted flower. The pods are long and very tight. Sesbania is ideally suited to wetlands and flourishes at higher altitudes, suffers from significant drought and salinity in the soil, and is grown on very low lands. It is not useful for drilling and is therefore produced only as GM. *Sesbania aculeata* has a solid, deep system of rooting and this was described as an advantage if a subsoil is to be opened.
2. **Green Leaf Manuring:** The green leaf manure refers to turning green leaves in the soil and tender green branches gathered from shrubs and trees cultivated on rivers, wastelands, and surrounding forest areas. *Leucaena leucocephala* is a thornless tree or shrub that can grow up to 7–20 m in altitude. Bipinnate leaves are 6–8 pairs with 11–23 pairs of leaflets 8–16 mm long. The inflorescence is a globular cream that produces a cluster of 13–18 mm, flat brown pods that contain 15–30 seeds.

13.2.4 EFFECT OF GREEN MANURING ON RICE

Manguiat et al. (1997) confirmed that immediate rice sowing after incorporating GM had no negative impact on upland rice growth and development. Chandra and Pareek (1998) noted that the introduction of different Sesbanian

GM species increased the dry weight of the plants at different intervals from 7.1 to 25.2%. All *Sesbania* species except 45 day-old *S. rostrata* produced significantly more plant dry matter than control at 51 DAT, however, different species did not differ significantly. *Sesbania rostrata* (60 day-old), however, recorded the highest plant dry matter at all growth stages. Total and effective tillers recorded at maturity showed the similar trend. Hemalatha et al. (2000) observed that the best plant height (97.61 cm), hill numbers of tillers (19.55), index of leaf sized (6.85 t), development of the dry matter (13848 kg/ha) and days to 50% flowering were registered in situ with the incorporation of dhaincha at 12.0 t ha^{-1} (101 days).

Singh et al. (2000) have recorded a substantial increase in root length density over control of FYM and GM. Mukherjee and Singh (2001) revealed a significant effect of *Sesbania* green manuring on plant height at 50 and 70 days after transplanting and at harvest. Summer green manuring of *Sesbania* before rice transplanting recorded significantly more number of tillers at 30 and 50 days after transplanting over residue incorporated, residue burnt and residue removed treatments. Ram et al. (2011) revealed that incorporation of 12.5 t ha^{-1} of *Sesbania* aculeate recorded the highest plant height (87.3 cm), number of tillers $hill^{-1}$ (15.4) and LAI (7.9). Green manuring of *Sesbania* equivalents to 45 kg N/ha which gave 20% more yield than the control and the response was 8.8 kg grain per kg N. Rice cv. Gayatri, rice Cuttack (*Sesbania aculeata*) and dhaincha were either transplanted to plots of pure dhaincha following the accumulation of water in various arrangements (parallel lines and mixed broadcasting) in dry soil, or rice seedlings. With dhaincha GMs compared with control the yield of both direct-seeded and transplanted crops has increased (without green manuring). They commented that the increase in yields under GM was the result of higher panicle weight, likely due to the combined supply of N after organic matter was decomposed through the dhaincha. Grain yield obtained with *Sesbania aculeata* was equivalent to 90 kg ha^{-1} in non-scented rice. The results of IARI experiments showed that GM in *Sesbania* increased the grain yield of rice by 0.4 t ha during the summer months (May-June). In a field experiment for 4 years with irrigated rice-wheat rotations, Aulakh et al. (2000) conducted a sandy loam soil to test the effects of cowpea (*Vigna unguiculata*) or Sesbania (*Sesbania aculeata*) on the cultivation productivity of GM. The pre-transplant rice grain yields ranged from 5.18 to 5.81 t ha^{-1} of GM ha^{-1} from 20 and 40 t.

There has been a rise of 3 Q ha^{-1} compared to beushening (beushening is crossing under lowland rice habitats in standing water when rice plants are planted). Green manuring of dhaincha (*Sesbania aculeata*) before transplanting rice gave higher the grain yield of rice than other treatments. Hemalatha et al. (2000) observed that in situ incorporation of dhaincha at 12 t ha^{-1} increased the grain yield by 18% and straw yield by 16% over no organic manure, owing to increase in growth and yield-attributing characters of rice. Mehla et al. (2000) reported mean grain yield of 6.89, 6.74, 6.16, and 5.43 t ha^{-1} with GM, FYM, ash, and control treatments, respectively. Ram et al. (2011) observed that Sesbania aculeata GM resulted in higher N, P, and K uptake (mg $plot^{-1)}$ by rice grain and straw as compared GM with *Sebania speciosa, Crotalaria juncea, Azolla microphylla*, cowpea, FYM, composted coir pith and paddy straw. Tiwari et al. (1980) observed that Sesbania green manure increased the N, P, and K contents in plants and their availability in soil. Saha et al. (2000) observed that green manuring registered significantly higher P uptake, which was 8.4 higher over fallow. Sesbania GM resulted in N and P uptake similar to 120 kg N + 13 kg P + 17 kg K ha^{-1} and 120 kg N + 26 kg P + 34 kg K ha^{-1}, respectively. Ram et al. (2011) studied the effect of *Sesbania aculeata, S. speciosa, Crotalaria juncea*, paddy straw, powdered FYM or composted raw coir pith, 28-day-old rice cv. IR 60 seedlings on nutrient uptake. The highest values for uptake of N (399.4 mg pot^{-1}), P (49.82 mg pot^{-1}), K (403.2 mg pot^{-1}), Zn (1059.9 μg pot^{-1}), Fe (18.28 mg pot^{-1}), Mn (6.69 mg pot^{-1}) and Cu (693.3 μg pot^{-1}) in IR 60 were recorded with the addition of *S. aculeata*.

13.2.5 GREEN MANURING + BIOFERTILIZERS

Microbial inoculants or biofertilizers are essential components of organic agriculture that help to feed crops by the nutrients they need. The microbes help to fix nitrogen in the environment, solubilize, and mobilize phosphorus, transfer small items such as zinc, copper, etc., to the plant, generate plant growth that promotes hormones, vitamins, and amino acids and regulates pathogenic fungi in plants, thereby improving soil health and increasing crop production. Long-term use has been made of biofertilizers such as *Rhizobium, Azotobacter, Azospirillum*, and blue-green algae (BGA). These species fix and supply atmospheric nitrogen to plants. Therefore, it is called biofertilizers. 20–30 kg N/ha/season is contributed by bacterial biofertilizers. Inoculant rhizobium is used in leguminous plants. *Azotobacter* can be used in crops such as wheat, maize, mustard, cotton, potatoes, etc. For sorghum, millets, corn, sugarcane, and wheat inoculants are primarily recommended. The BGA of the genus *Nostoc, Anabaena, Tolypothrix*, and *Aulosira* fix the atmospheric nitrogen and serve as inoculants for low and high ground paddy cultivation. However, the inoculants are most successful for cultivation in lowland rice and contribute 20–30 kg N per hectare per season with better grain quality. Azolla adds nitrogen up to 60 Kg/ha/Saison to Azolla in combination with water fern and enriches soils with organic matter. The term biofertilization or microbial inoculants can be described as preparations contending strains of microorganisms that can increase microbiological processes such as fixation of nitrogen, phosphates solubilization or mineralization, excretion from plant growth promotion of soil, compost, or other environments for the purpose of cellulose or lignin biodegradation (Gaur, 2006). The biofixing mechanism for nitrogen can be divided into 3 categories: (i) symbiotic system; (ii) Legume-Rhizobia symbiosis; and (iii) other symbiotic nitrogen fixation systems. An artificially prepared Rhizobium cultivation was used until the seed was seeded. A specific Rhizobium culture is needed for a particular legume crop with high infection, nodulation, fixation, and antibiotic resistance capacity (Bhattacharyya and Tandon, 2002). *Azotobacter* is non-symbiotic free-living soil bacteria which fixes nitrogen in cereals, vegetables, and flowers. Application is usually achieved by seed/seedling or soil treatment. It also reports its foliar submission. Both Azotobacter fertilizers based on carriers and liquids are available. The Azotobacter species are known to average 10 mg N/g sugar on nitrogen free medium in pure culture. The most effective strains of Azotobacter will have to oxidize approximately 1000 kg of organic matter in order to repair 30 kg N/ha. This does not sound practical for our very low active carbon soils. Furthermore, soil has a wide range of other bacteria, all competing for activated carbon. The blue-green algae (BGA) are frequently referred to as cyanobacteria or cyanophyte that are a phylum of bacteria which obtain energy from photosynthesis. The name “cyanobacteria” is taken from the color of the bacteria, cyan (blue); no use or production of cyanide is made by the bacteria. Cyanobacteria have been found to show fossil traces from around 3.8 billion years, which certainly prove that Blue-green algae are among the earliest forms of life on earth. Filamentous, photosynthetic aerobic N fastening species are blue-green algae. Over 100 BGA species are known to correct N. These can provide 25–30 kg N ha^{-1} as biofertilizer for humidity rice (paddy). They also separate hormones such as IAA, GA, and improve the structure of the soil by generating polysaccharides that help attach soil particles to better soil aggregation. For growth and N-fixation, BGA requires all plant nutrients. The optimum BGA temperature is about 30–35°C and its growth decreases at low temperatures. The optimal pH is between 7.5 and 10.0 for BGA growth in cultivation media and is around

6.5–7.0 (Kumar and Shivay, 2008). The cultivation method of rice (*Oryza sativa* L.)-wheat (*Triticum aestivum* L.) is spread out over 5 Asian countries, namely India, Pakistan, Nepal, Bangladesh, and China (approximately 28.8 million hectares) (Prasad, 2005; Amanullah and Inamullah, 2016). A rice-wheat crop system accounts for approximately a quarter of South-East Asia's total food grain output. This means the contribution of the rice-wheat cultivation method to the country's food demands. However, the rice-white cultivation system is considered the cornerstone of food self-sufficiency, which faces a sustainability challenge because of the current processing methods using chemical fertilizers and pesticides indiscriminately (Amanullah and Inamullah, 2016; Amanullah et al., 2019a, c; Amanullah and Khalid, 2020). Many studies have concluded that low productivity mainly concerns dryland farming management practices (Tawaha and Turk, 2001; Turk and Tawaha, 2002b; Abebe et al., 2005a–d; Abera et al., 2005; Assefa et al., 2005; Lee et al., 2005a–d; Nikus et al., 2005a, b; Mesfine et al., 2005; Sulpanjani et al., 2005a–c; Tawaha et al., 2005a, b; Yang et al., 2005; Zheng et al., 2005; Assaf et al., 2006). The consequences of the modern rice-wheat production system with unbalanced and harmful use of chemical fertilizers and pesticides include concerns such as declining factor productivity, deprivation of soil-based organic carbon and mineral nutrients (Imranuddin et al., 2017; Shah Khalid, 2017; Amanullah et al., 2019), water extraction and salinization, increasing nitrate concentrations on wells, etc. The adverse effects of these chemicals on the soil structure, microflora, water quality, food, and forage are clearly evident (Hidayatullah et al., 2013; Hidaytullah, 2015; Khalid et al., 2018a). According to a global survey conducted by Germany's Ockologie and Landbau (SOUL), only 5% of worldly cultivated food was developed in 2003. Austria's organic farming sector was the highest percentage, followed by Switzerland, Italy, Finland, Denmark, Sweden, and the Czech Republic. In recent years, awareness of improved food safety, health risks, and environmental concerns both at the international and national level has increased. Bio-farming is also favored due to increased demand from consumers for natural, high-quality, and ethical organic foods. Organically, it also produces strong yields.

13.2.6 EFFECT OF GREEN MANURING, FYM, AND BIOFERTILIZERS ON RICE

The application of FYM, *Eichhornia*, and *Azolla* compost yielded less grain and pain than 60 kg N ha^{-1} as urea. These suggests that organic materials may have a strong potential to be a source of alternative nutrients in a rice crop for the longer term (Flessa et al., 2002; Khan et al. 2004; Munda et al., 2018; Shrestha et al., 2018; Amanullah et al., 2019a, 2020; Rahman et al., 2019; Sharma et al., 2019). Dixit and Gupta (2000) have observed a growth in the economic yield by applying fertilizer manure alone or together at 10 t ha^{-1} and BGA (Cyanobacteria) inoculation BGA averaged 0.24 t ha^{-1} (7.5%) in grain yield, while the combined use of farm manure and BGA showed an improvement of 0.60 t ha^{-1} (19.2%). Dixit and Gupta (2000) also pointed out that content and uptake of N, P, and K showed increasing trends as a result of the application of FYM, and BGA inoculation either alone or in combination. Shanmugam and Veeraputhran (2000) revealed that application of either GM (*Sesbania aculeata* at 6.25 t ha^{-1}) or FYM at 12.5 t ha^{-1} combined with *Azospirillum* (2 kg ha^{-1}) significantly increased the growth attributes of rice. The FYM + SGM + BGA application was stated to have resulted in more N and P uptake of rice than FYM by Van Quyen and Sharma (2003). Highest N intake was achieved with the combined application of FYM + SGM + BGA + PSB, which was substantially higher than both inorganic and biologic combinations. The FYM + SGM + BGA + PSB combination however substantially increased P uptake by grain, paw, and grain + straw over FYM, SGM, and BGA combinations, while GM from Sesbania resulted in a comparable uptake in N and P of 120 kg N + 13 kg P + 17 kg K ha^{-1} and 120 kg N + 26 kg P +34 kg ha^{-1} respectively. Total rice N and P intakes were substantially higher than that found with FYM with sesbania GM. However, if FYM+SGM+BGA are added together they resulted in more N and P than FYM, SGM+FYM+BGA, in rice inoculation without significant impact on total N rice intake. In addition to all combines of inorganic and organic therapies, the FYM+SGM+BGA+PSB was added together to the N uptake. However, the combined grain, paw, and grain + straw increased considerably in P uptake over other FYM, SGM, and BGA combinations except FYM + SGM +FGA. Dixit and Gupta (2000) reported that higher number of tillers, number of total spikelets per panicle, test weight, grain, and straw yields, larger panicles and lesser unfilled spikelets in both years were recorded with application of Azolla along with 100% RDN. This was on par with the application of Azospirillum and BGA with 100% RDN. All the yield components, grain, and straw yields of rice due to inoculation of any one of the biofertilizers along with 80% RDN were statistically similar to that of 100% RDN alone without any biofertilizer inoculation, indicating a saving of 20% RDN (24 kg N/ha) due to application of any one of the three biofertilizers. The experiment was performed by Bhattacharya (2003), and the results demonstrated that the quality and compatibility of biofertilizers with inorganic fertilizers is gradually increasing. In the subsequent processes treated with the biologics, major differences in soil fertility status (available N, P, and K) and soil biota increased (the third season). Kumar et al. (2007) reported that approximately 50% of recommended dose of inorganic fertilizers could be saved by using 20 t ha^{-1} of FYM + 10 kg ha^{-1} of BGA [or cyanobacteria]. Supply of a portion of P and K along with secondary and micronutrients required by crops could help offset the negative nutrient balance and slow down nutrient depletion processes. Application of organic manures improved the physical, chemical, and biological properties of soil. Sharma and Namdeo (1999) reported that FYM + biofertilizers improved all the parameters of soil fertility over FYM alone as well over green manuring alone. Kumar et al. (2007) reported that the direct and residual application of GM increased the yield of hybrid rice by up to 1240 kg/ha (42.61%) during the 1st year and 1275 kg/ha (45.94%) during the 2nd year. The highest grain yield of hybrid rice (9300 and 8670 kg/ha during 1st and 2nd year, respectively) was recorded for 150% NPK + GM + biofertilizer. The increase in the yield of rice due to biofertilizer application ranged from 202 to 422 kg/ha during the 1st year, and from 290 to 388 kg/ha during the 2nd year. The greatest total removal of NPK (405.5 kg/ha and 394 kg/ha on the 1st and 2nd year in hybrid rice was registered for 150% NPK + GM + biofertilizer. The increase in nutrient use efficiency due to GM application was greater in the control and lower at higher rates of applied fertilizers for rice (0.7–43.9%, with a mean value of 17.4%, on the 1st year, and 10.68–44.36%, with a mean value of 16.61%, on the 2nd year). The percent increase in nutrient use efficiency due to biofertilizer application on the 1st and 2nd year reached 6.5 and 8.16% in rice.

13.3 CONCLUSION

The common organic amendments used in rice are FYM, vermicompost, green manuring, and biofertilizers. The combination and combined use of these various organic changes will meet the biorice nutrient requirement. Organic farming improves soil organic carbon, phosphorus content available and soil microbial population and enzymes, making it sustainable for the production of organic crops. The organic system becomes sustainable for the long term by protecting soils and increasing their productivity, ensuring future generations' productive potential.

KEYWORDS

- **biofertilizers**
- **biogas slurry**
- **compost**
- **farmyard manure**
- **green manures**
- **organic amendments**
- **vermicompost**

REFERENCES

Abebe, G., Assefa, T., Harrun, H., Mesfine, T., & Al-Tawaha, A. M., (2005d). Participatory selection of drought-tolerant maize varieties using mother and baby methodology: A case study in the semi-arid zones of the central rift valley of Ethiopia. *World Journal of Agricultural Sciences, 1*(1), 22–27.

Abebe, G., Hattar, B., & Al-Tawaha, A. M., (2005c). Nutrient availability as affected by manure application in cowpea [*Vigna unguiculata* (L.) Walp.] on calcareous soils. *Journal of Agriculture and Social Science, 1*(1), 1–6.

Abebe, G., Sahile, G., & Al-Tawaha, A. M., (2005a). Evaluation of potential trap crops on Orobanche soil seed bank and tomato yield in the central rift valley of Ethiopia. *World Journal of Agricultural Sciences, 1*(2), 148–151.

Abebe, G., Sahile, G., & Al-Tawaha, A. M., (2005b). Effect of soil solarization on Orobanche soil seed bank and tomato yield in the central rift valley of Ethiopia. *World Journal of Agricultural Sciences, 1*(2), 143–147.

Abera, T., Feyisa, D., Yusuf, H., Nikus, O., & Al-Tawaha, A. M., (2005). Grain yield of maize as affected by biogas slurry and N-P fertilizer rate at Bako, Western Oromiya, Ethiopia. *Bioscience Research, 2*(1), 31–38.

Amanullah, & Khalid, S., (2020). Agronomy-food security-climate change and the sustainable development goals. *Agronomy-Climate Change and Food Security*. Intech Open, London.

Amanullah, I., & Inamullah, X., (2016). Dry matter partitioning and harvest index differ in rice genotypes with variable rates of phosphorus and zinc nutrition. *Rice Sci., 23*(2), 78–87.

Amanullah, Iqbal, A., Khan, A., Khalid, S., Shah, A., Parmar, B., Khalid, S., & Muhammad, A., (2019a). Efficient management of phosphorus, organic sources, and beneficial microbes improve dry matter partitioning of maize. *Commun. Soil Sci. Plant Anal., 50*(20), 2544–2569.

Amanullah, Khalid, S., Imran, Khan, H. A., Arif, M., Al-Tawaha, A. R., Adnan, M., et al., (2019b). Organic matter management in cereals based system: Symbiosis for improving crop productivity and soil health. In: Lal, R., & Francaviglia, R., (eds.), *Sustainable Agriculture Reviews 29: Sustainable Soil Management: Preventive and Ameliorative Strategies* (pp. 67–92). Cham: Springer International Publishing.

Amanullah, Khalid, S., Khalil, F., & Imranuddin, (2020). Influence of irrigation regimes on competition indexes of winter and summer intercropping system under semi-arid regions of Pakistan. *Sci. Rep., 10*(1), 8129.

Amanullah, Khan, N., Khan, M. I., Khalid, S., Iqbal, A., & Al-Tawaha, A. R., (2019c). Wheat biomass and harvest index increases with integrated use of phosphorus, zinc, and beneficial microbes under semiarid climates. *J. Microbiol. Biotech. Food Sci., 9* (2), 242–247.

Amanullah, Khan, S. T., Iqbal, A., & Fahad, S., (2016). Growth and productivity response of hybrid rice to application of animal manures, plant residues, and phosphorus. *Frontiers in Plant Science, 7*, 1440.

Assaf, T. A., Hameed, K. M., Turk, M. A., & Tawaha, A. M., (2006). Effect of soil amendment with olive mill by-products under soil solarization on growth and productivity of faba bean and their symbiosis with mycorrhizal fungi. *World Journal of Agricultural Sciences, 2*(1), 21–28.

Assefa, T., Abebe, G., Fininsa, C., Tesso, B., & Al-Tawaha, A. M., (2005). Participatory bean breeding with women and smallholder farmers in eastern Ethiopia. *World Journal of Agricultural Sciences, 1*(1), 28–35.

Aulakh, M. S., Khera, T. S., Doran, J. W., Kuldip, S., & Bijay, S., (2000). Yields and nitrogen dynamics in a rice-wheat system using green manure and inorganic fertilizer. *Soil Sci. Soc. Am J, 64*(5), 1867–1876.

Badgley, C., Moghtader, J., Quintero, E., Zakem, E., Chappell, M. J., Aviles-Vazquez, K., Samulon, A., & Perfecto, I., (2007). Organic agriculture and the global food supply. *Rene. Agri. Food Sys., 22*(2), 86–108.

Bhattacharya, S., (2003). Effect of humic acid (earth) on the growth and yield of transplanted summer rice. *Environ. Ecol., 21*(3), 680–683.

Bhattacharyya, P., Dey, B. K., Nath, S., & Banik, S., (1986). Organic manures in relation to rhizosphere effect iii. effect of organic manures on population of ammonifying bacteria and mineralization of nitrogen in rice and succeeding wheat rhizosphere soils. *Zent. Mikro, 141*(4), 267–277.

Bridgit, T., & Potty, N., (2006). Effect of cultural management on root characteristics and productivity of rice in laterite soil. *J. Trop. Agric, 40*, 59–62.

Dixit, K., & Gupta, B., (2000). Effect of farmyard manure, chemical and biofertilizers on yield and quality of rice (*Oryza sativa* L.) and soil properties. *J. Indian Soc. Soil Sci., 48*(4), 773–780.

Dwivedi, D., & Thakur, S., (2000). Effect of organics and inorganic fertility levels on productivity of rice (*Oryza sativa*) crop. *Indian J. Agron., 45*(3), 568–574.

Flessa, H., Ruser, R., Dörsch, P., Kampb, T., Jimenez, M. A., Munchb, J. C., & Beese, F., (2002). Integrated evaluation of greenhouse gas emissions (CO_2, CH_4, N_2O) from two farming systems in southern Germany. *Agri. Eco. Envi. Beh., 91*(1–3), 175–189.

Gaur, A. C., (2006). *Handbook of Organic Farming and Biofertilizers*. Ambica Book Agency.

Ghimire, R., Norton, U., Bista, P., Obour, A. K., & Norton, J. B., (2017). Soil organic matter, greenhouse gases and net global warming potential of irrigated conventional, reduced-tillage and organic cropping systems. *Nutri. Cyc. Agroecology., 107*(1), 49–62.

Hidayatullah, A., Jan, A., & Shah, Z., (2013). Residual effect of organic nitrogen sources applied to rice on the subsequent wheat crop. *Int. J. Agron. Plant Prod, 4*, 620–631.

Hidaytullah, A., (2015). Sources, ratios, and mixtures of organic and inorganic nitrogen influence plant height of hybrid rice (*Oryza sativa* L.) at various growth stages. *EC Agric, 2*(3), 328–337.

Hobson, P., (2005). Book review: Organic recycling and biofertilization in South Asia. *Bioresour. Technol., 96*(3), 393.

Ibrahim, M., Arshad, M., & Tanveer, A., (2010). Variation in root growth and nutrient element concentration in wheat and rice: Effect of rate and type of organic materials. *Soil Envi., 29*(1), 47–52.

Imranuddin, Arif, M., Khalid, S., Nadia, Saddamullah, Idrees, M., & Amir, M., (2017). Effect of seed priming, nitrogen levels, and moisture regimes on yield and yield components of wheat. *Pure App. Biol., 6*(1), 369–377.

Khalid, S., Afridi, M. Z., Munsif, F., Imranuddin, & Nadia. (2018a). Effect of Sulphur foliar application on yield and yield components of *Brassica napus* L. *Int. J. Agric. Environ. Res, 2*, 232–236.

Khalid, S., Imranuddin, N., Nadeem, F., Saddamullah, A. M., & Ghani, F., (2017). Allelopathic effect of Parthenium liquid extract on mung bean germination ability and early growth. *Inter. J. Agron. Agri. Res., 11*(4), 31–36.

Khalid, S., Munsif, F., Nadia, I., Nadeem, F., Ali, S., Ghani, F., & Idrees, M., (2018b). Influence of source limitation on physiological traits of wheat. *Pure App. Biol., 7*(1), 85–92.

Khan, A. R., Chandra, D., Nanda, P., Singh, S. S., Ghorai, A. K., & Singh, S. R., (2004). Integrated nutrient management for sustainable rice production. *Arch. Agron. Soil Sci., 50*(2), 161–165.

Kumar, A., Tripathi, H., & Yadav, D., (2007). Correcting nutrient imbalances for sustainable crop production. *Indian J. Fer., 2*(11), 37.

Kumar, D., & Shivay, Y., (2008). *Definitional Glossary of Agricultural Terms* (p. 324). IK International, New Delhi.

Lee, K. D., Sulpanjani, Tawaha, A. M., & Min, Y. S., (2005a). Effect of phosphorus application on yield, mineral contents and active components of *Chrysanthemum coronarium* L. *Bioscience Research, 2*(3), 118–124.

Lee, K. D., Tawaha, A. R. M., & Supanjani, (2005c). Antioxidant status, stomatal resistance and mineral composition of hot pepper under salinity and boron stress. *Bioscience Research, 2*(3), 148–154.

Lee, K. D., Turk, M. A., & Tawaha, A. M., (2005b). Nitrogen fixation in rice based farming system. *Bioscience Research, 2*(3), 130–138.

Leifeld, J., Angers, D. A., Chenu, C., Fuhrer, J., Kätterer, T., & Powlson, D. S., (2013). Organic farming gives no climate change benefit through soil carbon sequestration. *Proc. Natl. Acad. Sci. USA, 110* (11), E984.

Lori, M., Symnaczik, S., Mäder, P., De Deyn, G., & Gattinger, A., (2017). Organic farming enhances soil microbial abundance and activity-A meta-analysis and meta-regression. *PLoS One, 12*(7).

Mäder, P., Fliessbach, A., Dubois, D., Gunst, L., Fried, P., & Niggli, U., (2002). Soil fertility and biodiversity in organic farming. *Sci. Cult, 296*(5573), 1694–1697.

Mesfine, T., Abebe, G., & Al-Tawaha, A. M., (2005). Effect of reduced tillage and crop residue ground cover on yield and water use efficiency of sorghum (*Sorghum bicolor* (L.) Moench) under semi-arid conditions of Ethiopia. *World Journal of Agricultural Sciences, 1*(2), 152–160.

Munda, S., Shivakumar, B. G., Rana, D. S., Gangaiah, B., Manjaiah, K. M., Dass, A., Layek, J., & Lakshman, K., (2018). Inorganic phosphorus along with biofertilizers improves profitability and sustainability in soybean (*Glycine max*)- potato (*Solanum tuberosum*) cropping system. *J. Saudi Soc. Agri. Sci., 17*(2), 107–113.

Nadeem, F., Khan, N. U., Imranuddin, Khalid, S., Azam, S., Saeed, B., Jah, T., et al., (2018). 4. Genotype × environment interaction studies in F5 populations of upland cotton under agro-climatic condition of Peshawar. *Pure App. Biol., 7*(3), 973–991.

Nikus, O., Abebe, G., Takele, A., Harrun, H., Chanyalew, S., Al Tawaha, A. M., & Mesfin, T., (2005b). Yield response of tef (*Eragrostis tef* (Zucc.) Trotter) to NP fertilization in the semi arid zones of the central rift valley in Ethiopia. *European Journal of Scientific Research, 4*(4), 49–60.

Nikus, O., Nigussie, M., & Al Tawaha, A. M., (2005a). Agronomic performance of maize varieties under irrigation in Awash valley, Ethiopia. *Bioscience Research, 2*(1), 26–30.

Prasad, R., & Misra, B. N., (2001). Effect of addition of organic residues, farmyard manure and fertilizer nitrogen on soil fertility in rice-wheat cropping system. *Arch. Agron. Soil Sci., 46*(5, 6), 455–463.

Prasad, R., (2005). *Rice-Wheat Cropping Systems* (pp. 255–339). Elsevier, New Delhi, India.

Rahman, F., Hossain, A., & Islam, M. R., (2019). Integrated effects of poultry manure and chemical fertilizer on yield, nutrient balance and economics of wetland rice culture. *Bang. Rice J, 22*(2), 71–77.

Ram, M., Davari, M., & Sharma, S., (2011). Organic farming of rice (*Oryza sativa* L.)-wheat (*Triticum aestivum* L.) cropping system: A review. *Inter. J. Agron. Plant Prod, 2*(3), 114–134.

Shah, K., Afridi, M. Z., Imranuddin, Nadia, Nadeem, F., Aamir, M., & Saddamullah, (2017). Effect of sulfur foliar fertilization on reproductive growth and development of canola. *Int. J. Agron. Agri. Res, 11*(3), 61–67.

Shanmugam, P., & Veeraputhran, R., (2000). Effect of organic manure, biofertilizers, inorganic nitrogen and zinc on growth and yield of rabi rice (*Oryza sativa* L.). *Madras Agric. J. 87*, 90–93.

Sharma, K., & Namdeo, K., (1999). Effect of biofertilizers and phosphorous on NPK contents, uptake and grain quality of soybean (*Glycine max* L. Merrill) and nutrient status of soil. *Crop Res, 17*(2), 164–169.

Sharma, S., Padbhushan, R., & Kumar, U., (2019). Integrated nutrient management in rice-wheat cropping system: An evidence on sustainability in the Indian subcontinent through meta-analysis. *Agronomy, 9*(2), 71.

Shrestha, S., Bhatta, B., Shrestha, M., & Shrestha, P. K., (2018). Integrated assessment of the climate and land-use change impact on hydrology and water quality in the Songkhram River Basin, Thailand. *Sci. Total Environ., 643,* 1610–1622.

Singh, S. K., Kumar, M., Singh, R. P., Bohra, J. S., Srivastava, J. P., Singh, S. P., & Singh, Y. V., (2018). Conjoint application of organic and inorganic sources of nutrients on yield, nutrient uptake and soil fertility under rice (*Oryza sativa*)-wheat (*Triticum aestivum*) system. *J. Indian Soc. Soil Sci., 66*(3), 287–294.

Sulpanjani, A., Yang, M. S., Tawaha, A. R. M., & Lee, K. D., (2005a). Effect of magnesium application on yield, mineral contents and active components of *Chrysanthemum coronarium* L. under hydroponics conditions. *Bioscience Research, 2*(2), 73–79.

Supanjani, Tawaha, A. M., Min, Y. M. S., & Lee, K. D., (2005b). Role of calcium in yield and medicinal quality of *Chrysanthemum coronarium* L. *Journal of Agronomy, 4*(3), 188–192.

Supanjani, Tawaha, A. M., Yang, M. S., & Lee, Y. D., (2005c). Calcium effects on yield, mineral uptake and terpene components of hydroponic *Chrysanthemum coronarium* L. *Research Journal of Agriculture and Biological Science, 1*(1), 146–151.

Tawaha, A. M., & Turk, M. A., (2001). Effects of dates and rates of sowing on yield and yield components of Narbon vetch under semi-arid conditions. *Acta Agronomica Hungarica, 49*(1), 103–105.

Tawaha, A. M., & Turk, M. A., (2002). Lentil (*Lens culinaris* Medic.) productivity as influenced by rate and method of phosphate placement in a Mediterranean environment. *Acta Agronomica Hungarica, 50*(2), 197–201.

Tawaha, A. M., Turk, M. A., Lee, K. D., Supanjani, Nikus, O., Al-Rifaee, M., & Sen, R., (2005a). Awnless barley response to crop management under Jordanian environment. *Bioscience Research, 2*(3), 125–129.

Tawaha, A. R. M., Turk, M. A., & Lee, K. D., (2005b). Adaptation of chickpea to cultural practices in a Mediterranean type environment. *Research Journal of Agriculture and Biological Science, 1*(2), 152–157.

Turk, A. M., & Tawaha, A. M., (2002a). Impact of seeding rate, seeding date, rate and method of phosphorus application in faba (*Vicia faba* L. Minor) in the absence of moisture stress. *Biotechnology, Agronomy, Society and Environment, 6*(3), 171–178.

Turk, A. M., & Tawaha, A. M., (2002b). Response of winter wheat to applied n with or without ethrel spray under irrigation planted in semi-arid environments. *Asian Journal of Plant Sciences, 1*(4), 464–466.

Turk, M. A., & Tawaha, A. M., (2001). Influence of rate and method of phosphorus placement to Garlic (*Allium sativum* L.) in a Mediterranean environment. *Journal of Applied Horticulture, 3*(2), 115–116.

Turk, M. A., Hameed, K. M., Aqeel, A. M., & Tawaha, A. M., (2003c). Nutritional status of durum wheat grown in soil supplemented with olive mills by-products. *Agrochimica, 209*–219.

Turk, M. A., Tawaha, A. M., Samara, N., & Latifa, N., (2003b). The response of six row barley (*Hordeum vulgare* L.) to nitrogen fertilizer application and weed control methods in the absence of moisture stress. *Pakistan Journal of Agronomy, 2*(2), 101–108.

Turk, M. A., Tawaha, A. M., Taifor, H., Al-Ghzawi, A., Musallam, I. W., Maghaireh, G. A., & Al-Omari, Y. I., (2003a). Two row barley response to plant density, date of seeding and rate and method of phosphorus application in the absence of moisture stress. *Asian Journal of Plant Science, 2*(2), 180–183.

Van, Q. N., & Sharma, S., (2003). Relative effect of organic and conventional farming on growth, yield and grain quality of scented rice and soil fertility. *Arch. Agron. Soil Sci., 49*(6), 623–629.

Yang, M. S., Tawaha, A. M., & Lee, Y. D., (2005). Effects of ammonium concentration on the yield, mineral content and active terpene components of *Chrysanthemum coronarium* L. in a hydroponic system. *Research Journal of Agriculture and Biological Science, 1*(2), 170–175.

Zheng, W. J., Tawaha, A. M., & Lee, K. D., (2005). *In situ* hybridization analysis of mcMT1 gene expression and physiological mechanisms of Cu-tolerant in *Festuca rubra* cv merlin. *Bioscience Research, 1*(1), 21–26.

CHAPTER 14

PROSPECTS OF ORGANIC AGRICULTURE IN FOOD QUALITY AND SAFETY

AKBAR HOSSAIN,[1] DEBJYOTI MAJUMDER,[2] SHILPI DAS,[3,4] APURBO KUMAR CHAKI,[4,5] MST. TANJINA ISLAM,[6] RAJAN BHATT,[7] and TOFAZZAL ISLAM[8]

[1]*Bangladesh Wheat and Maize Research Institute, Dinajpur–5200, Bangladesh*

[2]*Uttar Banga Krishi Viswavidyalaya, Cooch Behar, West Bengal, India*

[3]*Bangladesh Institute of Nuclear Agriculture, Mymensingh–2202, Bangladesh*

[4]*School of Agriculture and Food Sciences, University of Queensland, QLD–4072, Australia*

[5]*On-Farm Research Division, Bangladesh Agricultural Research Institute (BARI), Gazipur, Dhaka, Bangladesh*

[6]*Department of Agronomy, Hajee Mohammad Danesh Science and Technology University, Dinajpur–5200, Bangladesh*

[7]*Regional Research Station-Kapurthala, Punjab Agricultural University, Ludhiana, Punjab–144601, India*

[8]*Institute of Biotechnology and Genetic Engineering (IBGE), Bangabandhu Sheikh Mujibur Rahman Agricultural University, Gazipur–1706, Bangladesh*

ABSTRACT

Innovative farming practices in the traditional agriculture have significantly been progressed in the last 1,000 years. Most of these aboriginal agricultural methodologies were not finely honed. Concurrent improvement of biochemistry and engineering, farming practices were also rapidly improved during the early 40 years of the 20th Century. After the industrial revolution, the application of farm machineries, irrigation, and synthetic fertilizers and pesticides in crop production have dramatically changed the traditional farming practices. Although these modern approaches significantly increased the yield of crops to feed the increasing population, the degradation of environmental resources, including soils, has become apparent. Increasing environmental pollution and residual effects of the used synthetic chemicals pose a serious threat to human life and biodiversity. Considering these burning issues, researchers explored alternative sustainable techniques or solutions to alleviate these deleterious effects of modern farming, while preserving the maximum productivity. Therefore, an organic farming system (OFS) concept was started in the 1940s for avoiding the deleterious effects of the application of synthetic fertilizers and pesticides to the crop fields. The OFS was established as a substitution of the rapidly changing modern agricultural system in the early 20th century. In the OFS, agricultural inputs and practices are generated from the biological origins such as organic manure, compost, non-GM elite cultivars, crop rotation and companion planting. Instead of synthetic chemicals, pests are managed through integrated pest management (IPM) approach. The OFS aims to maintain agroecosystem by encouraging interior self-regulation for using synthetic agricultural inputs. The understanding of OFS revealed that the system is environment-friendly, and good for biodiversity and sustainability. Therefore, the environment-friendly OFS approaches have become a subject for the improvement of sustainable agriculture by many countries and organizations. This chapter highlights the current concept, status, economic benefits, and challenges for sustainable OFS intensification. The generated knowledge of OFS is thought to be useful for the sustainable production of safe and nutritional food in the era of global climate change.

14.1 INTRODUCTION

After the industrial revolution, various inorganic methods were introduced in modern agriculture for resolving the increasing demand of food and nutrition for the increasing population of the world. At the same time, the invention of the gasoline engine leads to improve the numerous farm apparatuses. A concurrent improvement of biochemistry and engineering has impacted on introduction of various farm machineries and synthetic chemicals in agriculture. The application of machineries and synthetic chemicals ultimately replaced the use of animal powers and organic inputs in the agriculture. The primitive agriculture was the innovative type of farming which has been experienced for 1000 of years. All aboriginal agricultural systems are now measured to be "organic farming system (OFS)" though that period there were no recognized non-living approaches (McConnell et al., 2017). Since, farming was started to upgrade through the introduction of new farming procedures and machinery, while several serious short and longer-term side effects had appeared as a result of the excessive and imbalanced use of synthetic fertilizers and pesticides. Therefore, soon it was well-understood that modern production systems were not environment-friendly and had severe adverse effects on human health and the environment. Considering these burning issues, researchers started to explore alternative sustainable techniques or solutions to alleviate these hazardous effects of modern farming to preserve the potential of maximum productivity of the farming system.

The OFS concept was started in practice in the 1940s for avoiding the harmful effects of modern farming using artificial fertilizers and pesticides (Colorado State University, 2020). The OFS is well-defined by the use of plant nutrition of biological origin, including compost, manure, green manuring compost, and bone meal, etc., and simultaneously producing crops in farmland with a prioritization on procedures such as non-GMO elite varieties, crop rotation and companion planting. The system varies basically in soil fertility and productivity, diseases, and pests' management, nutritional quality and yield stability as compared to conventional farming systems. Besides these, organic pests and diseases management, diversified cropping and raising of insect killers are inspired. The OFS focuses on the versatility and buffering limit in the ranch environment by animating inner self-guideline through useful agrobiodiversity in or more the dirt, rather than outer guideline through substance protectants. Natural norms are intended to permit the utilization of normally happening substances while precluding or carefully restricting manufactured substances (McEvoy, 2018). The OFS was established as a substitution of the modern agricultural system which initiated during the early in the 20th century in response to promptly altering agricultural activities (Coleman, 1995; Arsenault, 2014; USDA, 2016). In the OFS, agricultural inputs such as fertilizers and pesticides are generally obtained from the organic origins also must involve crop rotation and companion planting methods. The OFS aims to maintain agroecosystem by encouraging interior self-regulation for using synthetic agricultural inputs (Halberg, 2006; Strochlic and Sierra, 2007).

It revealed that OFS is friendly to the environment, biodiversity, and sustainability of the resource utilization. Therefore, various countries and organizations have started to adopt, improve, and extend the use of OFS for promoting sustainable agriculture. This chapter reviews and updates our understanding of the current concept, status, economic benefits, and challenges for sustainable intensification of OFS. The knowledge of eco-friendly OFS is described in relation to its usefulness for the safe production of nutritional food in the era of climate change.

14.2 PRESENT STATUS OF ORGANIC FARMING ACROSS THE GLOBE

Since 1990, the marketplace for an organic and additional foodstuff has full-grown quickly, attaining $63 billion globally in 2012 (Helga et al., 2013) and is now experienced in >120 countries (Willer and Yussefi, 2007). This need for organic products has motivated a comparable intensification in naturally accomplished farmland that raised from 2001 to 2011 at a compounding proportion of 8.9% yearly (Paull, 2011a, b). As of 2018, about 71.5 Mha land globally were cultivated under OFS naturally, demonstrating about 1.5% of total farmland worldwide (FiBL, 2020) (Figures 14.1–14.3). According to the latest FiBL organic agriculture survey data from 186 countries (conducted at the end of 2018) revealed that the growth rate of organic agriculture was 2.9% or 2 Mha compared to 2017. Presently, specialized OFS is occupied about 70 million ha land worldwide, where country Australia stands for 54% of global authorized organic land with the country recorded >35 Mha verified OFS land (Paull, 2018, 2019). As a result of the huge area of organic farmlands are practiced in Australia, half of the global OFS lands are situated in Oceania (36.0 Mha) (Figures 14.1–14.3). Considering the country based on OFS lands, Australia has the first (about 35.7 Mha), secondly Argentina (3.6 Mha), and then China (3.1 Mha) in 2018 (Paull and Hennig, 2018).

In the case of continents under OFS land, Australia is the 1st (about 35.7 Mha), Europe is the 2nd (15.6 Mha), third in Latin America (8 Mha). Asia is in 4th position (4.1 Mha), and North America is the 5th position (about 1.4 Mha) and Africa is the least (1.2 Mha) (Colom-Gorgues et al., 2009; Paull and Hennig, 2018). In the year 2018, the international market for organic food and products exceeded 97 billion euros (BES) for the first time, whereas the USA contributes 40.6 BEs, next place was occupied for Germany (10.9 BEs) and France (9.1 BEs). In the meantime, the organic market in French was grownup more than 15%. Consumers in Denmark and Switzerland have already spent 312 Euros per capita in 2018 for organic food. Similarly, Denmark spent 15% on organic food of its total food market. Additionally, as an Asian country, India is the maximum number of organic food producers (1,149,000), then place was occupied for African countries Uganda (210000) and Ethiopia (204000).

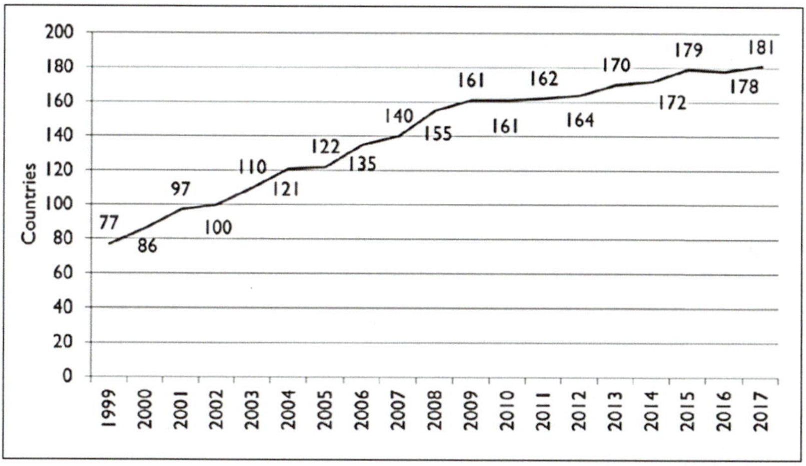

FIGURE 14.1 Development of organic agriculture in several countries globally from 1999 to 2017.

Source: Research Institute of Organic Agriculture (FiBL, 2020).

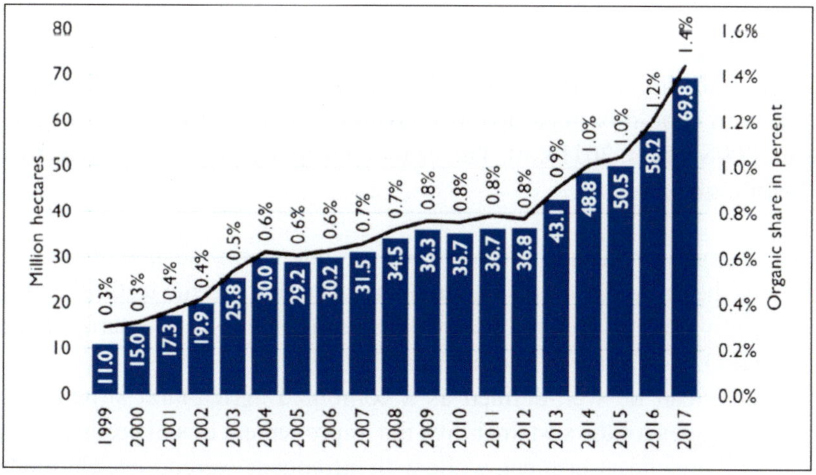

FIGURE 14.2 Trends (from 1999 to 2017) of organic agricultural land worldwide.

Source: Research Institute of Organic Agriculture (FiBL, 2020).

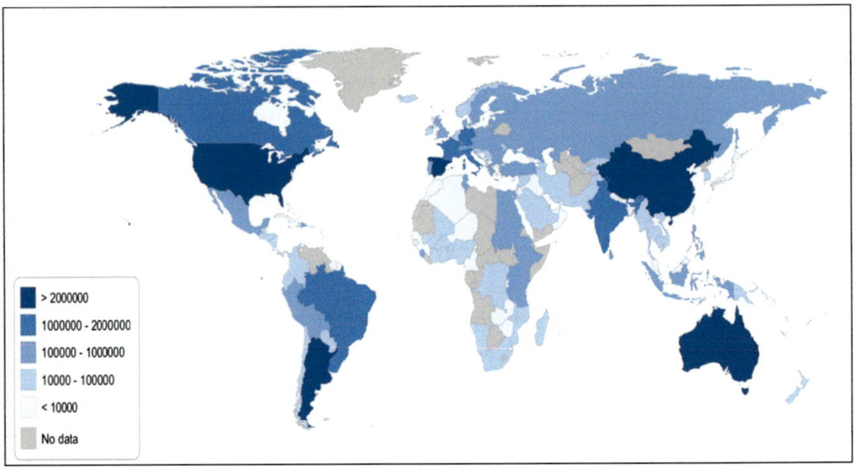

FIGURE 14.3 Area (ha) under organic farming across the globe.

Source: Research Institute of Organic Agriculture (FiBL, 2020).

14.3 CONCEPT, HISTORY, AND REGULATION OF ORGANIC AGRICULTURE

14.3.1 CONCEPT OF ORGANIC FARMING

The practice of term "organic" promoted by Howard and Rodale, who mentions more precisely to usage of organic agricultural status in soils, highlighted from the effort of soil researchers who stated the concept of 'humus farming' (Nayler, 2014). On the other hand, the term "organic" designates that a farm should be regarded as active organisms (Kirchmann and Bergstrom, 2008; Paull, 2011a, b). The exact concept of OFS is discussed by the following quotations:

- An organic farm is a farm whose arrangement is shaped in simulated of the construction of a natural structure that has the truthfulness, the individuality, and the benevolent requirement of a natural creature (IFOAM-Organics International, 2008).
- OFS can be well-defined as a combined agricultural structure that struggles for the sustainability, the improvement of soil productiveness and organic assortment, while with infrequent exemptions, elimination artificial pesticides, fertilizers, phytohormones, and GM crop cultivars (Martin, 2009; Gold, 2020).

- OFS is an assembly of a productive arrangement that withstands the healthiness of soils, agroecosystems, and anthropological activity. It trusts on environmental practices, biodiversity, and cycles adjusted to home-grown conditions, rather than the use of contributions with hostile effects. OFS associates custom, improvement, and discipline to profit the collective location and indorse reasonable associations and decent superiority of life for all involved (IFOAM-Organics International, 2008).
- OFS is an unconventional cultivated system that initiated initial of the 20th century in response to promptly altering agricultural practices. OFS approaches are globally synchronized and legitimately imposed by numerous countries, established in huge part on the ethics set by the IFOAM, which was established in 1972, an intercontinental umbrella organization for OFS (Paull, 2010).

14.3.2 HISTORY OF ORGANIC FARMING

The traditional agriculture was the innovative kind of farming that has been experienced for 1000 of years. Most of these traditional agricultural methodologies were underdeveloped and had unadorned adverse effects on human health and also on the surrounding environment. Since non-natural plant nourishments in agriculture were initially created at the time of mid-1900. Comparable progresses of inorganic pesticides were materialized in the 1940s, directed to the period being mentioned to as the 'era of pesticide' (Horne, 2008). As a concurrent improvement of biochemistry and engineering farming's during the initial 40-year of the 20-century, farms size and cropping were started bigger and specialized for using of efficient farm machinery and also for reducing the dependence on physical and animal employments in tillage, pesticides, and fertilizers applications. As an advancement of farm mechanization and excessive and imbalanced use of modern agricultural inputs including man-made agricultural stimulators, and the extreme events of climate change (Stinner, 2007), resulted several short and lengthier side effects on agroecological environment, leading to adverse effect on human health. Researchers also started to search techniques/solutions to alleviate these poisonous effects of modern farming, while preserving the maximum productivity. In the circumstances, farmers and policymakers in the emerging countries have adapted to contemporary biological approaches for commercial reasons (Paull, 2007). Therefore, an OFS concept was started in the 1940s for avoiding the poisonous effects that were emerged from the application of artificial fertilizers and pesticides. The OFS was established as a substitution of the agricultural system which initiated during the early in the 20th century in response to promptly altering agricultural activities. The historical trends of OFS have discussed in subsections.

14.3.2.1 PRE-WORLD WAR-II

Intentionally OFS started more or less concurrently in Europe and India. Historical evidence of the Pre-World War-II concept on OFS are discussed in the following several points:

1. In the late 1800s and early 1900s, scientists of soil environmental science started to search for techniques for remedying these adverse effects of synthetic agricultural inputs in modern agriculture, through preserving the maximum agricultural productivity.
2. OFS conception was established by Sir Albert Howard, F.H. King, Rudolf Steiner, and others in the early 19 Century, who presumed that the utilization of compost, organic manures, cover crops, crop rotation, and pest management through biological means ensued in a healthier farming system.
3. An American agronomist namely 'F. H. King' published his findings on 'Farmers of 40 Centuries' who also visited in China, Korea, and Japan during the year 1909 for lecturing on outdated agricultural system's nourishment, plowing, and wide-ranging of farming practices (King, 2004; Paull, 2011b).
4. From 1905 to 1924, the father of current OFS the British botanist Sir Albert Howard and his wife a plant physiologist namely Gabrielle worked as agricultural consultants in Pusa-Bengal, India, and exposed the primitive Indian agricultural activities and suggested several methodologies to respect them as loftier to their unadventurous agriculture (Conford, 2001; Heckman, 2007). As soon as Howard backed to UK in the initial of 1930s, he started to disseminate OFS in England and Europe (Lotter, 2003; Vogt, 2007; Kirchmann and Bergstrom, 2008).
5. In the year 1924, the first comprehensive system of OFS namely 'Development of biodynamic agriculture' was coined by Rudolf Steiner's (a German Scientist), exposed by a series of lectures with Steiner at a farm in Koberwitz (Kobierzyce now in Poland) (Paull, 2011a, b). In the lecture they revealed that when crops are cultivated in despoiled soil leads to deteriorate the productivity and

nutritional quality of crops and livestock as a consequence of excessive and imbalanced utilization of inorganic stimulants. Although the summary of these lecture series was first appeared as a publication in November 1924, but it was first translated to English in the year 1928 in 'The Agriculture Course' (Paull, 2011b).

6. A soil microbiologist namely Masanobu Fukuoka (Japan) first opened the contemporary agricultural movement. In 1937, he left his job as a scientist and backed to his family's farm in 1938, and dedicated his next 60 years of life for developing a fundamental no-till organic agricultural technique for increasing crop productivity. Now these techniques are known as natural, natural farming, 'do-nothing' farming or Fukuoka farming.
7. In July 1939, the term "OFS" was first coined by Walter James in his book 'Look to the Land' (Lord Northbourne), which was written in 1939 and was published in 1940) (Paull, 2011b, c, 2014). In the text of the book, Walter James demarcated all-inclusive, environmentally stable methodology to farming, known as "the farm as organisms" (Paull, 2006); establishing based on Steiner's farming moralities and approaches (Paull, 2011a, c, 2014).
8. In 1939, Lady Eve Balfour had been working on farming systems development since 1920 for generating data to test these beliefs at Haughley Experiment, Suffolk-England (Gordon, 2017). Lady Eve Balfour assumed that human healthiness and upcoming life are depending on how the soil was used for producing their foodstuffs. She also assumed that sketchy farming could harvest more healthy food (Gordon, 2017).

14.3.2.2 POST-WORLD WAR-II

Technological progress at the time of 2nd World War was enhanced for the post-war novelty in all sides of food production by ensuing in huge improvements in modernization and synthetic inputs. In particular, two chemicals were produced and used widely in agricultural production system. For example, ammonium nitrate was used as a cheap source of nitrogen for plant munitions. Besides the new pesticides, DDT was used widely for controlling disease-carrying insects:

1. In 1944, an intercontinental movement named the 'Green Revolution' was hurled in Mexico with a private subsidy from the America.

The concept of 'Green Revolution' inspired the expansion of hybrid crop cultivars, synthetic stimulators such as inorganic fertilizers and pesticides, irrigated crops, and also introduction of mechanization for intensive agricultural through reducing manual labors; as a result, a single farmer started to big areas of land and fields raised larger.

2. During the year 1950s, a topic of sustainable agriculture was started popularity in a scientific society, but researchers tend to spirit on evolving the new chemical methods as one of the principals of the ongoing 'Green Revolution,' for enhancing the agricultural productivity for a growing population. At that time, conversely, the hostile effects of "modern" agriculture sustained to spark an unimportant but increasing consciousness for OFS concept. For example, in the US, J. I. Rodale started to promote the word and approaches of OFS, mainly to clients by doing the advertising of organic horticulture initially.
3. During 1962, Rachel Carson, a prominent ecologist, published the book '*Silent Spring'* which recording the consequences of DDT and other insecticides on the surrounding environment (Paull, 2007). The US government in 1972 banned the usage of chemical DDT as per the suggestions of the book *Silent Spring*.
4. During 1970, the meaning of "eco-agriculture" was first specified by Charles Walters, in the Acres Magazine, in where Charles Walters suggested not to use 'synthetic stimulators' in the cultivation rather than organic agriculture; since these molecules have short and long-term side effects (ACRES-Magazine, 2020).
5. During the 1970s, scientists and policy were worried about environmental pollution as a consequence of intensive agricultural production systems and emphasized on OFS. One objective of the OFS effort was inspired for consuming locally produced food staffs through slogans like "Know Your Farmer, Know Your Food."
6. During 1972, the IFOAM was created in 'Versailles' (Paull, 2010), France for dissemination and discussion about the benefits of OFS.
7. During 1975, the book 'The One-Straw Revolution' was published by Fukuoka. In the book, Fukuoka discussed the robust impression in various fields of the modern farming across the globe. He stated that limited scale grain production should be accentuated carefully to maintain the equilibrium of the local agroecosystem through minimizing anthropological interference.
8. J. I. Rodale and his Rodale Press (now Rodale, Inc.) directed the technique to minimize the adverse effects of artificial agricultural

inputs at the time period of 1970s and 1980s. He also highlighted the benefits of OFS.

9. 'Oregon Tilth' started first for authorization of organic products and service in the USA during 1984 (Musick, 2008).
10. In the 1980s, the government forced to producers and consumers for authorization ruling of organic products. In the USA, the Organic Foods Production 'Act of 1990' rule was first published in the Federal Register in the year 2000 (USDA, 2000).
11. Since the early 1990s, due to increasing consumer request, OFS products in developed economies have been growing up about 20% annually.
12. The damage of Soviet financial livelihood resulting the failure of the Soviet Union in 1991; regulated to emphasizing on native farming production and the progress of an exclusive state-supported urban OFS program called 'Organopónicos.'

14.3.2.3 TWENTY-FIRST-CENTURY

All the way of history, for meeting the food demand of raising population, modern agriculture always emphasize the artificial agricultural inputs, not on OFS. After agricultural modernization, biotechnology, and genetic engineering approaches were introduced for the improvement of stress-tolerant crop cultivars (USEPA, 2020). Since, farming was started to upgrade through the introduction of new farming procedures and machinery, while several serious short and longer-term side effects were arisen due to the excessive and imbalanced use of synthetic stimulants including inorganic plant nutrients and pesticides. Considering these burning issues, researchers have started to search techniques or solutions to alleviate these poisonous effects of modern farming, while preserving the maximum productivity. Therefore, an OFS concept has been started for avoiding the deleterious effects that were emerged from the application of artificial fertilizers and pesticides. The OFS was established as a substitution of the agricultural system in response to promptly altering agricultural activities. Data in the year of 2018, revealed that approximately 71.5 Mha of land globally under OFS naturally (FiBL, 2020). According to the latest FiBL (2020) organic agriculture survey data from 186 countries (conducted at the end of 2018) revealed that the growth rate of organic agriculture was 2.0 Mha compared to 2017.

14.3.3 REGULATION OF ORGANIC FARMING

The products of OFS must be legally authorized by the governments of each country. Growers should be certified for their products and foods to be exclusively "organic," and there are exact organic criterions for foodstuffs. Organic authorization and fair-trade authorization are one of the eight successes of the millennium development goals (MDGs), that were time-honored in the Millennium Summit of the UN in the year 2000, since all UN member states were committed to assist for achieving the goal of MDGs by the year 2015 (Setboonsarng, 2015).

Considering the aspect, the body of 'Codex Alimentarius of the FAO' was established in November 1961. The Commission's major aims are to keep the healthily life of consumers and ensure quality food in the international food market. The Codex Alimentarius body is recognized by the WTO as an international authority for resolving arguments, regarding food security and end-user health-safeguard (WTO, 2020). Since, most nations have their own agendas, roles, and regulation for ensuring the quality and safety of organic products (Figure 14.4). The certifiers of the EU or the United States can examine and verify the other countries' growers and processors products when the country/organization want to export their product to these countries. In the EU, organic certification and examination are approved by authorized organic control bodies as per the standard levels of EU. In USA, OFS products has been certified and examined by the U.S. Department of Agriculture (USDA) for ensuring the National Organic Standards. In the USA, the Organic Foods Production 'Act of 1990' rule was first established and published in the Federal Register in the year 2000 (USDA, 2000).

14.4 TYPES, PRINCIPLES, COMPONENTS, AND BASIC METHODS OF ORGANIC AGRICULTURE

14.4.1 TYPES OF ORGANIC FARMING SYSTEM (OFS)

14.4.1.1 PURE ORGANIC FARMING SYSTEM

This includes the use of organic inputs while completely avoids the use of any types of chemicals. Further to become pure OFS, around 4–5 years required to diminish the residual effects of already applied fertilizers. Many agencies are there to provide certificates for the OFS. Generally, organic produce is somewhat expensive as compared to the conventional produce but superior from nutritional values (Brandt and Mølgaard, 2001). However, inhabitants, particularly of developed nations demanding it even after paying higher for it after considering its health benefits. However, in underdeveloped nations, OFS is still requiring to jump many hurdles (Meena et al., 2013).

FIGURE 14.4 Organic certification labels for different countries.

14.4.1.2 INTEGRATED ORGANIC FARMING SYSTEM

This permits the sustainable use of both organic and inorganic fertilizers to an extent at which it does not adversely affect our ecology. It is the kind of cultivation practices, where cultivating of crops is done solely utilizing natural resources (Meena et al., 2013). Legumes have a considerable scope under this farming (Drinkwater et al., 1998). Further, it could also involve the poultry, mushroom production, goat-rearing, and fishpond altogether for having regular incomes (Yadav, 2017).

14.4.2 BASIC PRINCIPLES OF ORGANIC FARMING SYSTEM (OFS)

There are several principles of OFS, which are discussed in the following sub-headings (Meena et al., 2013).

14.4.2.1 PRINCIPLE OF HEALTH

The OFS ultimately results in sustainable farming and reduces the carbon as well as water footprints. The OFS increases the product quality which further helps in attaining the overall sustainability. Good quality soils produce good quality products which further has a favorable effect on the consumers (Meena et al., 2013). Being having almost nil residual effects of chemicals, organic produce always improves the health status of the consumers. Organically produced products help to bring overall sustainability by improving the soil health, human health, etc., (Drinkwater et al., 1998; Brandt and Mølgaard, 2001).

14.4.2.2 PRINCIPLE OF ECOLOGY

The OFS has sustainable effects on our ecosystems by one or other way as it reduces the generation of GHGs, reduces carbon footprints, and expands the soil fertility and productivity by improving organic matters in the soils. It also has favorable effects on the beneficial insects which otherwise killed by the strong pesticides, which further controls the harmful insects. Further approaches such as *Trichoderma* has a significant role in controlling the pest population. The objective is to have a favorable effect on the ecological balance. Organic supervision must be reformed as per local agro-ecological and social conditions (Drinkwater et al., 1998). The OFS uses natural organic manures hence reduces the carbon as well as water footprints and makes the whole agricultural system ecologically sustainable (Meena et al., 2013).

14.4.2.3 PRINCIPLE OF FAIRNESS

Fair relationship between common environment and life opportunities must be there under the OFS (Brandt and Mølgaard, 2001), which involved well-being of every whosoever involved, viz. farmers, workers, processors, distributors, traders and consumers (Meena et al., 2013). Aim of the OFS is to produce a better-quality product which is good for society, the ultimate users (Drinkwater et al., 1998). Sustainable agriculture is a must for the upcoming generations (Brandt and Mølgaard, 2001). Fairness required in the production, transportation, handling, and trade of the product with a minimum of food loss as well as to the food wastage.

14.4.2.4 PRINCIPLE OF CARE

The OFS has a goal of improved health and wellness. Depending upon the demands both external as well as internal, organic produce produced under the principles of OFS, though it might be a bit costly as associated to the traditional systems. Precaution and responsibility of OFS must be critically followed for the optimum production (Meena et al., 2013).

14.4.3 COMPONENTS OF ORGANIC FARMING

Proper crop rotation, organic matter additions, nitrogen fixation, biogas, etc., are the main pillars for the OFS which must be followed while practicing the OFS. Among organic manures, farmyard manures (FYMs), vermicompost, green manuring, etc., are the key players for improving the soil health. OFS, in brief, involved the following pillars in subsections (Figure 14.5).

FIGURE 14.5 Essential components of organic farming system.

14.4.3.1 CROP ROTATION

Under this method, different crops with different rhizosphere and different nutrient mining capabilities are grown in sequence so as to grow crops sustainably with the least possible pressure on the natural resources. Introduced legumes in the crop rotation will improve the soil health and livelihoods of the farmers by one or other way. Further, fertilizer demands also reduced which reduced loads on the farmers' pockets. Hence proper crop rotation must be followed by adding legumes in the cereals.

14.4.3.2 CROP RESIDUE

Rice-wheat crop rotation followed on a significant portion of South Asia. Among their residues, management of the rice residues is really a challenging job as because of higher silica contents, and it is not used in the animal husbandry sector. Therefore, several methods viz., ZT sown wheat with a happy seeder, paralichar, phosphor-compost recommended in the region for the sustainable management of the rice residues. Under OFS, use of paralichar or phosphor-compost is highlighted.

14.4.3.3 ORGANIC MANURE

Use of organic manures viz., FYM, compost, vermicompost, etc., is recommended in the OFS for sustainably cutting of the chemical fertilizers. Further, organic manure slowly releases the nutrients and improves the soil health as manures act as the feed for the microorganisms. The use of only organic manures certainly cut down the potential yields, and that is why organic produce is somewhat costlier than the conventional produce because of its lower yields. Beside these above components, the following aspects are important for an OFS:

- Shifting of land from conventional use to organic use with almost no use of chemicals;
- Ensure biodiversity and sustainability of the system by overall ecological maintenance;
- Sustainable crop production with different approaches viz., crop rotation, management of crop residues, compost, manures, and managing pests and diseases through biological means;

- Control of insect-pest population through the biological control system;
- Use of organic concept in livestock by the use of organically produced feed.

14.4.4 ORGANIC FARMING METHODS

As the accumulation of organic matter is a precedence for OFS, hence, chemical fertilizers are ignored. Generally, organic matter is applied in the soil in the form of the FYM, vermicompost, green manure (GM), compost, and so on. The USDA National Organic Standards instructed those organic manures must be incorporated in the soils within 3 to 4 months, based on soil and air temperature, moisture availability, etc. Well decomposed compost or FYM only be utilized in the soil to enhance the soil health that ultimately further adds organic matter into the soils, improves the physicochemical properties of the soils, and finally the potential yields and their quality standards. As nutrients in these organic manures are un-mineralized, hence the use of microbes is a must to make them available to the plants for meeting their requirements, hence their release is slow, and uses efficiency is relatively higher than that of the chemical fertilizers which easily released the nutrients quickly. For example, urea easily hydrolyzed by the urease enzyme to produce the CO_2 and ammonia gas in the atmosphere, which further causes the greenhouse gas (GHG) effect.

Green manuring practiced under OFS to add the organic matter in soil and hence improves the product quality. Green manuring also helpful for loading the soil with different macro and micronutrients. Further, minimum tillage should be practiced as under the intensively tilled plots, hidden organic matter oxidized by the microorganisms and produces the CO_2 as GHGs enhances the C-footprints which is certainly not required. Further, under the OFS, to get ride-off from the different insects and pests, instead of using different insecticides and pesticides, organically produced liquid formulations containing neem leaves or neem oil or other organic products like *Lassi* preferred. Further, some bacteria like *Bacillus thuringiensis*, also used at some site-specific locations. However, under the OFS, some inorganic pesticides such as sulfur (S) and copper (Cu) are also allowed, since they have no side effect on the food chain confirmed by several earlier findings (NPR, 2011). Cu sources pesticides are recycled to fight against fungal and bacterial diseases in plants; since it is not very toxic to human health. Further, OFS much depends on the biological approaches to control insect pests. Further, breeding and biotechnological approaches must be applied to develop new plant cultivars which respond better to the OFS, could tolerate different stresses, and resistant to the attack of different insect-pest and diseases. The very basic methods of organic agriculture are discussed in the following sub-headings.

14.4.4.1 CROP DIVERSITY

Diversification already advocated minimizing the water footprints in the rice-based cropping systems (Bhatt et al., 2020a, b). Further, multi-cropping could also be practiced for resolving the ever-increasing demands of the burgeoning population. However, growing of a single crop for a number of consecutive years removes the nutrients from a particular soil depth for fulfilling further crop nutrient demands. Hence diversification is an important method to be kept in the mind.

14.4.4.2 SOIL MANAGEMENT

Mostly under conventional crop establishment methods, soil health is declining day after day due to intensively practiced rice-wheat cropping rotation with conventional faulty methods, which led to declined soil health and yields (Drinkwater et al., 1998; Bhatt et al., 2020a, b). However, under OFS mostly nutrient requirements fulfilled through the organic inputs viz., FYM, and vermicompost or even through the green manuring, etc., which further improves the soil organic matter (SOM) status which further has its favorable effects in the soil physic-chemical and biological properties (Dobbs and Smolik, 1996). Therefore, OFS depends on the natural way of farming and avoid the use of chemical fertilizers (Brandt and Mølgaard, 2001).

14.4.4.3 WEED MANAGEMENT

Plants that are not required where they are not required are known as "weeds" and they fight with the focal plants for water, nutrients, and sunlight, and ultimately affected the overall land productivity (Drinkwater et al., 1998). Under OFS generally, we used the practice of mulching and cutting instead of using heavy weedicides, which further pollutes the underground water and thus a threat to the ecosystem.

14.4.4.4 INSECT-PEST CONTROL

Agricultural produce significantly affected by the attack of insect-pest and diseases, which needs to be controlled for overall sustainability (Drinkwater et al., 1998). However, instead of using the chemicals under OFS more stress given on the biological control or spray of natural chemicals like neem oil. Further, by doing so, we could also take care of the useful insects which otherwise affected by the chemicals of conventional farming.

14.4.4.5 LIVESTOCK

OFS methods also focused on the other sectors like dairy, piggery, etc., along with the agricultural sector to enhance the system sustainability. Further, this also led to regular incomes for the farmers.

14.4.4.6 GENETIC MODIFICATION

In general, OFS discourages the use of the engineered animals/plants for direct consumption by the humans as many complications are there regarding their residual effects on the human health (Dobbs and Smolik, 1996). Moreover, up to now in India, these are not advocated except Bt cotton, which is a non-food crop.

14.5 PROSPECTS OF ORGANIC AGRICULTURE FOR FOOD QUALITY AND SAFETY

Agricultural activity was started 10,000 years ago without the utilization of synthetic stimulators. Synthetic fertilizers and pesticides were started to use in the mid-1900. Initially, these artificial inputs were very inexpensive, influential, and transportation of bulk volume was easy. Comparable progresses were also happened for artificial pesticides in the 1940s, which leading to the period being mentioned to as the 'Pesticide Era' (Horne, 2008). Although application of artificial agricultural inputs significantly improved the farming productivity (Hole et al., 2005), while excessive and imbalanced application of these chemical fertilizers and pesticides generated a serious short and longer-term side effects on soil and surrounding environment, and also human health (Stinner, 2007). The green revolution has also contributed to several environmental problems (Singh, 2000) which have directly or indirectly influenced the human civilizations. In the present day the number of food-insecure people in India, Bangladesh. Pakistan, Nepal, and Sri Lanka are increasing not due to food availability but also due to food accessibility. Considering the burning issue, researchers began to seek solutions to alleviate these hostile effects of chemicals, while preserving the maximum productivity. In the last part of the 1800's and mid 1900's, researchers started to look for approaches to cure these results while as yet keeping up higher productivity. Therefore, an OFS concept was started in the 1940s for avoiding the poisonous effects that were emerged from the application of artificial fertilizers and pesticides (Pugliese, 2001; Vogt, 2007). Under such circumstances, there is an urgent need to provide access to food. As per the report provided (IFAD, 2007), it has been observed that OFS can purposively solve the problems by providing local food security by generating diverse products with low input costs as compared to conventional agriculture. In the OFS, agricultural inputs generally must be from biological origin also must involve organic manure, compost, free from GMO, crop rotation and companion planting, etc. The OFS aims to maintain agroecosystem by encouraging interior self-regulation for using synthetic agricultural inputs. Undeniably, profits of organic crops/products have been established to be around 25% inferior to modern farming (Stolze and Lampkin, 2009), but OFS is environmentally friendly, could improve nutritional quality through incorporating diversified crop cultivars (Rigby and Cáceres, 2001). A study conducted over 22 years in Rodale Institute test farm (Pimentel et al., 2005; Steffanie et al., 2018). A 30th-anniversary report was published by Rodale in 2012 (Rodale Institute, 2020). During the trial, he expressed that the harvest yields for corn and soybeans were at par in the animal based natural farming, vegetable-based cropping pattern and regular cultivating frameworks. It likewise found that fundamentally less fossil energy was exhausted for the production of corn in the Rodale Institute's organic leguminous based cropping practices and natural vegetable cultivation frameworks than in the traditional agronomic framework. Similarly, there was little distinction in energy input consumption among the various treatments for the production of soybeans. In the natural cropping practices, inorganic fertilizers and chemical pesticides were avoided (McBride and Greene, 2013).

Organic food production system is needed to spread globally as the currently practiced conventional agricultural production system poses a detrimental effect on the environment, causing soil, water, land degradation and also contributing to GHG emission leading to global warming. Organic cropping system not only increases food productivity in some areas but also helps in combating hunger where poverty exists providing both food security and health enrichment. Initiatives taken by some developing countries to slowly shift towards an organic package of practices through integrated nutrient management are praiseworthy. It is to be mentioned that the challenge in adapting organic cropping system is neither agronomical constraints nor economical boundaries but completely socio-political. Prospects and benefits of OFS are discussed in the following sub-sections.

14.5.1 IMPORTANCE OF ORGANIC FARMING ON FOOD SECURITY

During the last few years, per capita world food production has increased by almost 25%, and world food commodity rates have drastically decreased by 40% in actual terms. As per the study conducted by Pretty et al. (2001), during the early 1960s and mid-90's the mean cereal productivity increased significantly from 1.2 t ha^{-1} to 2.52 t ha^{-1} in emerging nations like India, Bangladesh, Nepal, Sri Lanka, Vietnam, and Malaysia. However, total cereal production has been increased dramatically from 420 to 1,176 million tons yearly.

Gradual progression towards achieving food sufficiency by reducing hunger has been unfortunately slowed down in recent years. For the period of the late '1990s and early 20th Century, the total figure of hungry/starved people in the developing nations has been reduced by just 1%, from 824 million to 815 million. However, in the sub-Saharan African (SSA) region, the scenarios are just the opposite, where there is a significant rise in hungry people since the 1990s by almost 20% (Von Braun, 2005). Thus, it is very much evident that humanity is far away from the reality in achieving the goal which has been set as per MDGs of reducing the number of hungry people around the globe by almost half. Yet paradoxically rather ironically during the same period, nearly about 1.2 billion people, particularly in first world nations, are over-eating, which increases obesity-related health issues (World Watch Institute, 2000).

In 2007, the United Nations Food and Agriculture Organization (FAO) reported that chemical fertilizers were needed to avoid hunger, particularly in Africa where fertilizers are presently used 90% less than in Asia (FAO, 2007). Likewise, NEPAD, an advancement association of African governments, reported that taking care of Africans and forestalling ailing health requires manures and upgraded seeds (Africa Fertilizer Summit, 2006). For example, in Malawi, the yield has been boosted up using seeds and fertilizers (FAO, 2007). FAO additionally calls for utilizing biotechnology, as it can help small and marginal producers (farmers) to improve their monetary returns and food security (FAO, 2013). According to a 2012 study in Science Digest, organic best management practices show an average yield of only 13% less than the conventional cultivation system (McGill University, 2012). On the planet's less fortunate countries (under-developed nations) where a large portion of the world's ravenous individuals live, and where individual farmers could not able to afford costly input materials, adoption of natural and organic input materials enhances yields almost by 93% overall and hence could be a significant piece of alternative way for achieving food security (World Watch Institute, 2006). Many researchers, policymakers, and experts opined that OFS would not only increase the world's food demand but might be the only technique to eliminate malnourishment (World Watch Institute, 2006).

This is well established fact that there is an increasing need for capital and chemical inputs that have deleterious consequences in the sustainability of agriculture. There have been many arguments in the adaptation of conservation agriculture, recycling of nutrients, biodiversity, and synergistic effect between crops, animals including livestock and soils, etc., regeneration, and conservation of natural resources. However, these strategies should accord with policy, capacities, and interest with the small and medium farmers, who otherwise will not be enthusiastic enough in adapting so.

14.5.2 THE SYNERGISTIC WELFARES OF ORGANIC FARMING SYSTEM (OFS)

Organic methods reduce the production cost of food crop (Marshall, 1991; Brown, 2008). During 2000, uncompensated costs for 1996 reached nearly 2,343 million British pounds or £208 per ha. (£84.20/ac) (Pretty et al., 2000). An investigation of practices in the US distributed in 2005 presumed that cropland costs the economy roughly 5 to 16 billion dollars (\$30–96/ha–\$12–39/ac), while livestock production costs 714 million dollars (Tegtmeier and Duffy, 2005). Assessing the welfares and also the constraints of OFS is rather very multifaceted. The effect of transforming to OFS will largely depend on the initial interest and capabilities of the growers and farming communities, their inherent skills, and the natural means obtainable to them. However, several major relevant potentials prospects have been worked out:

- OFS can enhance agricultural output, particularly in regions where growers are prone or subject to the vulnerability of food shortages.

- OFS can enhance income generation and/or revenues to the farm employment. This goal can be attained either through reaping the better yields, superior prices due to higher demand, lower costs (for inputs) or incorporations of these three parameters.
- The divergence of the agricultural production system, which is intrinsic in OFS, decreases the risk of crop loss by any natural calamities or anthropogenic sources due to various stress and synergistic financial and food security restraints. The system consensuses powerfully with the risk diversion approaches which have been embraced by low income-generating farming families.
- OFS foodstuffs are safe, diversified, and nutritional.
- OFS bypasses the hidden risks related with the drastic and vigorous use of toxic chemicals.
- OFS plays a vital role in sustaining natural resource conservation such as decreasing water demand and run-off, soil erosion, enhancing biodiversity, natural resource conservation.
- OFS plays a dynamic role for creating awareness regarding the importance of sustainable food production and consumption amongst the farmers and consumers and also the importance of clean and safely produced in order to safeguard the environment.
- OFS has significantly contributed to women upliftment in farming societies as compared to additional relegated groups, as well as ensuring innovative employment prospects for the poor landless.
- OFS identifies the importance of outdated and aboriginal understanding and thus helps in incorporating this knowledge in its production system, thus growing social capability and self-value.
- OFS has proved to be sustainable in the long run. It increases the capabilities of a production system towards climate resilience such as drought, flooding, or other extreme events.
- OFS has a perfect cut role to encounter a variety of world ecological policies and aims, comprising strategies to combat desertification, preserving resource biodiversity, and bypassing the unfriendly effects of global warming.

14.5.3 FOOD QUALITY IMPROVEMENT UNDER ORGANIC PRODUCTION SYSTEMS

The quality of food products is judgmental and shows temporal and spatial variation. There are no specific criteria for classifying food superiority.

Hence, the definition of food quality is highly flexible. Initially, there were some fixed parameters for assessing food quality. Nowadays, a more holistic approach is undertaken for defining food quality. Kahl et al. (2010) analyzed the present condition/values of organic food quality as compared to the potential quality that has been claimed in the superior world. A rational gap between consumer demands regarding food quality of an individual to that which can be guaranteed by the government as per policies, regulatory law enforcement was also worked out. A study by Byerlee and Alex (2005) showed that diversification and optimization of farm productivity, reduction of inputs for farm operations, resulting towards the development of households,' a market-orientated scheme for earning extra income; organic production systems which have eventually resulted in alleviating hunger and poverty. It is proven that improved income allows individual farmers to buy food to a greater extent which would otherwise lead to a hunger crisis.

Incurred higher returns coming from the marketing of organic commodities often leads to seasonal or permanent diversification of cropping systems from principle staple crop production towards high value lucrative high-value commodities such as exotic vegetables, fruits, etc., depending upon the farmers' investment potentiality, agroecological conditions and local demands. Although in many instances, principle food systems will continue to dominate the market (for example, rice dominated 50% of Asia's cultivated lands) and off-farm activities will generate extra revenue, however, on the other hand, organic diversification contributes to higher net earnings from land and labor savings. However, the adapting diversified crop cultivation from conventional to organic practices requires strong will and capital for establishment. Assured land, water source, sound technological knowledge is some of the important prerequisites for investments in organic diversification and commercialization as compared to any other forms of agriculture.

Rural Schools and homestead gardens provide chemical-free nutritional organic products that improve the health of rural children and households. These systems significantly contribute to food accessibility, the safety of children and improving the health and nutritional status of the families. In certain instances, poorly known less popular varieties have proved to be an important source of income generation opportunities through the marketing of processed specialty foods (e.g., *Chenopodium quinoa*) or certain medicinal plants like Aswagandha, Cinchona, etc., or aromatic (Cinnamon) or dye plants, which have high very high demand locally and in international export markets.

14.5.4 CONTRIBUTION OF ORGANIC AGRICULTURE TO TRANSITIONAL FOOD CRISIS

Small and marginal farmers are often associated with pre- and post-production risks factors, and thus, they are very much vulnerable to losses. For them, sustainable food with assured income is more important than yield. Thus, organic production system shows less variation in harvests and divergence and hence, is one of the best-assured alternatives in cases of a sole crop disappointment, climatic adversities, or even socio-economic crisis. Under changing climatic scenarios with a rise in the number of risky weather events, snow balling the flexibility of agroecosystems to weather abnormalities has become an urgent need of present and future, especially in countries which are agriculturally highly dependent. A thorough correlation of energy effectiveness in grain production, produced yield, and animal husbandry concluded that OFS had a better return for each unit of energy consumed over the huge dominating animals and crop-based cropping system (Dalgaard et al., 2001). Contradictory results on the profitability of OFS have also been reported (Dalgaard et al., 2001). It has commonly been tracked down that the work input per unit of yield was higher for organic means of agricultural practices over conventional agricultural framework system (Pimentel et al., 1983).

14.5.5 ORGANIC AGRICULTURE CONTRIBUTION TO HEALTHY DIETS

About 58% of deaths in the world are caused of non-transmissible diseases like cardiac arrest, diabetes, kidney ailments due to malnutrition or imbalanced diets. In China alone, around 8.1% of the same households have both underweight and an overweight member. Recent food patterns, however, have immensely proved to be beneficial in combating malnutrition; specialized agricultural systems focusing on enhancing nutritional contents of some staple crops through biotechnological interventions have been a success in some countries. Low dietary contents and micronutrients viz. vitamins, iron, and iodine have affected more than half the population in developing nations like India, Bangladesh, Sri Lanka, Nepal, etc. This is indeed a key community health anxiety which has been spoken through providing external addition and biofortification in some important cereal crops like black rice, Golden rice, High nutritional enriched fruits, and vegetables but in small scale which have failed to target the majority of the population till date.

Encouraging a miscellaneous indigenous food source may be available to poor families, which has established to be a low cost and effective solution to combat undernourishment. The feasibility of an OFS is highly dependable to a different agroecosystem both spatially as well as temporally. Crop diversification with the organic package of practice with even low economic value have high nutrient enrichment is beneficial for the household in improving the overall health. The reintroduction, assortment, and genetically enrichment of location specific reformed cultivars makes an irreplaceable involvement to "hidden hunger," or nutritional micronutrient insufficiencies. Several consumers' survey reports have clearly depicted that consumers of organic products have a healthier dietary status, particularly owing to selections of "minor" legumes that contribute to improved foods.

While there may be some alterations in the quantities of nutrients and anti-nutrients when progressively produced food and unadventurously produced food products are compared, the variable idea of food creation and taking care of makes it hard to sum up outcomes, and there is lacking proof to make asserts that natural food is more secure or more grounded than ordinary food (Bourn and Prescott, 2002; Blair, 2012; Smith-Spangler et al., 2012; Barański et al., 2014).

14.5.6 ORGANIC AGRICULTURE SYSTEM INCREASES THE PROFITABILITY

In the USA, OFS has been revealed 2.7 to 3.8 times extra cost-effective to modern conventional farming (CT) (The Hindu, 2010; Gurung, 2011). Worldwide, according to Metadata analysis across five continents in the year 2015, OFS was found 22 to 35% more income than CT (Hindu Business Line, 2010; Times of India, 2010). Another observation found that on an inclusive scale, 5–7% price premiums were needed to break even with CT (Lotter, 2003; Crowder and Reganold, 2015). Martin and Kim (2008) found that organic food is profitable, since organic products could be sold at a relatively higher rate as compared to organically produced food.

14.5.7 ORGANIC FARMING SYSTEM (OFS) INCREASES EMPLOYMENT OPPORTUNITY OF LABORS

Organic production system is labor-intensive than CT (FAO, 2020), indicating that OFS provides more jobs per unit area than conventional systems

(Green and Maynard, 2006). The 2011 United Nations Environment Program (UNEP) Green Economy Report suggests that an increase in investment in OFS is anticipated to lead to growth rate in labor employment about 60% compared with CT. The report also highlighted that investment in OFS could create 47 million additional jobs compared with CT over the next 40 years (UNEP, 2011), particularly for emerging nations. A large part of the development in ladies work support in agribusiness and allied sectors are outside the male dominated field of CT. Predominant associates in OFS are 21% ladies, instead of 14% in conventional mode of farming sectors.

14.5.8 ORGANIC FARMING SYSTEM (OFS) CONSERVES SOIL PHYSIOLOGICAL AND CHEMICAL PROPERTIES

The OFS can build up SOM better than CT system, which suggests long-term yield benefits from OFS (USDA-ARS, 2007). Researchers claimed that naturally conserved soil has higher fertility and productivity (Johnston, 1986) leads to higher water retention (Kirchmann et al., 2007). USDA's Agricultural Research Service has observed that organic manure applications in tilled-OFS are better at constructing conserves soil physiological and chemical properties than no-till (Hepperly et al., 2008; Paulson, 2008).

Scientists at Oxford University examined 71 peer-reviewed scientific papers and looked into contemplates and observed that natural items are once in a while are harmful rather detrimental for the climate (University of Oxford, 2004) and generally, natural items required less energy, yet more land (University of Oxford, 2004). Per unit of item, natural produce creates higher nitrogen filtering, nitrous oxide outflows, smelling salts emanations, eutrophication, and fermentation potential than customarily developed produce (Meleca, 2008; Tuomisto et al., 2012) and that OFS can diminish petroleum derivative discharges (UNEP, 2011; Rodale Institute, 2014). Several experts in the field of OFS techniques accept that the expanding land for cultivating natural/organically produced food might actually annihilate the rainforests and crash numerous environments (Goldberg, 2007; Leonard, 2007).

14.5.9 ORGANIC FARMING SYSTEM (OFS) IS A BIODIVERSITY-FRIENDLY PRACTICE

A wide scope of living beings profits by OFS, however, it is indistinct whether natural techniques give more prominent financial advantages than traditional coordinated agro-ecological projects (Hole et al., 2005). OFS is regularly introduced as a greater biodiversity-accommodating practice, however, the over-simplification of the valuable impacts of OFS is bantered as the impacts show up frequently species-and setting reliant and flow research have featured the need to measure the overall impacts of neighborhood and scene scale the board on farmland biodiversity (Henckel, 2015). The protection of characteristic assets and biodiversity is a center rule of natural creation. Three expansive administration rehearses (preclusion/diminished utilization of synthetic pesticides and inorganic manures; thoughtful administration of non-trimmed environments; and safeguarding of blended cultivating) that are generally characteristic (yet not selective) to OFS are especially helpful for farmland natural life. Utilizing rehearses that draw in or present gainful creepy crawlies, give territory to birds and vertebrates, and give conditions that expansion soil biotic variety serve to supply crucial biological administrations to natural creation frameworks. Benefits to guaranteed natural tasks that carry out these kinds of creation rehearses include: (1) diminished reliance on external ripeness inputs; (2) decreased bug the executives costs; (3) more solid wellsprings of clean water; and (4) better fertilization (USDSA, 2018). Practically all non-crop, normally happening species saw in similar homestead land practice reads show an inclination for OFS both by plenitude and variety (Hole et al., 2005; Gabriel et al., 2006). A normal of 30% more species possesses natural homesteads (Bengtsston et al., 2005). Birds, butterflies, soil microorganisms, scarabs, worms, (Blakemore, 2000), arachnids, vegetation, and well evolved creatures are especially influenced. The absence of herbicides and pesticides improve biodiversity wellness and populace thickness (Gabriel et al., 2006). Many weed species draw in useful bugs that improve soil characteristics and rummage on weed bothers (van Elsen, 2000). Soil-bound creatures frequently advantage in view of expanded microorganisms populaces because of regular compost like fertilizer, while encountering decreased admission of herbicides and pesticides (Hole et al., 2005). Expanded biodiversity, particularly from helpful soil microorganisms and mycorrhizae have been proposed as a clarification for the significant returns experienced by some natural plots, particularly considering the distinctions found in a 21-year examination of natural and control fields (Fließbach et al., 2006). Biodiversity from OFS gives cash-flow to people. Species found in natural homesteads upgrade maintainability by diminishing human info (e.g., manures, pesticides) (Perrings et al., 2006). The USDA's agricultural marketing service (AMS) distributed a Federal Register notice on 15

January 2016, reporting the national organic program (NOP) last direction on Natural Resources and Biodiversity Conservation for Certified Organic Operations. Given the expansive extent of regular assets which incorporates soil, water, wetland, forest, and untamed life, the direction gives instances of practices that help the hidden protection standards and exhibit consistence with USDA natural guidelines (USDSA, 2018). The last direction furnishes natural certifiers and homesteads with instances of creation rehearses that help preservation standards and consent to the USDA natural guidelines, which expect activities to keep up or improve common assets (USDA, 2018). The last direction likewise explains the job of guaranteed tasks (to present an OSP to a certifier), certifiers (guarantee that the OSP depicts or records rehearses that clarify the administrator's checking plan and practices to help regular assets and biodiversity protection), and assessors (on location investigation) in the execution and confirmation of these creation rehearses (USDSA, 2018).

14.5.10 CAPACITY BUILDING IN DEVELOPING COUNTRIES

The OFS can donate to environmental sustainability, particularly in low incoming countries (Hine and Pretty, 2006). Organic principles of OFS employ of native assets (e.g., local seed varieties, manure, etc.). Local and international markets for organic foodstuffs express the excellent opportunities to improve the income of producers and exporters' (Lockie, 2006). For expansion and improvement of OFS, International Federation of OFS Movements accommodated more than 170 free handbooks and 75 training prospects online in the year 2007 (Niggli et al., 2008). United Nations Environmental Program (UNEP) and the United Nations Conference on Trade and Development (UNCTAD) specified that OFS can be more encouraging to food safety in Africa than CT, and more sustainable in the long-term and intensive production systems (Howden, 2008; UNEP-UNCTAD, 2008). The estimation of OFS in the accomplishment of the MDGs, especially in destitution decrease endeavors even with environmental change, is appeared by its commitment to both pay and non-pay parts of the MDGs (World Bank, 2008; Markandya et al., 2015; Setboonsarng, 2015). A few overviews and studies have endeavored to inspect and analyze regular and natural frameworks of cultivating and have tracked down that natural strategies, while not without hurt, are less harmful than ordinary ones since they diminish levels of biodiversity not exactly customary frameworks do and utilize less energy and produce less waste when determined per unit region (Stolze et al., 2000; Hansen et al., 2001).

14.6 CHALLENGES OF ORGANIC AGRICULTURAL SYSTEM

Despite the fact that we need to recall that the world is still having a shortage of good quality cultivable agricultural lands. Cultivating as of now involves 37% of the world's territory zone and the majority of the great quality land is as of now trimmed. By the mid-20th century we need to have to increase farm outputs about 2-fold for feeding 9.7 billion people. The natural homesteads as of now lose about a portion of their yield potential since they will not utilize nitrogen compost and the more viable manufactured pesticides. They are as of now suing regular ranchers for "dust contamination" from biotech seeds. Presently they are beginning to lobby for "calamity installments" on landslides and weeds. It has been proposed that OFS may profit farmland biodiversity more in scenes that have lost a huge piece of its previous scene heterogeneity.

In spite, it is a reality that the natural development faces a few obstacles as it extends globally. A new audit of OFS recorded a few difficulties confronting OFS (Halberg et al., 2005; Kristiansen et al., 2006; Smith et al., 2010) including biological equity, creature government assistance, reasonable exchange, store network improvement, profitability impediment and local transformation and worldwide harmonization for norms. The most important challenges for the sustainability of OFS under the growing food demand of increasing population as well as in the modern era of changing climate are discussed in subsections.

14.6.1 GENERAL CONSUMERS HAVE SEVERAL STANDARDS FOR QUALITY ASPECTS THAT ARE DIFFICULT FOR ORGANIC GROWERS TO MEET

Lettuce with holes in it or apples with a bit of scab are always passed over by shoppers, although nutrition and flavor quality might be excellent. Consumers have been trained to seek out food with Barbie-doll features. Organic growers have higher rates of unmarketable blemished product which often limits sales revenue.

14.6.2 PROFITABILITY IS LOW BECAUSE FOOD PRICES ARE LOW AND THE LAND IS EXPENSIVE

The vast majority of the farm producers we know have a normal day to day employment to help the agricultural production which they do tirelessly on evenings and throughout the week. Regardless of increment openness and several benefits which have been acquired as of late, actually most of them are yet not productive organizations.

14.6.3 ORGANIC FARMING ON COMMERCIAL LEVEL IS DIFFICULT

Numerous organically harvested crops which are part of the monocropping system, unlike conventional crops however, utilize naturally enlisted pesticides and composts. It is basic for natural cultivators to splash pesticides much more as often as possible as their customary partners to maintain control pest and disease infestations. Natural techniques are considerably more successful on a limited scale than on the commercial level.

14.6.4 THERE ARE MANY CONTRADICTORY CONCEPTS REGARDING ORGANIC FOOD PRODUCTS/CONSUMABLES

Many consumers buy organic because it seems like the ethical choice. But how can big businesses (like Wal-Mart, General Mills, and Kellogg) grow organically, and be any better than the produce grown in your own town? Is organic really synonymous with pure? How do ethics of shopping for organics compare with shopping local or fair-trade? Perhaps we are ready for a new standard. How about farm-direct?

14.6.5 ORGANIC CERTIFICATION IS EXCLUSIVE

Numerous small producers do not legitimize the incurred cost for obtaining organic certification. Some utilization strategies that are very appropriate for obtaining organic tags, yet does not fulfill the underlying criteria for organic tagging. On the off chance that you purchase directly from the producers shops or village markets, you can converse with the producers regarding the agronomic practices for producing the stuffs.

14.7 REGIONAL SUPPORT OF ORGANIC FARMING ACROSS THE GLOBE FOR PROVIDING SAFETY AND NUTRITIONAL FOOD

The Chinese government, especially the local government, has been providing various supports for the development of OFS since the 1990s. OFS has been perceived by neighborhood governments for its potential in advancing economic provincial turn of events (Aijuan, 2015). It is normal for nearby governments to encourage land access of agribusinesses by arranging land renting with neighborhood ranchers. The public authority likewise sets up show natural nurseries, gives preparing to natural food organizations to pass affirmations, and sponsors natural certificate expenses, bug repellent lights, and natural manure, etc. The public authority has likewise been assuming a functioning part in showcasing natural items through getting sorted out natural food exhibitions and marking upholds (Scott et al., 2015).

In India, in 2016, the northern territory of Sikkim accomplished its objective of changing over existing homestead grounds to 100% OFS (Hindu Business Line, 2010). Different provinces of India, including Kerala (Martin, 2016), Mizoram, Goa, Rajasthan, and Meghalaya, have likewise announced their expectations to move to completely natural development (Indian Express, 2014; Paull, 2017). The South Indian state Andhra Pradesh is likewise advancing OFS, particularly zero budget natural farming (ZBNF) which is a type of regenerative agribusiness (Naresh et al., 2018). The Dominican Republic has effectively changed over a lot of its banana yield to natural (Paull, 2017). The Dominican Republic represents 55% of the world's confirmed natural bananas (Paull, 2017).

In Thailand, the Institute for Sustainable Agricultural Communities (ISAC) was set up in 1991 to advance OFS (among other feasible horticultural practices). The public objective by means of the National Plan for OFS is to achieve, by 2021, 1.3 million rai of naturally cultivated land. Another objective is for 40% of the produce from these farmlands to be burned through locally (City Life-Chiang Mai (Thailand), 2017). Much advancement has been made (City Life-Chiang Mai (Thailand), 2017). For instance, numerous natural ranches have grown, developing produce going from mangos-teenager to stinky bean. A portion of the homesteads have likewise settled schooling communities to advance and share their OFS methods and information. For expanding the showcasing offices for OFS items, the public authority has been set up in excess of 18 natural business sectors (ISAC-connected) in Chiang Mai Province of Thailand.

14.8 CONCLUSIONS

To satisfy the growing demand of food for the increasing population in the world, modern agricultural practices such as inorganic synthetic chemicals and various farm machineries were introduced in crop production. However, these new farming inputs and machineries have been found to cause serious short and longer-term side effects to the environment, human health, and biodiversity. Furthermore, these practices also cause degradation of soil health and other natural resources and thus unsuitable for sustainable agriculture. To overcome these challenges, OFS has been introduced in the early 20th century as a suitable alternative to the modern agriculture for the sustainable production of safe food without affecting the environment and human health. In the OFS, agricultural inputs are generally obtained from of the organic origins and should involve diversified crops with crop rotation and companion planting for maintaining agroecosystem through encouraging interior self-regulation for using synthetic agricultural inputs. Discussion in this comprehensive review revealed that the OFS system is environment-friendly and good for human health and biodiversity. The information discussed in this review should be useful for the safe and sustainable production of crops using OFS in the era of changing climate. Although the OFS system works well in some developed countries as their population is declining day by day, advanced research is needed to improve the productivity OFS to ensure the food and nutritional security of the ever-increasing population in the developing countries from the decreasing areas of cultivable land.

KEYWORDS

- **environmentally friendly**
- **food quality**
- **food safety**
- **integrated pest management**
- **millennium development goals**
- **national organic program**
- **organic agriculture**

REFERENCES

ACRES-Magazine, (2020). *What is Eco-Agriculture?* Acres, USA. https://www.acresusa.com/pages/what-is-eco-agriculture (accessed on 13 July 2021).

Africa Fertilizer Summit, (2006). *The New Partnership for Africa's Development (NEPAD), 9–13 June 2006, Abuja, Nigeria.* https://web.archive.org/web/20100304023531/http://www.africafertilizersummit.org/ (accessed on 13 July 2021).

Aijuan, C., (2015). *China's Path in Developing Organic Agriculture: Opportunities and Implications for Small-scale Farmers and Rural Development.* PhD Thesis, University of Waterloo.

Arsenault, C., (2014). *Only 60 years of Farming Left if Soil Degradation Continues.* Scientific American. https://www.scientificamerican.com/article/only-60-years-of-farming-left-if-soil-degradation-continues/ (accessed on 13 July 2021).

Barański, M., Srednicka-Tober, D., Volakakis, N., Seal, C., Sanderson, R., Stewart, G. B., Benbrook, C., et al., (2014). Higher antioxidant and lower cadmium concentrations and lower incidence of pesticide residues in organically grown crops: A systematic literature review and meta-analyses. *The British Journal of Nutrition, 112*(5), 794–811. doi: https://doi.org/10.1017/S0007114514001366.

Bengtsston, J., Ahnström, J., & Weibull, A., (2005). The effects of organic agriculture on biodiversity and abundance: A meta-analysis. *Journal of Applied Ecology, 42*(2), 261–269. doi: https://doi.org/10.1111/j.1365-2664.2005.01005.x.

Bhatt, R., Hossain, A., & Hasanuzzaman, M., (2020b). Adaptation strategies to mitigate the evapotranspiration for sustainable crop production: A perspective of rice-wheat cropping system. *Agronomic Crops: Management Practices* (Vol. 2, pp. 559–582). Springer Nature, Switzerland. doi: https://doi.org/10.1007/978-981-32-9783-8.

Bhatt, R., Hossain, A., & Singh, P., (2020a). Scientific interventions to improve land and water productivity for climate smart agriculture in South-Asia. *Agronomic Crops: Management Practices* (Vol. 2, pp. 499–558). Springer-Nature, Switzerland. https://doi.org/10.1007/978-981-32-9783-8.

Blair, R., (2012). *Organic Production and Food Quality: A Down to Earth Analysis.* Wiley-Blackwell, Oxford, UK.

Blakemore, R. J., (2000). Ecology of earthworms under the 'Haughley Experiment' of organic and conventional management regimes. *Biological Agriculture and Horticulture, 18*(2), 141–159. doi: https://doi.org/10.1080/01448765.2000.9754876.

Bourn, D., & Prescott, J., (2002). A comparison of the nutritional value, sensory qualities, and food safety of organically and conventionally produced foods. *Critical Review in Food Science and Nutrition, 42*(1), 1–34. doi: https://doi.org/10.1080/10408690290825439.

Brandt, K., & Mølgaard, J. P., (2001). Organic agriculture: Does it enhance or reduce the nutritional value of plant foods? *The Journal of the Science of Food and Agriculture, 81*, 924–931.

Brown, K., (2008). *New Zealand's Ministry of Agriculture and Forestry Sector Performance Policy.* MAF Policy, Ministry of Agriculture and Forestry, Wellington, New Zealand. https://web.archive.org/web/20081015111550/http://www.maf.govt.nz/mafnet/rural-nz/sustainable-resource-use/organic-production/organic-farming-in-nz/org10005.htm (accessed on 13 July 2021).

Byerlee, D., & Alex, G., (2005). *Organic Farming: A Contribution to Sustainable Poverty Alleviation in Developing Countries.* German NGO Forum on Environment and Development, Bonn, Misereor, Naturland, EED, NABU and WWF.

City Life-Chiang Mai (Thailand). (2017). *Understanding the complexities of organic farming in Thailand: Does organic farming exist in Thailand.* https://www.chiangmaicitylife.com/clg/business/agriculture/understanding-the-complexities-of-organic-farming-in-thailand/ (accessed on 13 July 2021).

Coleman, E., (1995). *The New Organic Grower: A Master's Manual of Tools and Techniques for the Home and Market Gardener* (2nd edn., pp. 65, 108).

Colom-Gorgues, A., (2009). The challenges of organic production and marketing in Europe and Spain: Innovative marketing for the future with quality and safe food products. *Journal of International Food and Agribusiness Marketing, 21*, 166–190. doi: https://doi.org/10.1080/08974430802589675.

Colorado State University, (2020). *Some Pesticides Permitted in Organic Gardening.* Colorado State University Cooperative Extension, Fort Collins, Colorado, USA. https://denver.extension.colostate.edu/programs/horticulture/ (accessed on 13 July 2021).

Conford, P., (2001). *The Origins of the Organic Movement.* Glasgow, Great Britain: Floris Books.

Crowder, D. W., & Reganold, J. P., (2015). Financial competitiveness of organic agriculture on a global scale. *Proc. Natl. Acad Sci. USA, 112*(24), 7611–7616.

Dalgaard, T., Halberg, N., & Porter, J. R., (2001). A model for fossil energy use in Danish agriculture used to compare organic and conventional farming. *Agriculture, Ecosystems, Environment, 87*(1), 51–65. doi: https://doi.org/10.1016/S0167-8809(00)00297-8.

Dobbs, T. L., & Smolik, J. D., (1996). Productivity and profitability of conventional and alternative farming systems: A long-term on-farm paired comparison. *Journal of Sustainable Agriculture, 91*(1), 63–79.

Drinkwater, L. E., Wagoner, P., & Sarrantonio, M., (1998). Legume-based cropping systems have reduced carbon and nitrogen losses. *Nature, 396*, 262–265.

FAO, (2007). *Organic Agriculture Can Contribute to Fighting Hunger.* Food and Agricultural Organization of United Nations, Rome, Italy. http://www.fao.org/newsroom/en/news/2007/1000726/index.html (accessed on 13 July 2021).

FAO, (2013). *Overcoming Smallholder Challenges with Biotechnology.* Food and Agricultural Organization of United Nations, Rome, Italy. http://www.fao.org/news/story/en/item/202820/icode/ (accessed on 13 July 2021).

FAO, (2020). *Organic Agriculture.* FAO-Working Group on Organic Agriculture. FAO, Rome, Italy. http://www.fao.org/organicag/oa-faq/oa-faq5/en/ (accessed on 13 July 2021).

FiBL, (2020). *Global Organic Area Continues to Grow-Over 71.5 Million Hectares of Farmland are Organic.* FiBL and IFOAM. https://www.fibl.org/en/info-center/news/global-organic-area-continues-to-grow-over-71-5-million-hectares-of-farmland-are-organic.html (accessed on 13 July 2021).

Fließbach, A., Oberholzer, H., Gunst, L., & Mäder, P., (2006). Soil organic matter and biological soil quality indicators after 21 years of organic and conventional farming. *Agriculture, Ecosystems and Environment, 118*(1–4), 273–284. https://doi.org/10.1016/j.agee. 2006.05.022.

Gabriel, D., Roschewitz, I., Tscharntke, T., & Thies, C., (2006). Beta diversity at different spatial scales: Plant communities in organic and conventional agriculture. *Ecological Applications, 16*(5), 2011–21.

Gold, M. V., (2020). *What is Organic Production?* National Agricultural Library, USDA. https://www.nal.usda.gov/afsic/organic-productionorganic-food-information-access-tools (accessed on 13 July 2021).

Goldberg, B., (2007). *The Hypocrisy of Organic Farmers*. AgBioWorld. http://www.agbioworld.org/biotech-info/articles/biotech-art/hypocrisy.html (accessed on 13 July 2021).

Gordon, I., (2017). *Reproductive Technologies in Farm Animals* (p. 10). CABI.

Green, M., & Maynard, R., (2006). The employment benefits of organic farming. *Aspects of Applied Biology, 79*, 51–55.

Gurung, B., (2011). *Sikkim Races on Organic Route*. Telegraph India. https://www.telegraphindia.com/states/west-bengal/sikkim-races-on-organic-route/cid/478489 (accessed on 13 July 2021).

Halberg, N., (2006). *Global Development of Organic Agriculture: Challenges and Prospects* (p. 297). CAB International, Wallingford.

Halberg, N., Alrøe, H. F., & Kristensen, E. S., (2005). Synthesis: Perspectives for organic agriculture in a global context. In: Halberg, N., Alrøe, H. F., Knudsen, M. T., & Kristensen, E. S., (eds.), *Global Development of Organic Agriculture: Challenges and Promises* (pp. 344–368). CAB International, Wallingford.

Hansen, B., Alrøe, H. J., & Kristensen, E. S., (2001). Approaches to assess the environmental impact of organic farming with particular regard to Denmark. *Agriculture, Ecosystems and Environment, 83*(1, 2), 11–26. doi: https://doi.org/10.1016/S0167-8809(00)00257-7.

Heckman, J., (2007). *A History of Organic Farming: Transitions from Sir Albert Howard's War in the Soil to the USDA National Organic Program*. The Weston A. Price Foundation, Washington. https://www.westonaprice.org/health-topics/farm-ranch/a-history-of-organic-farming-transitions-from-sir-albert-howards-war-in-the-soil-to-the-usda-national-organic-program/ (accessed on 13 July 2021).

Helga, W., Julia, L., & Robert, H., (2013). *The World of Organic Agriculture: Statistics and Emerging Trends 2013*. Research Institute of Organic Agriculture (FiBL) and the International Federation of Organic Agriculture Movements (IFOAM).

Henckel, L., (2015). Organic fields sustain weed meta-community dynamics in farmland landscapes. *Proceedings of the Royal Society B, 282*(1808), 20150002. doi: https://doi.org/10.1098/rspb.2015.0002.

Hepperly, P., Jeff, M., & Dave, W., (2008). *Developments in Organic No-till Agriculture* (pp. 16–19). Acres USA: The Voice of Eco-Agriculture.

Hindu Business Line, (2010). *Sikkim 'Livelihood Schools' to Promote Organic Farming*. https://www.thehindubusinessline.com/todays-paper/tp-economy/Sikkim-lsquolivelihood-schools-to-promote-organic-farming/article20031763.ece (accessed on 13 July 2021).

Hine, R., & Pretty, J., (2006). *Capacity Building Study 3: Organic Agriculture and Food Security in East Africa*. United Nations Conference on Trade and Development (UNCTAD) and United Nations Environment Programme (UNEP). http://citeseerx.ist.psu.edu/viewdoc/download?doi=10.1.1.535.1631&rep=rep1&type=pdf (accessed on 13 July 2021).

Hole, D. G., Perkins, A. J., Wilson, J. D., Alexander, I. H., Grice, P. V., & Evans, A. D., (2005). Does organic farming benefit biodiversity? *Biological Conservation, 122*(1), 113–130. doi: https://doi.org/10.1016/j.biocon.2004.07.018.

Horne, P. A., (2008). *Integrated Pest Management for Crops and Pastures* (p. 2). CSIRO Publishing.

Howden, D., (2008). *Organic Farming Could Feed Africa*. The Independent. http://www.panna.org/sites/default/files/imported/files/IndependentOrganicFarmsCouldFeedAfrica20081022.pdf (accessed on 13 July 2021).

IFAD (International Fund for Agricultural Development), (2007). *Organic Agriculture and Poverty Reduction in Asia (2005)*. China and India Focus.

IFOAM-Organics International, (2008). *Concept of Organic Agriculture*. https://www.ifoam.bio/why-organic/organic-landmarks/definition-organic (accessed on 13 July 2021).

Indian Express, (2014). *CM: Will Get Total Organic Farming State Tag by 2016*. https://www.newindianexpress.com/cities/kochi/2014/nov/07/CM-Will-Get-Total-Organic-Farming-State-Tag-by-2016-679699.html (accessed on 13 July 2021).

Johnston, A. E., (1986). Soil organic-matter, effects on soils and crops. *Soil Use Management, 2*(3), 97–105. doi: https://doi.org/10.1111/j.1475-2743.1986.tb00690.x.

Kahl, J., Busscher, N., & Ploeger, A., (2010). Questions on the validation of holistic methods of testing organic food quality. *Biological Agriculture and Horticulture, 27*, 81–94.

King, F. H., (2004). *Farmers of Forty Centuries: Organic Farming in China, Korea, and Japan*. Courier Corporation.

Kirchmann, H., & Bergstrom, L., (2008). *Organic Crop Production-Ambitions and Limitations* (pp. 1–2). Springer Science & Business Media.

Kirchmann, H., Bergström, L., Kätterer, T., Mattsson, L., & Gesslein, S., (2007). Comparison of long-term organic and conventional crop-livestock systems on a previously nutrient-depleted soil in Sweden. *Agronomy Journal, 99*(4), 960–972. doi: https://doi.org/10.2134/agronj2006.0061.

Kristiansen, P., Taji, A., & Reganold, J., (2006). *Organic Agriculture: Opportunities and Challenges* (pp. 421–441). Organic Agriculture: A Global Perspective. CSIRO Publishing, Australia.

Leonard, A., (2007). *Save the Rain Forest-Boycott Organic?* How The World Works. https://www.salon.com/2006/12/11/borlaug/ (accessed on 13 July 2021).

Lockie, S., (2006). *Going Organic: Mobilizing Networks for Environmentally Responsible Food Production* (p. 239). Wallingford: CABI.

Lotter, D., (2003). Organic agriculture. *Journal of Sustainable Agriculture, 21*(4), 59. doi: https://doi.org/10.1300/J064v21n04.

Markandya, A., Setboonsarng, S., Qiao, Y. H., Songkranok, A. R., & Stefan, S., (2015). The costs of achieving the millennium development goals through adopting organic agriculture. In: Setboonsarng, S., & Markandya, A., (eds.), *Organic Agriculture and Post-2015 Development Goals: Building on the Comparative Advantage of Poor Farmers* (pp. 49–78). Manila: ADB.

Marshall, G., (1991). Organic farming: Should government give it more technical support? *Review of Marketing and Agricultural Economics, 59*(3), 283–296.

Martin, A., & Kim, S., (2008). *Sticker Shock in the Organic Aisles*. New York Times. https://www.nytimes.com/2008/04/18/business/18organic.html?pagewanted=all&_r=0 (accessed on 13 July 2021).

Martin, H., (2009). *Introduction to Organic Farming*. Ontario Ministry of Agriculture, Food and Rural Affairs. http://www.omafra.gov.on.ca/english/crops/facts/09-077.htm#define (accessed on 13 July 2021).

Martin, K. A., (2016). *State to Switch Fully to Organic Farming by 2016*. https://www.thehindu.com/todays-paper/tp-national/tp-kerala/state-to-switch-fully-to-organic-farming-by-2016-mohanan/article6517859.ece (accessed on 13 July 2021).

McBride, W. D., & Greene, C. R., (2013). Organic data and research from the ARMS survey: Findings on competitiveness of the organic soybean sector. *Crop Management, 12*(1), 1–11.

McConnell, D. J., Dharmapala, K. A. E., & Attanayake, S. R., (2017). *The Forest Farms of Kandy: And other Gardens of Complete Design*. Routledge.

McEvoy, M., (2018). *Organic 101: Allowed and Prohibited Substances*. USDA Blog. https://www.usda.gov/media/blog/2012/01/25/organic-101-allowed-and-prohibited-substances (accessed on 13 July 2021).

McGill University, (2012). *Can Organic Food Feed the World? New Study Sheds Light on Debate Over Organic vs. Conventional Agriculture*. Science Daily. www.sciencedaily.com/releases/2012/04/120425140114.htm (accessed on 13 July 2021).

Meena, R. P., Meena, H. P., & Meena, R. S., (2013). *Organic Farming: Concept and Components* (Vol. 1, No. 4, pp. 1–14). Popular Kheti.

Meleca, (2008). *The Organic Answer to Climate Change*. https://web.archive.org/web/20081211111856/http://www.organicguide.com/community/education/the-organic-answer-to-climate-change/ (accessed on 13 July 2021).

Musick, M. (2021). *WA Tilth Association History*. Washington Tilth. Retrieved from: http://www.tilthalliance.org/about/abriefhistoryoftilth (accessed on 13 July 2021).

Naresh, R. K., Vivek, M. K., Kumar, S., Chowdhary, U., Kumar, Y., Mahajan, N. C., Malik, M., et al., (2018). Zero budget natural farming viable for small farmers to empower food and nutritional security and improve soil health: A review. *Journal of Pharmacognosy and Phytochemistry, 7*(2), 1104–1118.

Nayler, J., (2014). *Second Thoughts About Organic Agriculture*. Soil and health library. https://web.archive.org/web/20140801061226/http://www.soilandhealth.org/01aglibrary/Second.Thoughts.pdf (accessed on 13 July 2021).

Niggli, U., Slabe, A., Schmid, O., Halberg, N., & Schlüter, M., (2008). *Vision for an Organic Food and Farming Research Agenda 2025: Organic Knowledge for the Future* (p. 48). IFOAM-EU and FiBL. https://orgprints.org/13439/1/niggli-etal-2008-technology-platform-organics.pdf (accessed on 13 July 2021).

NPR, (2011). *Public Health: Organic Pesticides: Not an Oxymoron*. NPR Health Newsletter. https://www.npr.org/sections/health-shots/2011/06/18/137249264/organic-pesticides-not-an-oxymoron (accessed on 13 July 2021).

Paull, J., & Hennig, B., (2018). Maps of organic agriculture in Australia. *Journal of Organics, 5*(1), 29–39.

Paull, J., (2006). The farm as organism: The foundational idea of organic agriculture elementals. *Journal of Bio-Dynamics Tasmania, 83*, 14–18.

Paull, J., (2007). Rachel carson: A voice for organics-the first hundred years. *Journal of Bio-Dynamics Tasmania*, (86), 37–41.

Paull, J., (2010). From France to the world: The international federation of organic agriculture movements (IFOAM). *Journal of Social Research and Policy, 1*(2), 93–102.

Paull, J., (2011a). The Betteshanger summer school: Missing link between biodynamic agriculture and organic farming. *Journal of Organic Systems, 6*(2), 13–26.

Paull, J., (2011b). Biodynamic agriculture: The journey from Koberwitz to the World, 1924–1938. *Journal of Organic Systems, 6*(1), 27–41.

Paull, J., (2013a). Breslau (Wrocław): In the footsteps of Rudolf Steiner. *Journal of Bio-Dynamics Tasmania, 110*, 10–15.

Paull, J., (2013b). Koberwitz (Kobierzyce): In the footsteps of Rudolf Steiner. *Journal of Bio-Dynamics Tasmania, 109*, 7–11.

Paull, J., (2014). Lord Northbourne, the man who invented organic farming, a biography. *Journal of Organic Systems, 9*(1), 31–53.

Paull, J., (2017). Four new strategies to grow the organic agriculture sector. *AGROFOR-International Journal, 2*, 61–70.

Paull, J., (2019). Organic Agriculture in Australia: Attaining the Global Majority (51%). *Journal of Environment Protection and Sustainable Development, 5*(2), 70–74.

Paulson, T., (2008). *The Lowdown on Topsoil: It's Disappearing: Disappearing Dirt Rivals Global Warming as an Environmental Threat*. https://www.seattlepi.com/national/article/The-lowdown-on-topsoil-It-s-disappearing-1262214.php?source=mypi (accessed on 13 July 2021).

Perrings, C., Louise, J., Kamaljit, B. S., Lijbert, B., Brush, S., & Tom, G., (2006). Biodiversity in agricultural landscapes: Saving natural capital without losing interest. *Conservation Biology, 20*(2), 263–264.

Pimentel, D., Berardi, G., & Fast, S., (1983). Energy efficiency of farming systems: Organic and conventional agriculture. *Agriculture, Ecosystems and Environment, 9*(4), 359–372.

Pimentel, D., Hepperly, P., Hanson, J., Douds, D., & Seidel, R., (2005). Environmental, energetic, and economic comparisons of organic and conventional farming systems. *BioScience, 55*(7), 573–582.

Pretty, J. N., Morison, J. I. L., & Hine, R. E., (2003). Reducing food poverty by increasing agricultural sustainability in developing countries. *Agriculture, Ecosystems and Environment, 95*, 217–234.

Pretty, J., Brett, C., Gee, D., Hine, R. E., Mason, C. F., Morison, J. I. L., Raven, H., Rayment, M. D., Van, D. B. G., et al., (2001). An assessment of the total external costs of UK agriculture. *Agricultural Systems, 65*(2), 113–136. doi:10.1016/S0308-521X(00)00031-7.

Pugliese, P., (2001). Organic farming and sustainable rural development: A multifaceted and promising convergence. *Sociologia Ruralis, 41*(1), 112–130.

Rigby, D., & Cáceres, D., (2001). Organic farming and the sustainability of agricultural systems. *Agricultural Systems, 68*(1), 21–40.

Rodale Institute, (2014). *Regenerative Organic Agriculture and Climate Change: A Down-to-Earth Solution to Global Warming*. https://rodaleinstitute.org/wp-content/uploads/Regenerative-Organic-Agriculture-White-Paper.pdf (accessed on 13 July 2021).

Rodale Institute, (2020). *The Farming Systems Trial Rodale 30 Year Report*. https://rodaleinstitute.org/wp-content/uploads/fst-30-year-report.pdf (accessed on 13 July 2021).

Scott, S., Si, Z., Schumilas, T., & Chen, A., (2018). *Organic Food and Farming in China: Top-down and Bottom-up Ecological Initiatives*. Routledge.

Setboonsarng, S., (2015). Can ethical trade certification contribute to the attainment of the millennium development goals? A review of organic and fair-trade certification. In: Setboonsarng, S., & Markandya, A., (eds.), *Organic Agriculture and Post-2015 Development Goals: Building on the Comparative Advantage of Poor Farmers* (pp. 79–103). https://www.adb.org/sites/default/files/publication/161042/organic-agriculture-post-2015-development-goals.pdf (accessed on 13 July 2021).

Singh, R. B., (2000). Environmental consequences of agricultural development: Case study from the green revolution state of Haryana, India. *Agriculture, Ecosystem and Environment, 82*, 97–103.

Smith, H. G., Dänhardt, J., Lindström, Å., et al., (2010). Consequences of organic farming and landscape heterogeneity for species richness and abundance of farmland birds. *Oecologia, 162*, 1071–1079. https://doi.org/10.1007/s00442-010-1588-2.

Smith-Spangler, C., Brandeau, M. L., Hunter, G. E., Bavinger, J. C., Pearson, M., Eschbach, P. J., Sundaram, V., et al., (2012). Are organic foods safer or healthier than conventional alternatives?: A systematic review. *Annals of Internal Medicine, 157*(5), 348–366. doi:10.7326/0003-4819-157-5-201209040-00007.

Steffanie, S., Zhenzhong, S., Theresa, S., & Aijuan, C., (2018). *Organic Food and Farming in China: Top-down and Bottom-up Ecological Initiatives* (1st edn., p. 236). Routledge.

Stinner, D. H., (2007). The science of organic farming. In: Lockeretz, W., (ed.), *Organic Farming: An International History* (pp. 40–72). Oxford shire, UK: CAB International.

Stolze, M., & Lampkin, N., (2009). Policy for organic farming: Rationale and concepts. *Food Policy, 34*(3), 237–244.

Stolze, M., Piorr, A., Häring, A. M., & Dabbert, S., (2000). Environmental impacts of organic farming in Europe. *Organic Farming in Europe: Economics and Policy* (Vol. 6). Universität Hohenheim, Stuttgart-Hohenheim.

Strochlic, R., & Sierra, L., (2007). *Conventional, Mixed, and "Deregistered" Organic Farmers: Entry Barriers and Reasons for Exiting Organic Production in California* (pp. 1–43). California Institute for Rural Studies. http://ccwiki.pbworks.com/f/CAStudy-Barriers-Organic-CIRS-2007.pdf (accessed on 13 July 2021).

Tegtmeier, E. M., & Duffy, M., (2005). External costs of agricultural production in the United States. *The Earthscan Reader in Sustainable Agriculture.* https://www.organicvalley.coop/why-organic-valley/ (accessed on 13 July 2021).

The Hindu, (2010). *Sikkim to Become a Completely Organic State by 2015.* https://www.thehindu.com/sci-tech/agriculture/Sikkim-to-become-a-completely-organic-state-by-2015/article15908694.ece (accessed on 13 July 2021).

Tuomisto, H. L., Hodge, I. D., Riordan, P., & Macdonald, D. W., (2012). Does organic farming reduce environmental impacts?: A meta-analysis of European research. *Journal of Environmental Management, 112*, 309–320.

UNEP, (2011). *Towards a Green Economy: Pathways to Sustainable Development and Poverty Eradication.* www.unep.org/greeneconomy (accessed on 13 July 2021).

UNEP-UNCTAD, (2008). *Organic Agriculture and Food Security in Africa.* United Nations. https://unctad.org/en/docs/ditcted200715_en.pdf (accessed on 13 July 2021).

University of Oxford, (2004). *Organic Farms Not Necessarily Better for Environment.* http://people.forestry.oregonstate.edu/steve-strauss/sites/people.forestry.oregonstate.edu.steve-strauss/files/Organic%20farms%20not%20necessarily%20better%20for%20environment%20-%20University%20of%20Oxford_2012.pdf (accessed on 13 July 2021).

USDA, (2000). *National Organic Program.* Federal Register/Vol. 65, No. 246/Rules and regulations. https://www.govinfo.gov/content/pkg/FR-2000-12-21/pdf/00-32257.pdf (accessed on 13 July 2021).

USDA, (2016). *USDA List of Allowed and Prohibited Substances in Organic Agriculture.* USDA. https://www.ecfr.gov/cgi-bin/text-idx?c=ecfr&SID=9874504b6f1025eb0e6b67cadf9d3b40&rgn=div6&view=text&node=7:3.1.1.9.32.7&idno=7 (accessed on 13 July 2021).

USDA-ARS, (2007). *Organic Farming Beats No-Till?* Agricultural Research Magazine, USDA- Agricultural Research Service. https://www.ars.usda.gov/news-events/news/research-news/2007/organic-farming-beats-no-till/ (accessed on 13 July 2021).

USDSA, (2018). *Guidance Natural Resources and Biodiversity Conservation* (pp. 1–9). Agricultural Marketing Service. https://www.ams.usda.gov/sites/default/files/media/NOP%205020%20Biodiversity%20Guidance%20Rev01%20%28Final%29.pdf (accessed on 13 July 2021).

USEPA (United States Environmental Protection Agency), (2020). Organic Farming. https://www.epa.gov/agriculture/organic-farming (accessed on 13 July 2021).

Van, E. T., (2000). Species diversity as a task for organic agriculture in Europe. *Agriculture, Ecosystems and Environment, 77*(1, 2), 101–109. doi:10.1016/S0167-8809(99)00096-1.

Vogt, G., (2007). Chapter 1: The origins of organic farming. In: Lockeretz, W., (ed.), *Organic Farming: An International History* (pp. 9–30). CABI Publishing, UK.

Von, B. J., (2005). *The World Food Situation: An Overview*. Presentation to the CGIAR annual meeting, Marrakech (Morocco).

Willer, H., & Yussefi, M., (2007). The current status of organic farming in the world-focus on developing countries. In: *International Conference on Organic Agriculture and Food Security* (pp. 12, 13). Rome, Italy.

World Bank, (2008). *Global Monitoring Report-2008: MDGs and the Environment: Agenda for Inclusive and Sustainable Development.* Washington, DC: World Bank.

World Watch Institute, (2000). *State of the World 2000*. Norton Publishing, New York.

World Watch Institute, (2006). *Can Organic Farming Feed Us All*? (Vol. 19, No. 3) World Watch Magazine. https://web.archive.org/web/20140209135230/http://www.worldwatch.org/node/4060 (accessed on 13 July 2021).

WTO (World Trade Organization), (2020). *Agreement on the Application of Sanitary and Phytosanitary Measures*. World Trade Organization. https://www.wto.org/english/docs_e/legal_e/15sps_01_e.htm (accessed on 13 July 2021).

Yadav, A. K., (2017). *Organic Agriculture-Concept, Scenario, Principals and Practices*. National Project on Organic farming, Department of Agriculture and Cooperation, Government of India.

CHAPTER 15

ORGANIC FOODS IN SUB-SAHARAN AFRICA: HEALTH IMPACT, FARMERS' EXPERIENCES, AND INTERNATIONAL TRADE

OSEBHAHIEMEN ODION IKHIMIUKOR,[1]
OLUWADAMILOLA MATHEW MAKINDE,[2]
CHIBUZOR-ONYEMA IHUOMA EBERE,[2] TOBA SAMUEL ANJORIN,[3]
and FAPOHUNDA STEPHEN OYEDELE[2]

[1]*Environmental Microbiology and Biotechnology Laboratory, Department of Microbiology, University of Ibadan, Nigeria*

[2]*Department of Microbiology, School of Science and Technology, Babcock University, Ilishan-Remo, Nigeria*

[3]*Department of Crop Protection, Faculty of Agriculture, University of Abuja, Nigeria*

ABSTRACT

The organic food industry is a rapidly growing sector in Africa, with strong links to economic and socio-cultural development. Organic food production involves an ecosystem-friendly approach that harnesses biodiversity, the biogeochemical cycles as well as the soil's biological activity in the production of fresh food void of chemical agricultural inputs. This chapter aims at discussing organic foods vis-à-vis its health and economic impact on consumers and producers within Africa with focus on creating awareness on practices and regulations guiding it. Agricultural practices in Africa relies heavily on chemical inputs to boost food production. Toxic concentrations of these chemicals cause an imbalance in the soil's natural ecosystem and may eventually affect food quality to be unsuitable for international trade. Despite efforts that gave birth to the strategic plan (2015–2025) for the development of organic agriculture in Africa, the continent scores low (26%) of the total number of producers of organic food worldwide. It is therefore the intention of the authors to bring to the consciousness of the continent these statistics as well as provide ways by which these statistics can be improved to guarantee safety, food security and foreign exchange for Sub-Saharan Africa (SSA).

15.1 INTRODUCTION

The Sub-Saharan region of Africa (SSA) accommodates over 950 million people. The population in SSA is projected to increase by 22% by 2050 (FAOSTAT, 2015). Adequate nourishment of the ever-increasing population in SSA is a challenge. Despite a reduction in the rate of undernourishment from 33% between 1990 and 1992 to 23% between 2014 and 2016, this region is still rated as highly undernourished among developing countries (FAOSTAT, 2015). Food insecurity in this region has been linked with peculiar challenges, one of which is low agricultural productivity. Close to 70% of the African population are involved with one form of agriculture or the other, as highlighted by AGRA (2017), and reports show that the major farming systems in this region are crop production and livestock management (AGRA, 2016). Current agricultural practices in SSA fail in achieving food sufficiency and security. From an agricultural perspective, there is a drastic need for changes in the food system. There is the need to produce more food at affordable prices, ensuring better livelihoods to farmers and the people while reducing the environmental cost of agriculture (Agama, 2015).

Organic farming is described by the Directorate-General for Agriculture and Rural Development of the European Commission (2009) as a variant of agriculture that utilizes eco-friendly, traditional methods in the cultivation, control of pest, and animal husbandry without the use of chemical farm inputs. It is simply characterized by the exclusion of chemicals (i.e., fertilizers or pesticides) and total reliance on natural processes or products to ensure food production (IFOAM, 2010). This system of farming aimed at producing food with minimal harm to ecosystems, animals, or humans, is the most prominent of the alternatives and is often proposed as a solution for more sustainable agriculture (Agama, 2015).

Organic farming draws its origins from concepts established over a century ago, borne out of a desire for efficient nutrient utilization on the farm (Lockeretz, 2007). In the 1940s, the term organic farming was coined to depict the farm as a biological system and not on the basis of the artificial inputs employed in food production today (Kuepper, 2010). This biological system comprises the minerals within the soil, its organic matter, soil microflora as well as the soil's flora and fauna and how they interact to create a stable system. In this biological system, resources internally generated as well as the biogeochemical cycle are of paramount importance (Letourneau and Bothwell, 2008). Biogeochemical cycling of resources in the soil's ecosystem has been significantly influenced by climate change which consequently affects organic farming. Evolving issues over the years, especially those of environmental concern such as climate change, have affected the way organic farming is perceived by placing a huge premium on enhancing the cultivation of crops using biological and eco-friendly inputs (USDA/AMS, 2000).

Sub-Saharan Africa (SSA) is faced with numerous challenges of which poor policies on agriculture leads to malnutrition and loss of productivity (Makinde et al., 2017). These poor policy decisions have also taken its toll on organic farming in recent years. For instance, the amount of arable land devoted to the practice of organic farming, despite huge growth in research and awareness created concerning its benefits, can be greatly expanded (De Ponti et al., 2012; Reganold and Wacher, 2016). In relation to the total amount of land the world over devoted to organic farming (43.7 million hectares) as at 2014, Africa had the lowest amount (3%, 1.3 million hectares) despite its vast amount of suitable land for agriculture which is reported to be 1031 million hectares (FAO, 2003). Uganda leads the continent with a meager 240,000 hectares of arable land devoted to organic farming (Willer and Lernoud, 2016).

In 2014, organic farming in SSA encouraged the cultivation of cash crops like coffee and olives (Willer and Lernoud, 2016). Each of these crops occupied 47% and 19% respectively of the total landmass available. The major cash crop cultivated on the continent was coffee covering about 241,500 hectares (Willer and Lernoud, 2016). Organic crops cultivated by organic farming on the continent include: oilseeds, cotton, cocoa, tea, fruits, medicinal and aromatic plants, olives, sesame, cereals, oils, nuts, spices, and vegetables. About 123,000 and 68,000 hectares of land were devoted to the cultivation of oilseeds and cotton, respectively (Willer and Lernoud, 2016). The statistics of organic farming in SSA as it concerns food security and economic enhancement is quite poor. It is, therefore, the intention of the authors to bring the Sub-Saharan populace to the consciousness of the poor state of regional agriculture, its implications and possible ways for improvement to guarantee safety, food security, and foreign exchange.

In SSA, where the focus is on improved nutrition, food security and sustainable agriculture, organic farming in the form of organoponic systems (which is adaptable to urban areas) can be explored to ensure high yield and healthy food products. Organoponics is eco-friendly and improves the microbial diversity of the soil which in turn influences soil fertility positively through the use of organic control systems (Prain, 2006). This form of organic farming has been successful in Venezuela and Cuba with the active support of their various governments (Orsini et al., 2013) and possesses huge potential if adopted to provide food for the people of SSA. It is encouraged that government at all levels on the continent show strong political will to support the furtherance of organic farming in ensuring food security within the continent.

15.2 AWARENESS PROFILE AND INTERNATIONAL TRADE

Organic food production in SSA is largely practiced by small scale farmers who are of the low-income class (Kisaka-lwayo and Obi, 2014). Despite the growing popularity and inherent potentials of organic food production in SSA, there are concerns over its ability to sustain and ensure food security in the region (Issaka et al., 2016).

Data on the production of organic foods in SSA is primarily collated by agencies of government as well as private sector organizations who are stakeholders in the business of organic farming. These efforts have improved the quality and availability of data in recent years within SSA. The data revealed that Uganda has the highest number of organic farms in Africa to the tune 187,893 farms, closely followed by Ethiopia and Tanzania with 100,000 and 85,366 farms, respectively (Willer and Lernoud, 2016).

Although Ivory Coast, Ghana, and Nigeria are the leading producers of cocoa beans in the world, only 36% of organic cocoa is produced in SSA, notably from DR Congo and the United Republic of Tanzania (FAOSTAT, 2015). Nigeria, a leading producer of pulses and vegetables, is reported to have little or no data on its organic cultivation of these crops (Rahmann et al., 2015).

The biggest African clients are in Europe, where it has its largest market. In 2009, Uganda was reported to earn as much as 36.87 million US Dollars from export as well as programs sponsored by international agencies. Organic foods generated reported a profit of €10 million through exports from Tanzania to the EU and the America (Kledal and Kwai, 2010).

15.3 NUTRITIONAL AND HEALTH IMPACTS

15.3.1 HIGHER YIELD

Organic food production places a huge premium on the quality and yield of food crops. According to De Ponti et al. (2012), crops whose growth is properly managed under the organic method produce higher yields than those produced by chemical-driven agriculture. With the use of chemical agricultural inputs, yield was 6–11% less than organic production. The soil ecosystem also contributes to getting the best in organic farming (Lockeretz et al., 1981; Lotter et al., 2003). The ability of organic soil to retain water over a long period of time confers on it this advantage (Niggli, 2014).

15.3.2 REDUCED PESTICIDE RESIDUES

Organic farming has been confirmed through scientific reports to have little or no pesticide residues in the crops (Baker et al., 2002; Pussemier et al., 2006; Smith-Spangler et al., 2012; Barański et al., 2014). A scary dimension was introduced when Curl et al. (2003); and Lu et al. (2006), reported that children who consumed conventionally produced foods had significantly high concentrations of pesticide residues in their urine.

15.3.3 ENHANCED NUTRIENT QUALITY

In terms of nutrition, existing literature has proven organic foods to surpass conventionally produced foods (SAOS, 2000; Brandt and Mølgaard, 2001; Worthington, 2001). Despite little or no premium placed on nutrients obtained from organic foods, they are reported to possess higher concentrations of vitamin C, antioxidants, and omega-3 fats compared to conventionally produced foods (Brandt et al., 2011; Smith-Spangler et al., 2012).

With increasing production, marketing, and consumption of conventional foods, the occurrence of pesticides, fertilizers, and other contaminants in foods generally has also increased, leading to the preference of organic foods by individuals in recent times (Hurtado-Barroso et al., 2017). A number of factors have been linked to the preference in organic foods, for example, their environmental friendliness and health benefits (Dettmann and Dimitri, 2009; Padilla et al., 2013). Similarly, some studies have linked the consumption of organic foods to health and lifestyle indicators, more physical activity, and lower body-mass index (BMI) than those who seldom or do not consume organic foods (Dimitri, 2009; Eisinger-Watzl et al., 2015; Brantsaeter et al., 2017). Many studies have shown that there is no significant difference between organic foods and conventional foods in terms of its core nutrient levels of carbohydrates, vitamins, and minerals (Williams, 2002; Matallana González et al., 2010), however, studies have also shown low levels of nitrate in the former than the latter (Worthington, 2001; Matallana González et al., 2010). This is a desirable quality as nitrates have been shown to be associated with a high risk of gastrointestinal cancer and increased risks of methemoglobinemia in infants (Forman and Silverstein, 2012). Higher vitamin C concentration has been observed in organic leafy vegetables like spinach, lettuce, and chard (21 out of 36 studies) (Williams, 2002). Some other studies have postulated health benefits due to antioxidants effects from organic foods as a result of their higher total phenol content as when compared to conventional foods (Asami et al., 2003).

Differences in nutritional content in food has been argued to be a result of many factors, including geographic location, soil quality, climate condition, maturity, and time of harvest and storage conditions as well. Considering all these factors, concluding on nutritional differences may not be definitive with a report by Dangour (2009) highlighting high levels of nitrogen from conventionally produced foods when compared to organic foods, which in turn revealed high levels of phosphorus.

The health benefits of organic foods could also be viewed from its compositional studies. The composition of dairy products, for example, is dependent upon various factors like genetic variability and cattle breed. However, milk produced both organically and conventionally has shown to contain the same protein, vitamin, lipid, and trace elements. In terms of microbial load, there is no evidence to prove that organic milk possesses high levels of contamination by pathogenic bacteria compared to conventionally produced milk, however, conventionally produced milk could contain antibiotic-resistant bacteria due to the use of antibiotics in animal husbandry (Forman and Silverstein, 2012). Hormones like estradiol and progesterone were lower in concentration in conventionally produced milk than in organic foods.

An animal-based study described by Huber (2010), showed higher growth rate on animals which fed on conventional foods as against those fed organic foods, however, immune responsiveness and recovery was observed to be higher among those on organic foods. A number of studies have been made to buttress the health impact of organic foods as compared to conventional foods and they include:

- Reduction of allergies among children (5–13) through the consumption of organic foods according to the cross-sectional PARSIFAL study by Alfvén et al. (2006) and confirmed by Stenius (2011).
- Lowered risk of eczema at 2 years among babies who consumed organic daily products according to the KOALA study by Alim et al. (1999); and Kummeling (2008).
- Lower risk of being overweight (28% in women and 27% decrease in men) or obese among organic food consumers according to the Nutrient-Sainte study (Mie and Wivstad, 2015).
- A study in the United Kingdom reported by Bradbury et al. (2014), described that there was no association between preference for organic foods and all forms of cancer; however, 9% risk increase was reported for organic food preference and breast cancer.

Health implications resulting from pesticide ingestion ranges from neurotoxicity to its effect in human infertility through endocrine system disruption (Laster and Rea, 1983). The latter has been subjected to studies that came up with evidences of hormone synthesis disruption in male mice by the herbicide glyphosate. It was also shown that exposure either by ingestion or occupationally to xeno-estrogen pesticides can affect the human male offspring's reproductive system (North and Golding, 2000). Savona (1998) also reported that a reduction in the consumption of conventionally produced foods by women presenting with signs of hormonal instability aids them in regaining hormonal balance.

It was established that exposure to a combination of pesticides causes an increased incidence of birth defects (Garry et al., 1996). The association between exposures to pesticides and the risk of cancer has also been considered. And it was shown that Canadian farmers with exposures to pesticides had higher incidences of cancer (Wiggle et al., 1990). Correlations between organic food consumption and sperm health was also established to show that men who were involved with organic farming and consumed organic foods had higher sperm concentrations than control groups (Abell, 1994; Chaudhary et al., 2015).

There is a paucity of data on the health impact of consumption of organic foods as compared to conventionally produced foods in SSA, however, emphasis elsewhere has been on the occurrence of pesticides and their levels of occurrence in foods generally and this has raised some concerns on food safety (Betarbet et al., 2000; Trewavas, 2001). The health impact of organic foods is not conclusive, however, the negative health consequences of pesticides either individually or in combination give organic foods an edge over conventional foods.

15.4 MICROBE-ORGANIC FOOD RELATIONS

Soil which is home to a wide variety of microbes, is notably the first requirement in organic farming. Soil health, is therefore of paramount essence as its unique features-one of which is compost, forms the foundation of organically grown food. Organic farming being an alternative to the use of synthetic pesticides and fertilizers for optimal crop yield and reduced environmental pollution, has shown huge success and growing awareness throughout the globe. One of the pivotal pillars in its success is the significant role played by the microbial population in the soil. Microorganisms play important roles in terms of pest and disease control as well as sustainable fertility management. Microbial biomass present in soil comprises of fungi, protozoa, algae, yeast, eubacteria, actinomycetes, and archaea. Their populations, however, vary due to factors such as soil texture and structure, pH, air/moisture content, temperature, and organic matter (Zarb et al., 2005). Their various contributions to organic farming are highlighted in subsections.

15.4.1 SOIL ENRICHMENT

Microorganisms enrich soil through various ways, including their possession of the enzyme urease, which helps in the degradation or hydrolysis of organic nitrogen (Zarb et al., 2005). In addition, microorganisms have the ability to melt and take up phosphate which is usually found in low concentrations in soils where conventional farming is practiced. Microbes known for phosphate uptake include vesicular-arbuscular mycorrhizas (VAM), Ecto-mycorrhizas, some fungi (*Fusarium*, *Aspergillus*, and *Penicillium*) and some bacteria (*Bacillus* and *Pseudomonas*) (Wang et al., 2019).

15.4.2 SOIL STABILITY

Microorganisms also play important roles in soil formation and stability by binding soil aggregates with hyphae and exudates secreted by them.

This is exemplified in the stimulated attachment of *Pasteuria penetrans* (a parasite of plant-parasitic nematode) by a microbial population (including mycorrhizal and nematophagous fungi) to the nematodes thereby reducing nematode infection (Ozturk et al., 2017).

15.4.3 ANTAGONISM AND ANTIBIOSIS

Soil organisms also play antagonistic roles against other organisms. For example, over 100 species of fungi are known to trap and prey on nematodes (Zarb, 2005), while many other fungi are hyperparasitic to other fungi (Toppo and Naik, 2015). Microorganisms such as streptomycetes, filamentous bacteria, which are known soil saprophytes are also associated with antibiotic and extracellular hydrolytic enzymes. The microbial secretions according to Samac et al. (2003) significantly contribute to a disease management system due to their ability to colonize plants and bring about a decrease in damage from a wide range of pathogens.

The plant growth-promoting rhizobacteria (PGPR) has the ability to suppress plant pathogens through antibiosis, hormone production, and competition with other pathogens for basic resources. Some of these bacteria are also known to promote root and shoot growth, bring about nodule formation and mycorrhizal establishment (Nkebiwe et al., 2016; Thonar et al., 2017; Mpanga et al., 2018).

15.4.4 DISEASE AND PEST CONTROL

Microorganisms control plant pests and disease through a number of ways which are discussed in subsections.

15.4.4.1 COMPETITION

This is achieved when high compost or organic matter is introduced into soil, thereby causing a rise in carbon source availability leading to increased indigenous microbial activity. With a raise in native microbial biomass, competition for basic resources with pathogens is subsequently increased, ultimately leading to outright stamping out of pathogens within an organically treated farm field (Mohammadi, 2013).

15.4.4.2 PROTECTION

Studies have highlighted the use of microbes to protect seedlings against a damping off disease caused by a *Pythium* species (Carr, 2016). Disinfecting and subsequently inoculating seeds can help avoid or prevent seed-borne diseases such as *Fusarium* (Nebert et al., 2016). Specific organisms such as *Trichoderma*, *Pseudomonas* are known to protect seeds and seedlings from various diseases (Trabelsi and Mhamdi, 2013; Mpanga et al., 2018).

In addition, some organisms have introduced certain structural changes in their roots, thereby creating physical barriers to pathogen entry. Some of these structural changes as reported by Eulenstein et al. (2016) occurs when associations between soil fungi and plant roots (as Arbuscular mycorrhiza) protects host plant from drought while enhancing nutrient uptake and plant growth even under water stress conditions (Tauschke et al., 2015; Lopez-Raez, 2016).

Microorganisms also have high tolerance for heavy metals (HMs), thereby conferring protection to the plants from the uptake of HMs by way of adsorption or various forms of fungal metabolism. Compost, as a major ingredient in organic farming, can serve as a means of inoculation of soil microbiota which in turn carry out unique activities for desired results. The use of compost extract can inhibit weed seed germination, thereby reducing weed competition (Mohammadi, 2013). Generally, organic farming system (OFS) maintains organic matter levels through practices such as mixed farming, rotations, recycling compost, farmyard and green manures (GMs). Proprietary cultures in the form of root dips, feed additives and sprays are also encouraged to help support the presence of soil microbiota, thereby leading to improved soil structure and stability.

15.5 AFRICAN EXPERIENCE: AN UPDATE

The producers of organic foods have various motivating factors for engaging in organic farming. Since the ultimate aim of food production is consumption, most farmers thrive in organic farming based on the preference of consumers for their products. Most consumers patronize organic farmers due to one or more of the following reasons (SSNC, 2014):

- Food is produced with less or no agrochemicals;
- Food production does not lead to residue contamination;
- Organic farming leaves a cleaner environment with a richer wildlife;

- It is a practice which is kind to animals;
- Organic foods boost consumer health.

Before products are labeled as organic, farmers need to meet up with specifications which are recognized either locally or internationally. Many organic farmers' associations exist in various countries, and these countries ascertain help in proper marketing and distribution of organic products. Countries which hold organic products in high regard have local markets in strategic places for the sale and distribution of organic produce.

While these efforts help in boosting organic farming, certain challenges exist among farmers who choose to engage in organic farming irrespective of the choice of products cultivated. Some of the challenges faced by organic farmers according to Saleki and Saleki (2012); and SSNC (2014) include:

- High price differences between organic and conventional produce. Organic foods are notably more expensive in most countries, including those within Sub-Saharan Africa. This gives a high percentage of the consumers a choice in terms of purchasing power.
- Lack of trust for some organic foods despite their labels.
- Lack of substantive awareness about organic farming, organic foods and organic food markets.

Despite the difficulties and challenges faced in organic farming, farmers are rather encouraged by the reason described above, thus ensuring its continuity by upcoming generations of farmers.

With extreme poverty ravaging a vast majority of the citizenry in SSA, organic food production provides a sure way of ameliorating this problem as well as providing a wide range of benefits to farmers. Some of the benefits enjoyed range from additional source of income from sales of excess food, extra cost of employing chemically produced fertilizers and chemicals is saved, fixed prices when international trade is involved as well as additional value for processed products of organic origin exported and consumed from and within the continent respectively (Twarog, 2006; UNCTAD, 2006, 2008). In terms of earnings from the export of products, organic farming has been reported to be a significant source of foreign exchange and net farm earnings compared to conventional farming. Other significant areas such as the income of organic farmers and their families are reported to witness significant increase to the tune of 87% with a resultant decrease in poverty amongst farmers and increased food supply within the continent (UNCTAD, 2008).

Organic foods are produced via a systemic production process that combines both traditional and scientific knowhow to ensure food safety and availability. The level of awareness on organic foods has increased in the developing countries of the world, which includes Africa, due to the huge benefits it provides ranging from economic, environmental, and socio-cultural (EU Commission, 2012).

According to IFOAM (2012), Africa has an estimated 570,000 producers of organic food utilizing 3% (1 mio ha) of the world's certified agricultural land. Uganda (0.23 mio ha), Tunisia (0.18 mio ha) and Ethiopia (0.14 mio ha) are also reported to be the countries with the most organic land on the continent.

In economic terms, prices of organic food products within developing countries including those from Africa; do not command premium prices compared to those exported which command relatively high prices especially those renowned for exceptional quality barring the challenges associated with logistics such as transportation, ensuring quality is not compromised as well as proper certification for the organic food products (Codex Alimentarius Commission and FAO/WHO, 2007; EU Commission, 2012).

15.5.1 ECONOMICS OF PRODUCTION AND AFFORDABILITY

According to FAO (2003), the cost of organic food products is observed to be high when compared to food products from chemical laden agriculture due to the fact that demand for organic food products is quite high due to its immense benefits compared to its supply. Also, due to the high labor cost involved per unit of organic food produced, and the diversity of products making economics of scale unachievable, production cost of organic food tends to increase with this having a huge impact on the price of the final product (Post and Schahczenski, 2012). In addition, higher costs are incurred during postharvest handling of small amounts of organic food products due to the fact that they must be adequately separated from conventional food products when it comes to transportation and processing, thus contributing to the final cost which is borne by the consumer. In addition, inefficient marketing and distribution systems for organic products due to its relatively low volume also contribute to the high prices incurred by the consumers.

Despite the increasing demand for organic agricultural products, the effective utilization of research findings from science and technology, which translates into an increase in production, should help reduce the cost of price of organic food products all through the production chain. This is realizable because the prices of organic food products are not only made up of the cost of production but also other factors unique to it and not considered when prices of chemical laden (conventional) food products are set. Factors such as curbing environmental pollution via commercial cultivation of organic foods ensure the effective maintenance of soil fertility (Peterson et al., 2012). However, the higher prices placed on these products helps to make up for the low income made from crop rotation periods employed by chemical laden agriculture. Also, animal products are required to meet high standards, making the welfare of these animals of high priority and thus impacting on the price. In addition, farmers are saved from the health impact of improper handling of pesticides, thus averting spending on their health (Dalton, 2008). Finally, employment within the farm is created, and producers gain maximum value for their products.

15.6 REGULATIONS

Organic food production is expected to be devoid of chemical inputs. However, farmers are allowed to apply only environmentally friendly chemical pesticides especially such as complexes of copper, sulfates, and chlorides, plant oils like neem, and the bacteria-*Bacillus thuringiensis* can also be used in biocontrol. Despite the safety concerns about some of the pesticides allowed in organic farming, the foods are expected to have no residues in them. Where they occur, incidence rates and levels are usually lower than when found in non-organic or conventional foods (Woese et al., 1997; Heaton, 2001; Williams, 2002). Washing and peeling fruit and vegetables are approved for precautionary purposes prior to eating (Buffin, 1997). For regulatory purposes, maximum residue level (MRLs) and acceptable daily intakes (ADIs) are set as Legal Limits for individual pesticides in food produces. Regulation of organic foods on the continent is subject to international regulations provided by the Codex Alimentarius Commission and FAO/WHO (2007). They are responsible for setting up guidelines for the production, processing, labeling, and marketing of organic food for producers as well as ensure that consumers are safe and not defrauded (EU Commission, 2012).

Guidelines specifically made by Codex Alimentarius for organic foods carry no legal weight as international standards however they serve as recommendations. These recommendations are encouraged for use by developing countries, including organic food producing countries in SSA (Codex Alimentarius Commission and FAO/WHO, 2007). Producers are heavily regulated by the private sector. Guidelines for producers was propounded by

IFOAM called the International Basic Standards for Organic Production and Processing, with the sole aim of regulating how organic food products are produced, processed, and handled.

According to EU Commission (2012), about 73 countries of the world have already implemented guidelines and legislation on organic foods. Of this total number, SSA has 11 countries that have either drafted regulations or fully implemented such regulations. Egypt, Morocco, South Africa, Zambia, and Zimbabwe are in the process of drafting regulations on organic food production while Tunisia and the Eastern African countries of Burundi, Kenya, Rwanda, Tanzania, and Uganda have fully implemented regulations and legislations guiding the production, processing, and handling of organic foods.

15.7 CONCLUSION

With the proven advantages ascribed to organic farming in SSA, there is the need for better policies by governments that will encourage the allocation of more agricultural land for its practice. This is due to the fact that this region is far behind other sub-regions in land allocation. Awareness and enlightenment programs need to be enhanced in order to achieve desired results on food safety. Improved healthy living and economic advantage through local and international trade are directly associated with the consumption of organic food. As the sub-region is economically challenged, a conscious effort should be put in place to encourage the development of local and national organic food production and markets. It is factual that organic agriculture has a high potential to contribute immensely to food security, employment, and increased incomes generation to the Sub-Saharan populace.

KEYWORDS

- **conventional farming**
- **ecosystem**
- **International trade**
- **organic farming**
- **pesticides**
- **regulation**
- **Sub-Saharan Africa**

REFERENCES

Abell, A., Ernst, E., & Bonde, J. P., (1498). High sperm density among members of organic farmers' association. *The Lancet, 1994*, 343.

Agama, J., (2015). Africa: Latest development in organic agriculture in Africa. In: Willer, H., & Lernoud, J., (eds.), *The World of Organic Agriculture. Statistics and Emerging Trends.* FiBL-IFOAM Report-Organics International, Bonn.

AGRA, (2016). *Africa Agriculture Status Report 2016: Progress Towards Agricultural Transformation.* Nairobi, Kenya: Alliance for a Green Revolution for Africa (AGRA).

AGRA, (2017). *Africa Agriculture Status Report: The Business of Smallholder Agriculture in Sub-Saharan Africa* (No. 5). Nairobi, Kenya: Alliance for a Green Revolution in Africa (AGRA).

Alfvén, T., Braun-Fahrlander, C., Brunekreef, B., Von, M. E., Riedler, J., Scheynius, A., Van, H. M., et al., (2006). Allergic diseases and atopic sensitization in children related to farming and anthroposophic lifestyle-the PARSIFAL study. *Allergy, 61*(4), 414–421.

Alim, J. S., Swartz, J., Lilja, G., Scheynius, A., & Pershagen, G., (1999). Atopy in children of families with an anthroposophic lifestyle. *Lancet, 353*(9163), 1485–1488.

Asami, D. K., Hong, Y. J., Barrett, D. M., & Mitchell, A. E., (2003). Comparison of the total phenolic and ascorbic acid content of freeze-dried and air-dried marionberry, strawberry, and corn grown using conventional, organic, and sustainable agricultural practices. *Journal of Agricultural and Food Chemistry, 51*(5), 1237–1241.

Baker, B. P., Benbrook, C. M., Groth, E. I. I. I., & Benbrook, K. L., (2002). Pesticide residues in conventional, integrated pest management (IPM)-grown and organic foods: insights from three US data sets. *Food Additives and Contaminants, 19*, 427–446.

Barański, M., Srednicka-Tober, D., Volakakis, N., Seal, C., Sanderson, R., Stewart, G. B., Benbrook, C., et al., (2014). Higher antioxidant and lower cadmium concentrations and lower incidence of pesticide residues in organically grown crops: A systematic literature review and meta-analyses. *British Journal of Nutrition, 112*, 794–811.

Betarbet, R., Shera, T. B., McKenzie, G., Garcia-Osuna, M., Panov, A. V., & Greenamyre, J. T., (2000). Chronic systemic pesticide exposure reproduces features of Parkinson's disease. *Nature Neuroscience, 3*(12), 1301–1306.

Bradbury, K. E., Balkwill, A., Spencer, E. A., Roddam, A. W., Reeves, G. K., Green, J., Key, T. J., Beral, V., Pirie, K., & Million Women Study Collaborators, (2014). Organic food consumption and the incidence of cancer in a large prospective study of women in the United Kingdom. *British Journal of Cancer, 110*(9), 2321–2326.

Brandt, K., & Mølgaard, J. P., (2001). Organic agriculture: Does it enhance or reduce the nutritional value of plant foods? *Journal of the Science of Food and Agriculture, 81*, 924–931.

Brantsæter, A. L., Ydersbond, T. A., Hoppin, J. A., Haugen, M., & Meltzer, H. M., (2017). Organic food in the diet: Exposure and health implications. *Annual Review of Public Health, 38*(1), 295–313.

Carr, R., (2016). *Deploying Microbes as a Seed Treatment for Protection Against Soil-Borne Plant Pathogen.* Rodale Institute.

Chaudhary, S. A., Manani, Y., Pithadiya, A., Masram, P., Joshi, K., & Rathia, S., (2015). Modern life-style: A threat for the fertility. *International Journal of Herbal Medicine, 3*(5), 47–51.

Codex Alimentarius Commission and the FAO/WHO, (2007). Food standard program: Organically produced foods, Rome.

Curl, C. L., Fenske, R. A., & Elgethun, K., (2003). Organophosphorus pesticide exposure of urban and suburban preschool children with organic and conventional diets. *Environmental Health Perspectives, 111*, 377–382.

Dalton, T. J., Parsons, R., Kersbergen, R., Rogers, G., Kauppila, D., McCrory, L., Bragg, L. A., & Wang, Q., (2008). *A Comparative Analysis of Organic Dairy Farms in Maine and Vermont: Farm Financial Information From 2004–2006.* University of Maine, Maine Agricultural and Forest Experiment Station Bulletin 851, Orono, ME.

Dangour, A. D., Dodhia, S. K., Hayter, A., Allen, E., Lock, K., & Uauy, R., (2009). Nutritional quality of organic foods: A systematic review. *American Journal of Clinical Nutrition, 90*(3), 680–685.

De Ponti, T., Rijk, B., & Van, I. M. K., (2012). The crop yield gap between organic and conventional agriculture. *Agricultural Systems, 108*, 1–9.

Dettmann, R. L., & Dimitri, C., (2009). Who's buying organic vegetables? Demographic characteristics of US consumers. *Journal of Food Products Marketing, 16*, 79–91.

Directorate-General for Agriculture and Rural Development of the European Commission, (2009). *What is Organic Farming?* http://ec.europa.eu/agriculture/organic-farming/what-organic_en (accessed on 13 July 2021).

Eisinger-Watzl, M., Wittig, F., Heuer, T., & Hoffmann, I., (2015). Customers purchasing organic food-do they live healthier? Results of the German national nutrition survey II. *European Journal of Nutrition and Food Safety, 5*, 59–71.

Eulenstein, F., Tauschke, M., Behrendt, A., & Monk, S., (2017). The application of mycorrhizal fungi and organic fertilizers in horticultural potting soils to improve water use efficiency of crops. *Horticulture, 3*, 8.

FAOSTAT, (2015). *The FAO Homepage*. FAO, Rome. http://www.fao.org/faostat/en/#data/QC (accessed on 13 July 2021).

Forman, J., & Silverstein, J., (2012). Organic foods: Health and environmental advantages and disadvantages. *Pediatrics, 130*, 1406–1415.

Garry, V. F., Schreinemachers, D., Harkers, M. E., & Griffith, J., (1996). Pesticide applicators, biocides and birth defects in rural Minnesota. *Environmental Health Perspectives, 104*, 394–399.

Heaton, S., (2001). *Organic Farming, Food Quality and Human Health* (p. 87). Soil Association Report, Bristol, UK.

Huber, M., Van De, V. L. P., Parmentier, H., Savelkoul, H., Coulier, L., Wopereis, S., Verheij, E., et al., (2010). Effects of organically and conventionally produced feed on biomarkers of health in a chicken model. *British Journal of Nutrition, 103*(5), 663–676.

Hughner, R. S., McDonagh, P., Prothero, A., Shultz, C. J., & Stanton, J., (2007). Who are organic food consumers? A compilation and review of why people purchase organic food. *Journal of Consumer Behavior, 6*, 94–110.

Hurtado-Barroso, S., Tresserra-Rimbau, A., Vallverdu-Queralt, A., & Lamuela-Raventos, R. M., (2017). Organic food and the impact on human health. *Critical Reviews in Food Science and Nutrition, 30*, 1–11.

IFOAM (International Federation of Organic Agriculture Movements), (2010). *Organic Food and Farming: A System Approach to Meet the Sustainability Challenge* (pp. 1–24). IFOAM EU-Group, Belgium.

Issaka, Y. B., Antwi, M., & Tawia, G. A., (2016). Comparative analysis of productivity among organic and non-organic farms in the West Mamprusi District of Ghana. *Agriculture, 6*(2).

Kisaka-lwayo, M., & Obi, A., (2014). Analysis of Production and consumption of organic products in South Africa. In: Pilipavicius, V., (ed.), *Organic Agriculture Towards Sustainability* (pp. 25–50). Intech Open, London.

Kledal, P. R., & Kwai, N., (2010). Organic food and farming in Tanzania. In: Helga, W., Youssefi-Menzler, M., & Sorensen, N., (eds.), *The World of Organic Agriculture: Statistics and Emerging Trends 2008* (p. 111). FiBL and IFOAM.

Kuepper, G., (2010). *A Brief Overview of the History and Philosophy of Organic Agriculture*. Kerr Center for Sustainable Agriculture.

Kummeling, I., Thijs, C., Huber, M., Van De, V. L. P. L., Snijders, B. E. P., Penders, J., Stelma, F., et al., (2008). Consumption of organic foods and risk of atopic disease during the first 2 years of life in the Netherlands. *British Journal of Nutrition, 99*(3), 598–605.

Laster, J., & Rea, W., (1983). Chlorinated hydrocarbon pesticides in environmentally sensitive patients. *Clinical Ecology, 2*, 3–12.

Letourneau, K. L., & Bothwell, S. G., (2008). Comparison of organic and conventional farms: Challenging ecologists to make biodiversity functional. *Frontier in Ecology and Environment, 6*(8), 430–438.

Lockeretz, W., (2007). *Organic Farming: An International History* (pp. 1–282). Cromwell Press, Trowbridge, UK.

Lockeretz, W., Shearer, G., & Kohl, D. H., (1981). Organic farming in the corn belt. *Science, 211*, 540–547.

Lopez-Raez, J. A., (2016). How drought and salinity affect arbuscular mycorrhizal symbiosis and strigolactone biosynthesis. *Planta, 243*, 1375–1385.

Lotter, D., Seidel, R., & Liebhardt, W., (2003). The performance of organic and conventional cropping systems in an extreme climate year. *American Journal of Alternative Agriculture, 18*, 146–154.

Lu, C., Toepel, K., Irish, R., Fenske, R. A., Barr, D. B., & Bravo, R., (2006). Organic diets significantly lower children's dietary exposure to organophosphorus pesticides. *Environmental Health Perspectives, 114*, 260–263.

Makinde, O. M., Ikhimiukor, O. O., & Fapohunda, S. O., (2017). Bioprocess engineering and genetically modified foods: Tackling food insecurity in Africa. *Brazilian Journal of Biological Sciences, 4*(8), 233–246.

Matallana, G. M. C., Martínez-Tomé, M. J., & Torija, I. M. E., (2010). Nitrate and nitrite content in organically cultivated vegetables. *Food Additives and Contaminants: Part B, 3*(1), 19–29.

Mie, A., & Wivstad, M., (2015). *Organic Food-Food Quality and Potential Health Effects*. A review of current knowledge, and a discussion of uncertainties. SLU, EPOK-Center for Organic Food and Farming. http://www.slu.se/Documents/externwebben/centrumbildningarprojekt/epok/Organic_food_quality_and_health_webb.pdf (accessed on 13 July 2021).

Mohammadi, G. R., (2013). Alternative weed control methods: A review. *Weed and Pest Control-Conventional and New Challenges*.

Mpanga, I. K., Dapaah, H. K., Geistlinger, J., Ludewig, U., & Neumann, G., (2018). Soil type-dependent interactions of P-solubilizing microorganisms with organic and inorganic fertilizers mediate plant growth promotion in tomato. *Agronomy, 8*(10), 213.

Nebert, L., Bohannan, B., Ocamb, C., Still, A., Kleeger, S., Bramlet, J., & Heisler, C., (2016). *Managing Indigenous Seed-Inhabiting Microbes for Biological Control Against Fusarium Pathogen in Corn*. University of Oregon, USA.

Niggli, U., (2014). Sustainability of organic food production: Challenges and innovations. *Proceedings of the Nutrition Society, 74*(1), 83–88.

Nkebiwe, P. M., Weinmann, M., & Müller, T., (2016). Improving fertilizer-depot exploitation and maize growth by inoculation with plant growth-promoting bacteria: From lab to field. *Chem. Biol. Technol. Agric., 3*, 15.

North, K., & Golding, J., (2000). A maternal vegetarian diet in pregnancy is associated with hypospadias. *BJU International, 85*, 107–113.

Orsini, F., Kahane, R., Nono-Womdim, R., & Gianquinto, G., (2013). Urban agriculture in the developing world: A review. *Agronomy for Sustainable Development, 33*, 695–720.

Ozturk, L., Avci, G. G., Behmand, T., & Elekcioglu, I. H., (2017). *Pasteuria penetrans*: A bacteria parasitizing plant-parasitic nematodes in vineyards and orchards of Thrace region. In: *6th Entomopathogens and Microbial Control Congress* (pp. 14–16). Gaziosmanpaşa University, Tokat, Turkey.

Padilla, B. C., Cordts, A., Schulze, B., & Spiller, A., (2013). Assessing determinants of organic food consumption using data from the German national nutrition survey II. *Food Quality Preference, 28*, 60–70.

Peterson, H. H., Barkley, A., Chacon-Cascante, A., & Kastens, T. L., (2012). The motivation for organic grain farming in the United States: Profits, lifestyle, or the environment? *Journal of Agricultural and Applied Economics, 44*(2), 137–135.

Ponisio, L. C., M'Gonigle, L. K., Mace, K. C., Palomino, J., De Valpine, P., & Kremen, C., (2015). Diversification practices reduce organic to conventional yield gap. *Proceedings of the Royal Society B*, (282), 1–7.

Post, E., & Schahczenski, J., (2012). *Understanding Organic Pricing and Costs of Production* (pp. 1–12). National Sustainable Agriculture Information Service. National Center for Appropriate Technology, IP441.

Prain, G., (2006). Integrated urban management of local agricultural development: The policy arena in Cuba. In: Veenhuizen, R. V., (ed.), *Cities Farming for the Future: Urban Agriculture for sustainable cities* (pp. 308–311). RUAF Foundation, IDRC and IIRR.

Pussemier, L., Larondelle, Y., Van, P. C., & Huyghebaert, A., (2006). Chemical safety of conventionally and organically produced foodstuffs: A tentative comparison under Belgian conditions. *Food Control, 17*, 14–21.

Rahmann, G., Olabiyi, T. I., & Olowe, V. I., (2015). Achieving social and economic development through ecological and organic agricultural alternatives. *Scientific Track Proceedings of the 3rd African Organic Conference.* Lagos, Nigeria. Food and Agricultural Organization of the United Nations and African Union Commission.

Reganold, J. P., & Wachter, J. M., (2016). Organic agriculture in the twenty-first century. *Nature Plants, 2*, 1–8.

Saleki, Z. S., & Saleki, S. M. S., (2012). The main factors influencing purchase behavior of organic products in Malaysia. *Interdisciplinary Journal for Contemporary Research in Business, 4*, 98–116.

Samac, D. A., Willert, A. M., McBride, M. J., & Kinkel, L. L., (2003). Effects of antibiotics-producing *Streptomyces* on nodulations and leaf spot in alfalfa. *Appl. Soil. Ecol., 22*, 55–66.

SAOS (Soil Association Organic Standard), (2000). *Organic Farming, Food Quality and Human Health: A Review of the Evidence* (pp. 1–88). Soil Association, Bristol.

Savona, N., (1998). Female health problems and their nutritional solutions. *Optimum Nutrition, 11*(3), 46–51.

Smith-Spangler, C., Brandeau, M. L., Hunter, G. E., Bavinger, J. C., Pearson, M., Eschbach, P. J., Sundaram, V., et al., (2012). Are organic foods safer or healthier than conventional alternatives? *Annals of Internal Medicine, 157*, 348–366.

SSNC (Swedish Society for Nature Conservation), (2014). *Organic Food and Farming for All* (pp. 1–30). SSNC, Stockholm.

Stenius, F., Swartz, J., Lilja, G., Borres, M., Bottai, M., Pershagen, G., Scheynius, A., & Alm, J., (2011). Lifestyle factors and sensitization in children- the ALADDIN birth cohort. *Allergy, 66*(10), 1330–1338.

Tauschke, M., Behrendt, A., Monk, J., Lentzsch, P., Eulenstein, F., & Monk, S., (2015). Improving the water use efficiency of crop plant by application of mycorrhizal fungi. In: Currie, L., & Burkitt, K. L., (eds.), *Moving Farm Systems to Improved Nutrient Attenuation* (pp. 1–8), Fertilizer and Lime Research Center, Massey University, New Zealand.

Thonar, C., Lekfeldt, J. D. S., Cozzolino, V., Kundel, D., Kulhánek, M., Mosimann, C., Neumann, G., et al., (2017). Potential of three microbial bio-effectors to promote maize growth and nutrient acquisition from alternative phosphorous fertilizers in contrasting soils. *Chem. Biol. Technol. Agric., 4*, 7.

Toppo, S. R., & Naik, U. C., (2015). Isolation and characterization of bacterial antagonist to plant pathogenic fungi (*Fusarium* spp.) from agro-based area of Bilaspur. *International Journal of Research Studies in Biosciences*, 6–14.

Trabelsi, D., & Mhamdi, R., (2013). Microbial Inoculants and their impact on soil microbial communities: A review. *Biomedical Research International, 2013*(836240), 1–11.

Trewavas, A., (2001). Urban myths of organic farming. *Nature, 410*, 409–410.

Twarog, S., (2006). Organic agriculture: A trade and sustainable development opportunity for developing countries. In: United Nations conference on trade and development (UNCTD). *Trade and Environment Review* (pp. 141–224). UN, New York.

UNCTAD (United Nations Conference on Trade and Development), (2008). *Certified Organic Export Production. Implications for Economic Welfare and Gender Equality Among Smallholder Farmers in Tropical Africa (UNCTAD/DITC/TED/2007/7).*

UNCTAD (United Nations Conference on Trade and Development), (2006). *Trade and Environment Review.* UN, New York (UNCTAD/DITC/TED/2005/12).

USDA/AMS. (2000). 7 CFR Part 205. National Organic Program; Final Rule. *Federal Register, 65(46)*, 80564.

Wang, J., Fu, Z., Ren, Q., Zhu, L., Lin, J., Zhang, J., Cheng, X., et al., (2019). Effects of arbuscular mycorrhizal fungi on growth, photosynthesis and nutrient uptake of *Zelkova Serrata* (Thunb.) Makino seedlings under salt stress. *Forests, 10*(2), 1–16.

Wiggle, D. T., Semenciw, R. M., Wilkins, K., Riedel, D., Ritter, L., Morrison, H., & Mao, Y., (1990). Mortality study of Canadian male farm operators: Non-Hodgkin's lymphoma mortality and agricultural practices in Saskatchewan. *Journal of the National Cancer Institute, 82*, 575–582.

Willer, H., & Lernoud, J., (2016). *The World of Organic Agriculture: Statistics and Emerging Trends.* Frick: Research Institute of Organic Agriculture (FiBL) and IFOAM-Organics International, Switzerland.

Williams, C. M., (2002). Nutritional quality of organic food: shades of grey or shades of green? *Proceedings of the Nutritional Society, 61*(1), 19–24.

Woese, K., Lange, D., Boess, C., & Werner, B. K., (1997). A comparison of organically and conventionally grown foods - results of a review of relevant literature. *Journal of the Science of Food and Agriculture, 74*(3), 281–293.

Worthington, V., (2001). Nutritional quality of organic versus conventional fruits, vegetables, and grains. *Journal of Alternative and Complementary Medicine, 7*, 161–173.

Zarb, J., Ghorbani, R., Koocheki, A., & Leifert, C., (2005). The importance of microorganisms in organic agriculture. *Outlooks on Pest Management, 16*(2), 52–55.

INDEX

D

E

F

G

H

X

Y

Z